Erratum

F. Aboitiz and D. Cosmelli (eds.) *From Attention to Goal-Directed Behavior.*
ISBN 978-3-540-70572-7
© Springer-Verlag Berlin Heidelberg 2009

Page IV

Legend to the cover illustration

From Attention to Goal-Directed Behavior

Francisco Aboitiz • Diego Cosmelli
Editors

From Attention to Goal-Directed Behavior

Neurodynamical, Methodological
and Clinical Trends

 Springer

Editors

Francisco Aboitiz
Pontificia Universidad Católica de Chile
Escuela de Medicina
Depto. de Psiquiatría
Centro de Investigaciones Médicas
391 Marcoleta Ave.
Santiago
Chile
faboitiz@uc.cl

Diego Cosmelli
Pontificia Universidad Católica de Chile
Escuela de Psicología
Vicuña Mackenna 4860
Macul, Santiago
Chile
dcosmelli@uc.cl

ISBN 978-3-540-70572-7   e-ISBN 978-3-540-70573-4

Library of Congress Control Number: 2008931164

© 2009 Springer-Verlag Berlin Heidelberg

Cover design: WMXDesign GmbH, Heidelberg, Germany

Printed on acid-free paper

9 8 7 6 5 4 3 2 1

springer.com

Contents

Contributors

F. Aboitiz
Departamento de Psiquiatría, Centro de Investigaciones Médicas, Escuela de
Medicina, Pontificia Universidad Católica de Chile, Marcoleta 391, Santiago,
Chile

A. Barnea
Bio Keshev Clinic, Kibbutz Givat Chaim Ichud, Israel

C. Bosman
F.C. Donders Center for Cognitive Neuroimaging, Radboud University,
Kapittelweg 29, 6525 EN Nijmegen, The Netherlands, and Departamento de
Psiquiatría, Centro de Investigaciones Médicas, Escuela de Medicina, Pontificia
Universidad Católica de Chile, Santiago, Chile

X. Carrasco
Servicio de Neurología y Psiquiatría Infantil, Hospital Luis Calvo Mackenna
and Departamento de Pediatría Oriente, Facultad de Medicina Medicina,
Universidad de Chile, Santiago, Chile, and Departamento de Psiquiatría, Centro
de Investigaciones Médicas, Escuela de Medicina, Pontificia Universidad Católica
de Chile, Santiago, Chile

D. Cosmelli
Escuela de Psicología, Pontificia Universidad Católica de Chile, Vicuña
Mackenna 4860, Macul, Santiago, Chile, and Laboratorio de Neurociencias
Cognitivas, Centro de Investigaciones Médicas, Escuela de Medicina, Marcoleta
391, 2° Piso, Pontificia Universidad Católica de Chile, Santiago, Chile

F. Daiber
Departamento de Psiquiatría, Centro de Investigaciones Médicas, Escuela
de Medicina, Pontificia Universidad Católica de Chile, Santiago, Chile

O. David
Inserm, U 836, Grenoble Institut des Neurosciences, CHU Grenoble, Bât Edmond
J Safra, Chemin Fortuné Ferrini, BP 170, 38042 Grenoble CEDEX 9, France, and
Université Joseph Fourier, Grenoble, France

F.J. Flores
Programa de Fisiología y Biofísica, Instituto de Ciencias Biomédicas, Facultad de
Medicina, Universidad de Chile, Casilla 70005, Santiago, Chile

P.A. Gaspar
Clínica Psiquiátrica Universitaria, Hospital Clínico, Universidad de Chile, Av.
La Paz 1003, Casilla 70014, Santiago, Chile, and Departamento de Psiquiatría,
Centro de Investigaciones Médicas, Escuela de Medicina, Pontificia Universidad
Católica de Chile, Santiago, Chile

M. Henríquez
Departamento de Psiquiatría, Centro de Investigaciones Médicas, Escuela de
Medicina, Pontificia Universidad Católica de Chile, Santiago, Chile

K. Herzberg
Department of Psychology, University of California, Los Angeles, Los Angeles,
CA 90095-1563, USA

A. Hill
Department of Psychology, University of California, Los Angeles, Los Angeles,
CA 90095-1563, USA

J.M. Hurtado
Departamento de Psiquiatría, Centro de Investigaciones Médicas, Escuela de
Medicina, Pontificia Universidad Católica de Chile, Marcoleta 391, Santiago,
Chile, and Inflexa Research, Santiago, Chile

J.-P. Lachaux
INSERM, U821, Brain Dynamics and Cognition, 69500 Lyon, France, Institut
Fédératif des Neurosciences, 69000 Lyon, France, Université Lyon 1, 69000
Lyon, France, and Centre Hospitalier Le Vinatier (Bât. 452), 95 Bd Pinel, 69500
Bron, France

Y.H. Li
Department of Psychology, University of California, Los Angeles, Los Angles,
CA 90095-1563, USA

V. López
Escuela de Psicología, Pontificia Universidad Católica de Chile, Av. Vicuña
Mackenna 4860, Macul, Santiago, Chile, and Laboratorio de Neurociencia

Cognitiva, Departamento de Psiquiatría, Centro de Investigaciones Médicas,
Escuela de Medicina, Pontificia Universidad Católica de Chile, Santiago, Chile

S.J. Luck
Center for Mind and Brain, University of California, Davis, 267 Cousteau Place,
Davis, CA, 95618, USA

B. Luna
Departments of Psychiatry and Psychology, Laboratory of Neurocognitive
Development, Western Psychiatric Institute and Clinic, University of Pittsburgh
Medical Center, Loeffler Building, 121 Meyran Avenue, Pittsburgh, PA 15213, USA

P.E. Maldonado
Programa de Fisiología y Biofísica, Instituto de Ciencias Biomédicas, Facultad de
Medicina, Universidad de Chile, Casilla 70005, Santiago, Chile

Y. Meltzer
Center for Learning Disabilities, Academic College of Tel-Hai, Israel

R. Ortega
Laboratorio de Neurociencia Cognitiva, Departamento de Psiquiatría, Centro de
Investigaciones Médicas, Escuela de Medicina, Pontificia Universidad Católica de
Chile, Santiago, Chile

J.P. Ossandón
Programa de Fisiología y Biofísica, Instituto de Ciencias Biomédicas, Facultad de
Medicina, Universidad de Chile, Casilla 70005, Santiago, Chile

T. Ossandón
INSERM, U821, Brain Dynamics and Cognition, 69500 Lyon, France, Institut
Fédératif des Neurosciences, 69000 Lyon, France, and Université Lyon 1, 69000
Lyon, France

A. Rassis-Ariel
Bio Keshev Clinic, Kibbutz Givat Chaim Ichud, Israel, and Department of
Psychology, College of Tel Aviv-Jaffa, Israel

P. Reyes
Instituto de Ciencias Biomédicas y Departamento de Ciencias Neurológicas,
Facultad de Medicina, Universidad de Chile, Avda. Independencia 1027,
Santiago, Chile

G. Rojas
Laboratorio de Imágenes Médicas, Servicio de Neurorradiología, Instituto de
Neurocirugía Asenjo, Santiago, Chile

S. Rotem
Nitzan, The Israeli Association for Children with Learning Disabilities, Tel Aviv,
Israel

P. Rothhammer
Departamento de Psiquiatría, Centro de Investigaciones Médicas, Escuela de
Medicina, Pontificia Universidad Católica de Chile, Santiago, Chile

S. Ruiz
Institute of Medical Psychology and Behavioral Neurobiology, Graduate School
of Neural & Behavioural Sciences, International Max Planck Research School,
Tübingen, Germany, and Departamento de Psiquiatría, Centro de Investigaciones
Médicas, Escuela de Medicina, Pontificia Universidad Católica de Chile,
Santiago, Chile

J.R. Silva
Departamento de Salud Mental y Psiquiatría, Facultad de Medicina, Universidad
de la Frontera, Temuco, Chile

A. Slachevsky
Instituto de Ciencias Biomédicas y Departamento de Ciencias Neurológicas,
Facultad de Medicina, Universidad de Chile, Avda. Independencia 1027,
Santiago, Chile

T. Womelsdorf
Donders Center for Cognitive Neuroimaging, Radboud University, Kapittelweg
29, 6525 EN Nijmegen, The Netherlands

E. Zaidel
Department of Psychology, University of California, Los Angeles, Los Angeles,
CA 90095-1563, USA

F. Zamorano
Departamento de Psiquiatría, Centro de Investigaciones Médicas, Escuela de
Medicina, Pontificia Universidad Católica de Chile, Santiago, Chile

Introduction

Attention is a key psychological construct in the understanding of human cognition, and the target of enormous efforts to elucidate its physiological mechanisms, as the wealth of literature—both primary and secondary—attests (for recent compilations see Itti, Rees, & Tsotsos, 2005; Paletta & Rome, 2008; Posner, 2004). But in addition to asking what attention actually is, decomposing and analyzing its varieties, or delimiting its neurobiological mechanisms and effects, in this volume we want to explore attention somewhat differently. We believe that a full-fledged theory of attention must consider its workings in the context of motivated, goal-directed, and environmentally constrained organisms.

That attention is related to goal-directed behavior is not news. What the contributions to this volume do suggest, however, is the existence of fundamental links between attention and two key processes that are crucial for *adapted* conduct: goal-directed behavior and cognitive control. Importantly, they show that these relations can be explored at multiple levels, including neurodynamical, neurochemical, evolutionary, and clinical aspects, and that in doing so multiple methodological challenges arise that are worth considering and pursuing. The reader will find here, therefore, a selection of contributions that range from basic mechanisms of attention at the neuronal level to developmental aspects of cognitive control and its impairments. Another trend that will become evident is that, in different ways, the authors stress the need to understand these issues as they unfold in natural behavior (both healthy and pathological), thus arguing for a more ecological approach to these questions.

An interesting dichotomy emerges from the reading of the chapters compiled here. Attention as a complex process seems to stand in the middle of the two somewhat opposing forces mentioned above: motivated, goal-directed behavior and cognitive control. While the former is necessary for the exploration of the world in search for reward, the latter is critical to restrain and tune action so that predicted outcome can be accurately gauged against potential risks. Attention indeed faces both ways: it is endogenously driven and exogenously called upon, and the balance of the two extremes appears crucial for survival. In other words, beneath these three themes—attention, goal-directed behavior, and cognitive control—lies the unifying thread of a precarious but fundamental equilibrium that is characteristic of animal life: the tension between self-motivated behavior and world-constrained opportunities for action. Both extremes are at work in adapted behavior. Importantly, a functional attentional system is critical to ensure that we can navigate an unpredictable world while keeping in accordance with experience-dependent goals, yet allowing for behavioral novelty through spontaneous exploration. As several of the authors show, unbalance in any of these three processes will inevitably lead to departures from what we usually understand as "normal" or "healthy" behavior. Interestingly, while sometimes this unbalance can be catastrophic, on other occasions it can result in the expression of alternative cognitive strategies that might even have a positive evolutionary value.

These considerations push us to propose the underlying thesis of this book: It is necessary to tackle the study of attention in the ecological context of motivated beings, while taking into consideration, from the very start, the variety of relations and intertwinement that exist with goal-directed behavior and cognitive control. We believe that, far from obscuring the object of study, this strategy holds the potential of making common mechanisms and complementarities appear—or begin to be intuited—for the benefit of a more comprehensive view of cognition. Moreover, we suggest that this can be done with the aid of numerous methodological advances that have been developed in the last few years, some of which are explicitly reviewed by our contributors.

Several concrete lines of inquiry seem to open up from here: It is evident that neurophysiological mechanisms of attention, goal-directed behavior, and cognitive control are not exhausted by "hot spots" of activity in the brain: dynamic neural connectivity, both at the functional and at the effective, anatomical level is critical; neurotransmitter "wet" dynamics are likewise a fundamental biological mechanism that cannot be disregarded; evolutionary and developmental processes are necessary to understand any supposedly stereotyped (mature) form of cognitive activity; clinical investigations of pathological conditions offer a unique window into how different cognitive processes interact and are interdependent or isolatable; finally, the ecological context cannot be sacrificed for the alleged benefit of analytical clarity because it obscures the way any given process is related to actual—adapted, flexible—behavior.

In the following paragraphs we provide a brief synopsis of the different contributions to this volume. We hope the reader will enjoy the diversity of approaches presented here; we invite him or her to look for the many (not so) hidden connections that have the potential of priming novel perspectives and empirical approaches to these rather fundamental issues of human cognition.

Attentional Networks: Basic Mechanisms and Methodological Issues

Conrado Bosman and *Thilo Womelsdorf* (Chap. 1) open this book with an in-depth analysis of how two basic mechanisms of attentional modulation of sensory response can proceed: selective gain of sensory spiking and selective synchronization within local neuronal groups. Bosman and Womelsdorf, however, transcend the traditional opposition that has been established between these two candidate mechanisms. They show how both can act in concert and in complement to bring about what they consider the ultimate consequence of attentional modulation: the selective gain control that underlies enhanced sensory representations. In this context, Bosman and Womelsdorf discuss three main themes: the importance of oscillatory activity and how phase relations between active neuronal groups are crucial for selection through rhythmic behavior; the relation between bottom-up and top-down mechanisms and how these might proceed through different frequency bands (especially in the gamma and beta range); and, finally, the importance of inhibitory networks for the establishment of gamma-band oscillatory activity and their role in bringing about effective gain modulation.

Jean-Philippe Lachaux and *Tomás Ossandón* (Chap. 2) present a review of one of the most rapidly growing domains of inquiry into the brain mechanisms underlying cognition in general and attention in particular: intracortical recordings in humans. They highlight the relevance it has had to demonstrate actual cortical oscillatory activity in the gamma band and precise synchronization among distant populations in the human brain. This chapter converges and agrees with several conclusions that Bosman and Womelsdorf put forth, such as the importance of the gamma band in selective attention. However, the approach here is different, with stress on the functional role synchronization might have in selective attention, as well as the methodological challenges that imply dealing with this question in human beings at the mesoscopic level. Lachaux and Ossandón present several examples in visual perception, attention to reading and to memory that show complex intracortical responses in the gamma band. Interestingly, they discuss the functional importance of the dynamic balance between synchronization and desynchronization in this frequency band. Finally they present the dynamical spectral imaging approach developed by their group, whereby ongoing spectral decompositions of actual brain signals can be obtained during attentional, or other, cognitive tasks.

Steven Luck (Chap. 3) proposes a highly original perspective in his chapter on the spatiotemporal dynamics of visual–spatial attention. He discusses the importance of understanding how attentional mechanisms are at play during actual fixation, a much neglected aspect in the study of attention at work in natural conditions (see also Chap. 4 by Maldonado et al.). Luck describes in detail the relation and dynamic interplay between covert attention and eye movements. In contrast to the traditional approach of studying covert attention by isolating it from the act of fixating, Luck argues that such shifts to peripheral regions of visual space and eye movements are intertwined at least on two functionally relevant levels: covert shifts of attention facilitate the ulterior foveation of regions that are likely to be relevant sources of information; consequently, covert attention is then critical to adjust dynamically around the foveated target so that distracting objects get effectively filtered out. The author highlights in his review the necessity to consider visual attention in a more ecological setting, yet shows how this is possible without losing the analytical detail that traditional approaches have provided.

Pedro Maldonado et al. (Chap. 4) tackle a complex issue in their contribution: the problem of natural vision. They challenge the classical receptive field notion by showing the importance of contextual, dynamical modulation in goal-directed behavior that dominates when animals find themselves in natural complex settings. Through a detailed analysis of the structure of natural images and the related neuronal responses, they argue that traditional receptive field explanations cannot account for the response of visual neurons in theses conditions. Consequently, Maldonado et al. stress the notion of "active vision" where goal-oriented behavior and environmental relevance meet, therefore questioning the traditional top-down/ bottom-up divide. In this context they argue for the need for an ecological approach to the study of attention, where the natural exploratory behavior of the organism is taken into account from the beginning.

In line with the call for a consideration of the more ecological aspects of attention, *Diego Cosmelli* (Chap. 5) undertakes a quite neglected theme in cognitive neuroscience: the ongoing stream of conscious experience. Cosmelli argues that a full-fledged theory of attention will have to consider attentional shifts as they take place during mind-wandering, internal discourse, and in general the restless movement of our focus of interest which characterizes our intimate daily experience. Drawing from the phenomenological tradition, he proposes that in addition to the traditional endogenous-voluntary/exogenous-reflexive distinction, the spontaneous nature of attentional shifts be considered a type in its own right – and studied accordingly. Cosmelli suggests starting from the foundations provided by traditional experimental psychology paradigms that target aspects such as vigilance changes and the process of attentional orienting. He then explores the recent reappraisal of stimulus-independent cognition and the default mode of brain function, and the opportunity these developments represent for a concerted approach to studying the workings of attention in the stream of consciousness.

In their chapter, *Rodrigo Ortega* and *Vladimir López* (Chap. 6) survey mechanisms of cross-modal integration and attention. This process is crucial for associating predicting stimuli or behavioral actions with outcomes, something on which our everyday behavior strongly relies. Contrary to early concepts of modular sensory processing, recent evidence using event-related potentials strongly indicates that stimuli presented in one modality may influence processing in other modalities. This mechanism may be crucial for establishing associations between different conditioned and unconditioned stimuli in everyday behavior.

Considering the conflicting findings of electrophysiological studies in brain lateralization, *Andrew Hill* et al. (Chap. 7) describe an experiment in which subjects are trained with an EEG biofeedback protocol, with electrodes located either at lateralized or at midline sites in the scalp. Subjects were assessed for attentional skills using the Lateralized Attentional Network Test, before and after the training procedure. They found that training with left-hemisphere electrodes improved attention in the right hemisphere and, vice versa, training in the right hemisphere improved attention in the left hemisphere. This suggests the existence of a contralaterally organized meta-cognitive control system. This work indicates that besides being lateralized, attentional networks are strongly dependent on behavior and on the valuation of behavioral outcomes.

In one of the most methodologically inclined chapters of this book, *Olivier David* (Chap. 8) presents the dynamic causal modeling framework and how it can be highly beneficial for the study of attentional networks. As David shows, this approach to brain imaging is founded on explicit prior probability distributions regarding brain function that take into consideration biophysical constraints. As

David suggests this hypothesis-driven approach allows for testing the validity of different neuropsychological models of attention through the study of event-related activity. This allows one to evaluate effective connectivity during cognitive paradigms, therefore producing a network picture that is beyond classical inverse-problem approaches. After introducing the basic mathematical concepts, David presents two examples from his own work to illustrate the framework: an emotional attention paradigm and a classic auditory oddball.

From Attention to Behavioral Control

In his contribution, *José Hurtado* (Chap. 9) deals with the contrast between goal-oriented and routine behaviors. He reviews the structural basis for a fronto-supervisory/posterior-sensorimotor dichotomy and how this might relate to the way executive attention comes into play when novel behavior must be deployed. Hurtado discusses two related concepts. Drawing from the representational properties of sensorimotor neuronal assemblies, he shows how intracortical high-order assemblies can act as "pointers" on low-level sensory assemblies to bias routine behavior. On the other hand, he presents the "scaffolding hypothesis," whereby the frontal system can match actions and their outcomes – and therefore their putative reward value – for the system. This is a clear instance of the frontoposterior dichotomy that implies a large network through which frontal (supervisory) systems recruit sensorimotor (procedural) ensembles to enact behavioral novelty. In turn, the sensorimotor–basal ganglia loop can stabilize behavioral sequences that are good predictors of reward through modulation of the corticothalamic system.

Francisco Aboitiz (Chaps. 10, 11) reviews the dopaminergic mechanisms of behavioral control and cognitive function. Dopamine regulates behavior and cognition at multiple levels, from behavioral activation, stimulus reward associations, and reward predictions to modulation of attentional mechanisms, working memory, and in general executive function. Aboitiz proposes an evolutionary framework to unify this complex signaling mechanism, which started from a simple system to invigorate behavior and acquired more complex functions as the capacity to associate predicting stimuli to reward or punishment became increasingly sophisticated, and predicting stimuli became increasingly separated in time and space from outcomes. In this context, cognitive and attentional function can be seen as subsidiary mechanisms of the elaboration and increasing complexity of goal-directed behavior.

Clinical and Developmental Issues

Andrea Slachevsky et al. (Chap. 12) review the neuroanatomy and functions of the human frontal lobe, mainly on the basis of analyses of brain-damaged patients. They highlight the mosaic nature of frontal lobe architecture, which corresponds to

xvi   Introduction

specific behavioral and cognitive subprocesses, in the final instance all related to behavioral control. Thus, frontal lobe function is far from being a strictly unitary process, but rather corresponds to a complex set of functions that concur to generate appropriate behavior according to contextual information, and allows the individual to modify behavior to obtain reward when it is not directly accessible.

Beatriz Luna (Chap. 13) describes brain maturation mechanisms during adolescence, placing strong emphasis on cognitive mechanisms of behavioral control that are proposed to be refined via myelination and synaptic pruning mechanisms. Cognitive development studies indicate that adolescence is characterized by improvements in existing abilities, including the speed and capacity of information processing, and in the ability to have consistent cognitive control of behavior. Temporoparietal association and prefrontal cortices, involved in sensorimotor integration mechanisms and crucial for control mechanisms, mature during this important period.

Ximena Carrasco et al. (Chap. 14) discuss the evidence for electrophysiological and genetic alterations in attention deficit–hyperactivity disorder (ADHD), and the emergent trend to assess genetically based variability in neurodynamical patterns using ADHD as a model. These authors pose the questions of whether ADHD reflects a variant within the spectrum of phenotypic variability more than a specific morbid condition, and that perhaps more than an attention deficit condition, ADHD relates to suboptimal function of behavioral control mechanisms.

Pablo Gaspar et al. (Chap. 15) analyze the operation of cognitive networks in schizophrenia. A current interpretation of the schizophrenic condition is that it reflects a disconnection pattern in neural networks involved in cognitive and emotional processes. Gaspar et al. expand this idea to the concept of "aberrant connectivity," highlighting imaging evidence that indicates that, in addition to decreased functional connectivity in some functions, there may be domains of abnormally increased connectivity. Overall, this evidence points to a generalized dysfunction in the mechanisms linking sensory stimuli, behavioral mechanisms, and valuation of behavioral outcomes.

References

Itti, L., Rees, G., & Tsotsos, J. (2005). *Neurobiology of attention*: Elsevier.
Paletta, L., & Rome (Eds.). (2008). *Attention in Cognitive Systems. Theories and Systems from an Interdisciplinary Viewpoint: 4th International Workshop on Attention in Cognitive Systems, WAPCV 2007 Hyderabad, India, January 2007, Revised Selected Papers*: Springer.
Posner, M. I. (Ed.). (2004). *Cognitive Neuroscience of Attention*. New York: The Guilford Press.

Acknowledgements

This publication is part of the activities of the Centro de Neurociencias Integradas CENI, Iniciativa Científica Milenio. This organization has supported part of the work presented in this book. We are also grateful to Leticia Clede for valuable assistance with the tedious editorial issues.

Part I
Attentional Networks: Basic Mechanisms and Methodological Issues

Chapter 1
Neuronal Signatures of Selective Attention – Synchronization and Gain Modulation as Mechanisms for Selective Sensory Information Processing

C. Bosman and T. Womelsdorf

Abstract Voluntary attention selectively enhances the influence of neuronal responses conveying information about relevant sensory attributes. This attentional modulation of sensory responses reflects selective gain changes of sensory spiking response and is accompanied by selective synchronization of those neuronal responses within a local neuronal group which convey information about the attended visual input. While these empirical signatures of selective attention in visual cortex are derived from different levels of analysis (single neurons vs. neuronal groups), they provide complementary insights into the selective modulation of sensory processing in the brain: neurons which are tuned to the spatial and featural attributes of the attended sensory inputs are those which most strongly synchronize within a local neuronal group, while likewise showing the strongest gain change of their spiking response. Both of these neuronal consequences of attention enhance the signal-to-noise ratio for the attended information, but their different roles for selective neuronal processing have not been reconciled. Here, we propose that rhythmic synchronization is the key mechanism underlying the selective gain control within sensory cortex during attentional processing. We derive this hypothesis from computational studies and from insights into the physiological mechanisms that impose synchronized oscillatory activity within neuronal groups. In summary, voluntary attention appears to recruit rhythmic neuronal activity to establish a selective gain pattern within visual cortex to ultimately enhance the representation of the attended sensory information.

C. Bosman
F. C. Donders Centre for Cognitive Neuroimaging, Radboud University Nijmegen, Kapittelweg 29, 6525 EN Nijmegen, The Netherlands, e-mail: conrado.bosman@fcdonders.ru.nl

F. Aboitiz and D. Cosmelli (eds.) *From Attention to Goal-Directed Behavior.*
© Springer-Verlag Berlin Heidelberg 2009

Introduction

Voluntary "top-down" attention is the key mechanism to select relevant subsets of sensory information among the multitude of environmental inputs that our sensors receive at any moment in time. Attentional selection ensures that our brain's limited processing resources are focused on behaviorally relevant input, while irrelevant and distracting sensory information is suppressed. The functional consequences of such prioritization of relevant inputs are manifold: Attended sensory information is processed more rapidly and accurately and with higher spatial resolution and sensitivity for fine changes, while nonattended information appears lower in contrast and is sometimes not perceived at all (Carrasco, Ling, & Read, 2004; Simons & Rensink, 2005).

These behavioral consequences of selective attention originate in highly selective changes of neuronal activity that evolve rapidly and span all levels of cortical information processing, from single neurons to local groups of neurons and integrated network interactions among distant neuronal groups. The dynamic restructuring of neuronal activity during selective attentional processing may be best illustrated by briefly surveying the response modulations as they evolve during typical attentional paradigms: These tasks typically begin with an instructional cue at the start of a trial or block of trials. The cue informs the subject about which stimulus attribute and/or motor output is behaviorally relevant and thus triggers voluntary attention to covertly shift toward the respective information. A critical prerequisite for attentional paradigms is that there are always alternative stimuli available within a trial, but that attention is covertly directed to only subsets of these stimuli.

On the basis of variants of this task design, neurons in various areas have been shown to change their firing pattern with a short delay following the cue. The top-down information about the behaviorally relevant visual field location can be read out as soon as 150 ms following cue onset from response changes of neurons in prefrontal cortex, the frontal eye field, intraparietal sulcus, and the superior colliculus (Buschman & Miller, 2007; Everling, Tinsley, Gaffan, & Duncan, 2002; Gottlieb, Kusunoki, & Goldberg, 1998; Ignashchenkova, Dicke, Haarmeier, & Thier, 2004; Inoue & Mikami, 2007; McPeek & Keller, 2004; Muller, Philiastides, & Newsome, 2005; Thompson, Biscoe, & Sato, 2005; Tomita, Ohbayashi, Nakahara, Hasegawa, & Miyashita, 1999). Thus, top-down information is available in the response patterns in a distributed neuronal circuitry. These top-down signals have been shown to directly influence downstream neurons in sensory cortices (Armstrong & Moore, 2007; Johnston, Levin, Koval, & Everling, 2007; Miller & D'Esposito, 2005). The top-down influence is thereby highly selective in enhancing and suppressing response strength and response synchronization of neurons activated by the attended stimulus (Gilbert & Sigman, 2007; Maunsell & Treue, 2006; Reynolds & Chelazzi, 2004; Womelsdorf & Fries, 2007). For example, the dynamical interplay of top-down information and attentional response modulation can be seen already in the earliest visual cortical processing stages (Khayat, Spekreijse, & Roelfsema, 2006), revealing an early attentional routing of sensory information transfer.

Dynamic top-down control of neuronal responsiveness in sensory areas is at the heart of attentional processing (Fries, Nikolic, & Singer, 2007; Gilbert & Sigman, 2007). It essentially shows that selective processing is accomplished by flexible changes of the interaction pattern among neurons conveying top-down information and neurons controlling behavioral and processing sensory inputs: Attention determines which of the anatomically possible connections are made efficient, while neuronal connections conveying irrelevant information are rendered ineffective. Although the mechanisms underlying selective changes in effective neuronal communication across cortical areas are of critical importance for attentional processing, only a few studies have investigated them directly (Buschman & Miller, 2007; Saalmann, Pigarev, & Vidyasagar, 2007; Womelsdorf et al., 2007). These studies generally point toward a critical role of rhythmic synchronization of neuronal responses to sharpen and stabilize communication between those neuronal groups processing relevant information. However, in contrast to the analysis of neuronal interactions, the most detailed characteristics of attentional modulation has been gathered on the basis of analysis of single neuron responses in isolated areas (Reynolds & Chelazzi, 2004; Treue, 2001). One major insight following these studies is that top-down information is imposing a highly selective gain change on neuronal responses in sensory cortices. Attentional gain enhances the responses of neurons, which convey information about the attended stimulus feature, and suppresses the influence of neurons tuned to stimulus attributes opposite to those of the attended feature, with negligible effects on neurons not tuned to the attended attribute (Martinez Trujillo & Treue, 2004; Treue & Martinez Trujillo, 1999).

This brief overview illustrates that there are two basic attentional signatures which evolve from different levels of analysis: Attentional response modulation of single neurons is best described as gain change, while attentional effects at the level of neuronal groups and interactions among them appear to critically rely on mechanisms of selective synchronization. In the following we will argue that results from both levels of analysis provide a common window on basic properties of neuronal circuitry. Consideration of recent empirical and theoretical insights about the physiological origins and functional relevance of rhythmic synchronization within neuronal circuitry, strongly suggests that mechanisms underlying selective response synchronization are likely the key to understanding selective sensory processing (Fries, 2005; Womelsdorf & Fries, 2007).

Principle Characteristics of Synchronization Relevant for Selective Attention

The last paragraph conclusion is inferred from various recent findings demonstrating the functional relevance of rhythmic synchronization and basic neuronal and network properties, which render rhythmic activity particularly suited to rapidly establish and sustain selective patterns of neuronal synchronization. Among these core findings are five main aspects, which will be briefly listed discussed and followed up in the subsequent sections of this review:

1. Firstly, synchronization of neuronal responses in various cortical regions has been shown to convey stimulus-specific information. In other words, the degree of synchronization within a local neuronal group is tuned to particular stimulus parameters so that changes in the strength of synchronization convey information about the relative weighting of these input parameters to upstream neurons. The degree of synchronization not only follows basic input properties such as stimulus contrast or coherence of motion signals (Henrie & Shapley, 2005; Lee, Simpson, Logothetis, & Rainer, 2005; Siegel, Donner, Oostenveld, Fries, & Engel, 2006), but is likewise tuned to stimulus orientation and spatial frequency (Frien, Eckhorn, Bauer, Woelbern, & Gabriel, 2000; Gray & Singer, 1989; Kayser & Konig, 2004; Kreiter & Singer, 1996; Siegel & Konig, 2003), the speed and direction of visual motion (Liu & Newsome, 2006), the spatial motor intentions and movement directions (Scherberger & Andersen, 2007; Scherberger, Jarvis, & Andersen, 2005), and objects defined by complex feature combinations (Kraskov, Quiroga, Reddy, Fried, & Koch, 2007; Kreiman et al., 2006).

2. Secondly, synchronization is modulated by selective attention (Womelsdorf & Fries, 2007). In particular, attending to one of two visual stimuli or to searching for particular stimulus features within the visual field induces selective synchronization among those neurons activated by the attended stimulus feature (Bichot, Rossi, & Desimone, 2005; Buschman & Miller, 2007; Fries, Reynolds, Rorie, & Desimone, 2001). The strength of synchronization thereby allows one to predict how efficient the attended stimulus is processed and how it is related to the perceptual accuracy during a shape discrimination task (Taylor, Mandon, Freiwald, & Kreiter, 2005; Womelsdorf, Fries, Mitra, & Desimone, 2006). Thus, synchronization carries information about which spatial location or stimulus feature is behaviorally relevant.

3. Thirdly, synchronization of neuronal inputs is a powerful mean to increase the response gain of the postsynaptic, "receiving" target neurons and could thus underlie the gain modulation during selective attentional processing (Fellous, Rudolph, Destexhe, & Sejnowski, 2003; Salinas & Sejnowski, 2001; Tiesinga, 2005). In particular, already moderate amounts of input synchronization can change the spiking response of a neuron manifold and enhance neuronal excitability. Notably, increases in the average input rate alone and in the absence of synchronization do not automatically increase response strength due to active synaptic mechanisms such as synaptic depression, which can actually lead to enhanced spiking thresholds and thus decreased spiking output of neurons (Tsodyks & Markram, 1997; Wespatat, Tennigkeit, & Singer, 2004). Moreover, enhanced spiking probability is promoted by rhythmic fluctuations at the postsynaptic membrane, with enhanced excitability periods corresponding to periods of maximum depolarization Taken together, the described impact of synchronous inputs and the neuronal preference to generate spikes at time windows, or particular phases, of the rhythmic membrane fluctuations complements each other. They provide powerful mechanisms to control neuronal sensitivity and

responsiveness and could thus underlie selective gain modulation during attentional processing.

4. The fourth key feature of neuronal synchronization pertains to its spatial and temporal selectivity. How are the subsets of neuronal connections that convey information about attended information made more effective, while connections conveying irrelevant information are rendered less effective? Recent empirical and modeling studies suggest that the critical key component for such a selective routing is the control of the phase of rhythmic synchronization of neuronal activity (Buia & Tiesinga, 2006; Fries, 2005; Fries et al., 2007; Mishra, Fellous, & Sejnowski, 2006; Schoffelen, Oostenveld, & Fries, 2005; Womelsdorf et al., 2007). Aligning the phase of rhythmic activity between neuronal groups promotes mutual interactions, because the time windows of excitability overlap. In contrast, neuronal groups that synchronize at random phase or at antiphase are less likely to influence each other. Such phase-dependency of neuronal interactions shows a high spatial and temporal resolution. In a recent study the pattern of neuronal interactions could be predicted by the phase of rhythmic synchronization between groups that were as close as 600 μm from each other (Womelsdorf et al., 2007). The generality of the mechanism described is highlighted by the fact that the same phase dependency predicted interaction patterns among distant groups of neurons from different cortical areas (Womelsdorf et al., 2007). Thus, rhythmic synchronization can be highly selective in determining the interaction patterns between groups of neurons and spans broad spatial scales. Thus, attentional top-down control could recruit these mechanisms to achieve neuronal coupling that is selective in time and space.

5. The fifth critical aspect of the hypothesized role of selective synchronization for attentional processes is its feasible implementation in neuronal circuitry. Importantly, recent computational studies have reproduced the neurophysiologically observed attentional gain changes on neuronal responses with model architectures, which explicitly acknowledge the complex interplay of inhibitory and excitatory neurons and the synaptic characteristics of spike generation (Bartos, Vida, & Jonas, 2007). Rhythmic synchronization emerges spontaneously in neuronal populations when excitatory drive diverges onto inhibitory interneurons, which then feed back their inhibition onto the excitatory cells. The inhibitory feedback thereby gates the excitatory activity by imposing synchronized inhibition onto their activity (Bartos et al., 2007; Borgers, Epstein, & Kopell, 2005; Whittington, Traub, Kopell, Ermentrout, & Buhl, 2000). Based on variations of this architecture, recent studies succeeded in reproducing the empirically observed attentional effects by inducing small biases of the phase of ongoing inhibitory synchronization (Mishra et al., 2006; Tiesinga, 2005; Tiesinga, Fellous, Salinas, Jose, & Sejnowski, 2004). Modulating the phase and precision of inhibitory synchronization in neuronal networks selectively facilitated the influence of groups of excitatory neurons that convey information about the attended information. Thus, modeling studies demonstrate in detail how (and potentially at which site) attention could act upon neuronal circuitry to rapidly establish, sustain, and switch between selective representations of the behaviorally

relevant information by considering the rich interplay of inhibitory and excitatory neuronal activity.

Taken together, the five aspects of selective synchronization described suggest that it is a pivotal mechanism that is invoked during selective attentional processing. The following sections will survey empirical evidence supporting this conclusion. We will begin by reviewing recent insights of attentional modulation within local neuronal groups and between groups of neurons from distant cortical areas. Later sections will outline in more detail the role of inhibitory circuitry in imposing gain changes in cortical neurons and groups of neurons.

Attention Affects Neurons Tuned Toward the Attended Stimulus Attribute

Top-down attention modulates the responses of neurons in all sensory areas studied so far (Maunsell & Cook, 2002). Studies from the past two decades have revealed various key characteristics of this selective response modulation. First, response modulation can proceed not only on spatially selected information (spatial attention), but can likewise be based on information of behaviorally relevant features (e.g., motion vs. color), and on parameters within a particular feature space (e.g., upward vs. downward motion directions, or red vs. green color) (Maunsell & Treue, 2006; Reynolds & Chelazzi, 2004). The general conclusion from studies of space-based and feature-based attention is that attention enhances the responsiveness of those neurons which are tuned for the attended stimulus attribute. This is most convincingly demonstrated for spatial attention, where attention typically enhances response strength and selective synchronization only of those neurons which overlap the spatial focus of attention, i.e., which prefer the spatial attribute of the attended stimulus (Connor, Preddie, Gallant, & Van Essen, 1997; Fries et al., 2001; Taylor et al., 2005; Womelsdorf, Anton-Erxleben, Pieper, & Treue, 2006; Womelsdorf, Fries et al., 2006). Similar to the spatial domain, attention in "feature space" to the color of a stimulus enhances responses of neurons in color-selective areas, while attention to the motion direction of an otherwise identical stimulus results in the strongest modulation in motion-selective areas, like such as the middle temporal (MT) area in the parietal cortex (Hopf et al., 2006; McMains, Fehd, Emmanouil, & Kastner, 2007; Schoenfeld et al., 2007).

Importantly, within a functionally specialized visual area such as area MT, attentional modulation reveals that attention not only enhances responses, but also suppresses responses of those neurons which are tuned toward features opposite to those of the attended stimulus attribute (Fig. 1.1. This finding has given rise to the feature-similarity gain hypothesis of attention (Treue & Martinez Trujillo, 1999), which holds that top-down information about the behaviorally relevant stimulus attribute imposes a unique gain factor on neuronal responses in sensory cortices. The strength and sign of attentional enhancement/suppression will thereby depend on the overlap of a single neuron's tuning preferences and the attended feature.

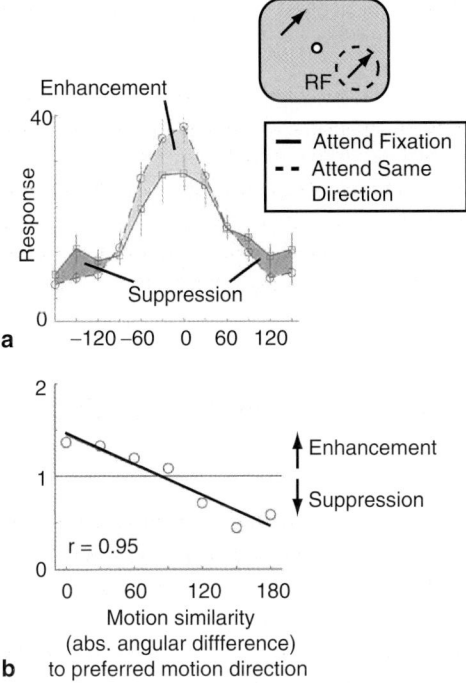

Fig. 1.1 Attention enhances and suppresses neuronal responses depending on the overlap of the attended stimulus feature (here, motion direction) and the tuning preference of single neurons in macaque middle temporal (MT) area. (**a**) The relation of average neuronal firing rate (*y*-axis) of a neuron to stimuli of different directions of motion (*x*-axis, *arrows* in the *inset* illustrate stimulus locations). The *solid line* shows the tuning curve of the neuron when the monkey attended the fixation spot (the *white circle* in the *inset*). The *dashed line* shows the response to identical motion stimuli within the neuronal receptive field (*RF*) when the monkey attended a second motion stimulus in the hemifield opposite the RF. The two stimuli (within and outside the RF) always moved in the same direction. The graph illustrates that attending to the preferred motion direction (here, 0°) enhances neuronal responses, while attending to antipreferred motion directions (here, 180°) suppresses the response. (**b**) Average result across the population of area MT neurons illustrating that the attentional effect (*y*-axis) varies as a function of the similarity of the attended motion direction with the feature preference of the neurons. (Adapted from Martinez-Trujillo & Treue 2004)

Recent evidence has demonstrated that the feature-similarity rule cannot only be traced back to firing rates, but is likewise evident in changes of response synchronization (Bichot et al., 2005). Attention to a particular feature selectively synchronized the responses of those sensory neurons which are tuned to the attended feature. To show this Bichot et al. (2005), recorded neuronal spiking responses and local field potentials (LFPs) in macaque visual area V4 while monkeys searched in multistimulus displays for a target stimulus defined either by color, shape, or both. When monkeys searched, e.g., for a red stimulus by shifting their gaze across stimuli on the display, the receptive fields of the recorded neurons could either encompass nontarget stimuli (e.g., of blue color), or the (red) target stimulus prior to the time when the monkey detected the target. The authors found that neurons synchronized to the LFP more strongly in response to their preferred stimulus feature when it was the attended search target feature rather than a distracter feature. Notably, the observed modulation depended on the overlap of feature preference and attended feature and was independent of the spatial location of attention. This feature-based modulation was also evident during a conjunction search task involving targets which were defined by two features: When monkeys searched for a target stimulus with a particular orientation and color (e.g., a red horizontal bar), neurons with tuning preference to one of these features synchronized their responses more strongly (Bichot et al., 2005). This enhancement was observed not only in response to the color-shape-defined conjunction target, but also in response to distracters sharing one feature with the target (e.g., red color). This finding is not only capable of explaining the increased difficulty and search times during conjunction versus simple search tasks. It also suggests that attention modulates neuronal synchronization in visual cortex gradually, and not according to a strict all-or-nothing principle as a function of the overlap of the neurons' tuning preference and the attended stimulus feature (Womelsdorf & Fries, 2007).

Attention Imposes a Gain Pattern on Sensory Neurons

The effects of attention on neuronal responses described generally reflect a modulation of the input–output relation of neuronal activity, which can be formalized as gain modulation. Existing evidence suggests, that selective attention utilizes a gain control mechanism that (1) enhances neuronal sensitivity to synaptic inputs and (2) can likewise act on the normalization stage of neuronal response functions and thereby reflect response, or output, gain control. These different scenarios emerge from the few studies which have investigated a broad range of input values during selective attention.

The most compelling evidence showing that attention imposes a multiplicative response gain arises from its effects on the tuning functions in extrastriate visual cortex in areas of the ventral, temporal, as well as the dorsal, parietal processing pathway. In particular, multiplicative scaling of tuning functions has been demonstrated for direction of motion in area MT of the macaque monkey, and for orientation tuning curves in area V4 (McAdams & Maunsell, 1999; Treue & Martinez Trujillo, 1999). For example, neurons in area MT respond to stimuli moving in varying directions of motion, with a gradual, bell-shaped response profile peaking at their preferred direction of motion. Attending to that preferred direction of motion outside the receptive field causes the tuning function to be scaled up multiplicatively, while attending to the antipreferred direction of motion causes the tuning function to be scaled down (Martinez Trujillo & Treue, 2004; Treue & Martinez Trujillo, 1999). This finding demonstrates that attention does not sharpen basic feature selectivity at the single-neuron level, i.e., by narrowing the range of motion directions that elicit responses, but rather by a change in the gain of the neuronal response. Importantly though, the multiplicative gain modulation of single-neuron responses translates into strongly enhanced feature selectivity of the population response of modulated neurons in area MT: Neurons with a tuning preference close to the attended direction of motion enhance their contribution to the population response profile, while neurons with preferences away from the attended feature reduce their contribution (Martinez Trujillo & Treue, 2004)

In addition to multiplicative gain modulation of featural input to a cortical neuron, attention interacts with other input, or bottom-up, aspects of visual stimuli. In particular, attention has been demonstrated to interact with the luminance contrast of the attended stimulus to effectively increase the effective contrast of the attended stimuli (Martinez Trujillo & Treue, 2002; Reynolds, Pasternak, & Desimone, 2000; Williford & Maunsell, 2006). Visual cortical neurons typically respond to stimuli of increasing contrast with a sigmoidal contrast-response function. Attention modulates this input–output function of cortical neurons by changing the gain of the input in various ways. Early studies suggested that attention modulated particularly stimuli of intermediate contrasts (Martinez Trujillo & Treue, 2002; Reynolds et al., 2000), which is evident in a leftward shift of the contrast-response function ("contrast gain"). A more recent study revealed a more complicated gain pattern across neurons, with response modulation of a larger proportion of neurons reflecting a rather constant gain factor modulating responses along the whole contrast-response function ("response gain"), or with an additional general increase of response strength ("activity gain") (Williford & Maunsell, 2006). Interestingly, a recent attention experiment based on human scalp recordings from visual cortex in response to flickering stimuli likewise suggests that attention boosts the amplitude of the steady-state population response even at highest contrasts and thus reflects response-gain modulation rather than a nonlinear contrast gain control (Kim, Grabowecky, Paller, Muthu, & Suzuki, 2007).

The findings surveyed demonstrate that the firing rate output of single neurons constrains which gain-control mechanism, or combination of mechanisms, provides adequate descriptions of the working principles of attention at that level of processing. The following sections will extend this level to include interactions among groups of neurons, by surveying evidence of selective synchronization during selective attentional processing across cortical areas, before the last sections describe the mechanistic aspects of gain control through rhythmic activity in inhibitory networks.

Attention Invokes Selective Long-Range Synchronization Between Cortical Areas

The evidence for attentional effects outlined following the feature-similarity principle and acting according to gain-control mechanisms were restricted to neuronal responses to local groups of neurons within functionally specialized sensory cortices. Few studies have gone beyond single-area recordings to reveal a more comprehensive picture of how gain-modulated neuronal responses at one level of the processing pathways affect and mutually interact with neuronal groups in other areas, which are likewise known to be attentionally modulated. Importantly, as insinuated in the "Introduction," selective interareal communication is the major characteristic of selective attention. Top-down information from higher-order frontal and parietal areas impinges on sensory cortex recruiting selective subsets of neuronal groups in distributed, functionally specialized visual areas. Similarly, distributed neuronal groups in sensory cortices need to be functionally linked according to the behavioral demands. Recent studies have begun to provide critical evidence that such a dynamic interareal communication between neuronal groups is likely mechanistically subserved by selective phase synchronization (Womelsdorf et al., 2007). In particular, neuronal groups were shown to strongly interact at those time periods when they precisely synchronized their rhythmic activity. At the same time, epoch interactions were reduced, or rendered negligible to neuronal groups, which were active at random phases or out of phase (Fig. 1.2a). These findings strongly support the notion that selective neuronal communication is critically promoted by selective synchronization (Fries, 2005). Selective synchronization could therefore be utilized by attention to dynamically link subsets of distributed neuronal groups conveying behaviorally relevant information (Womelsdorf & Fries, 2007).

Existing evidence supports this proposed role for selective long-range synchronization during attentional processing (Buschman & Miller, 2007; Saalmann et al., 2007; Schoffelen et al., 2005). Knowledge from earlier studies of awake cats demonstrated that nonselective states of expectancy of a behaviorally relevant stimulus (in, e.g., "go/no-go tasks") increase interareal beta-frequency synchronization among visual cortical and premotor regions (Roelfsema, Engel, König, & Singer, 1997; von Stein, Chiang, & König, 2000) and between thalamic and early visual cortical regions (Wrobel, Ghazaryan, Bekisz, Bogdan, & Kaminski, 2007).

Recent studies have critically extended these findings by showing that frontoparietal and intraparietal interactions are accompanied by synchronization at high beta frequencies (20–35 Hz) during task epochs requiring searching for and selecting behavioral relevant visual stimuli (Buschman & Miller, 2007; Saalmann et al., 2007). Figure 1.2b illustrates findings from a visual search task requiring monkeys to detect a search target that is salient and pops out among distracting stimuli ("bottom-up search"), or that is nonsalient by sharing features with distracting stimuli (Buschman & Miller, 2007). In contrast to bottom-up salient targets, the nonsalient target stimuli were detected more slowly, indicating that they require attentive search through the stimuli in the display before they are successfully detected ("top-down search"). Paralleling the difference in behavioral demands, the authors found a selective synchronization pattern among the LFPs in frontal and parietal cortex. While attentive "top-down search" enhanced specifically rhythmic synchronization at 20–35 Hz compared with the "bottom-up" search, the stimulus-driven "bottom-up" search enhanced interareal synchronization at higher frequencies (Fig. 1.2b). The pattern of results is

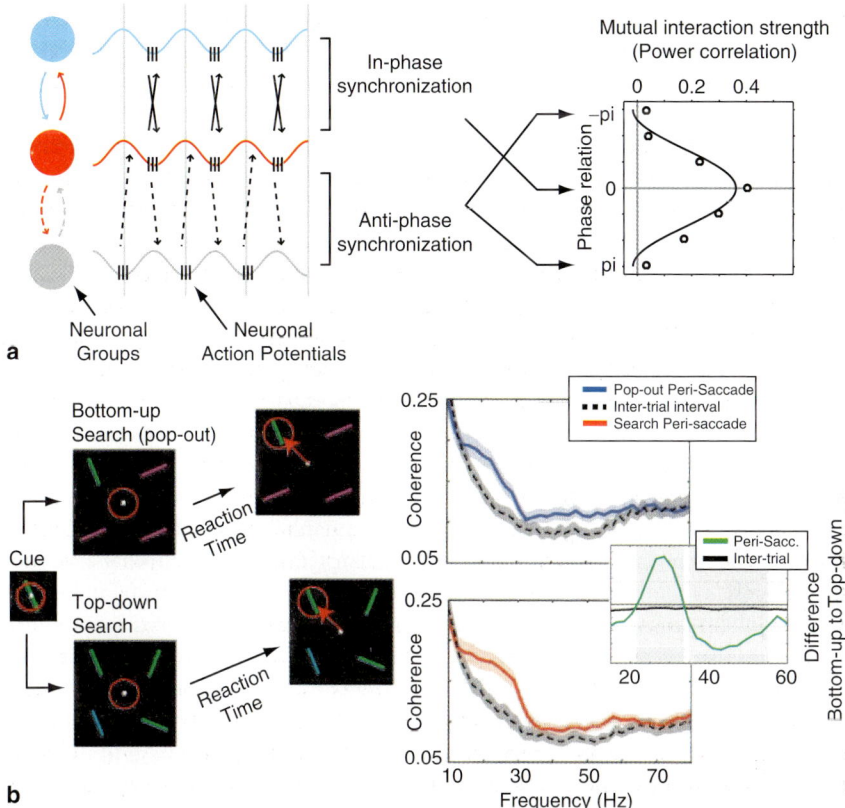

Fig. 1.2 Precise synchronization of rhythmic activities between neuronal groups determines the strength of their mutual interactions and is modulated by attention. (**a**) Rhythmic activity (local field potential oscillations with spikes at the troughs) of three neuronal groups (*circles*). Time windows of maximum excitability are either aligned in-phase (red and blue group) or are not aligned (red and grey group, here illustrated as groups with antiphase synchronization). The plot on the *right* shows increasing mutual interactions (*upper axis*, correlation of the power of the local field potential and the spiking response of the neuronal groups) during a period of in-phase synchronization and lower ones otherwise. (**b**) *Left panels* sketch the visual search tasks used by Buschman & Miller (2007). A cue instructed monkeys about the orientation and color of a bar that was the later search target in a multistimulus display during a bottom-up search task (both target color and orientation were unique) and during a top-down search task (target shared color or orientation with distracting stimuli). Monkeys covertly attended the multistimulus array and made a saccade to the target stimulus position as soon as they found it. The authors measured the coherence of the local field potential activity of neuronal groups in frontal cortex and the lateral intraparietal area. The line plots on the *right* show the coherence (*y*-axis) for different frequency bands (*x*-axis) in the bottom-up and top-down tasks, along with the coherence difference across tasks (*green line* in inset). The results show that attentional demand modulated long-range coherence at different frequency bands (relative to an intertrial baseline epoch) among the distant cortical sites. ((**a**) Adapted from Womelsdorf et al., 2007. (**b**) Adapted from Buschman & Miller, 2007)

most likely due to differences in task demands in both search modes and was unaffected by differences in reaction times, suggesting that interareal communication during top-down attentional control and bottom-up feedforward signaling is conveyed through rhythmic synchronization at different frequencies.

Importantly, the neurophysiological evidence surveyed on interareal synchronization during selective sensory processing is paralleled by a variety of human EEG and MEG studies, demonstrating a modulation of beta- and gamma-frequency oscillations and phase synchronization during tasks requiring selective attention, working memory, and speeded reaction times (Aboitiz, Lopez, & Montiel, 2003; Engel, Fries, & Singer, 2001; Jensen, Kaiser, & Lachaux, 2007; Varela, Lachaux, Rodriguez, & Martinerie, 2001; Womelsdorf & Fries, 2007). In this context, gamma-band synchronization appears to be mechanistically relevant to modulate visual cortex activity during visual attention. This modulation has relevant behavioral consequences. Thus, the understanding of the emergence of gamma-band phase synchronization within and across neuronal groups and its putative relation with changes in the gain of local circuits of neurons could provide insights into the mechanisms underlying selective attentional modulation in visual cortex. The following section will focus on these mechanisms that allow neuronal groups to rhythmically synchronize their activity. In that sense, it has been well studied that inhibitory interneuron networks have a prominent role in the generation of gamma oscillations, imposing a transient and recurrent inhibition to the excitatory drive of principal cells (Bartos et al., 2007; Jefferys, Traub, & Whittington, 1996). The next section will describe several models that can reproduce the observed changes in neuronal gain during visual attention, based on the dynamic interaction of inhibitory and excitatory neurons, by inducing small biases of the phase of ongoing inhibitory synchronization that restructures the balance of excitatory and inhibitory network interactions (Borgers et al., 2005; Mishra et al., 2006; Tiesinga et al., 2004; Tiesinga & Sejnowski, 2004).

Mechanisms of Attention

Selective Synchronization Via Inhibitory Networks

Oscillatory synchronization at gamma-band frequencies (30–90 Hz) appears to be a ubiquitous phenomenon; however, there is no single mechanism that accounts for the emergence of neuronal gamma-synchronized oscillations. At least two mechanisms have been suggested: (1) the spiking activity of fast rhythmic bursting (FRB) cells (also called "chattering cells"), which have been found mainly in neocortical regions, and (2) the activity of inhibitory interneurons, whose impact on network rhythms has been largely studies in hippocampal slices. We will briefly refer to both mechanisms.

FRB neurons comprise a unique class of cortical neurons that, during depolarization, intrinsically generate bursts of high-frequency action potentials (350–700 Hz), with interburst frequencies in the gamma-band range (Gray & McCormick, 1996; Nowak, Azouz, Sanchez-Vives, Gray, & McCormick, 2003). This type of cell has been described in visual cortex (Gray & McCormick, 1996), but also in sensory-motor and association cortices (Steriade, Timofeev, Durmuller, & Grenier, 1998). Early work by Gray and McCormick (1996) found stimulus-evoked gamma-frequency burst generated by FRB cells in the cat visual cortex. On the basis of these findings, Gray and McCormick (1996) have postulated that FRB cells acts as pacemakers, imposing a particular LFP modulation in the range of the gamma-band frequency on neighboring pyramidal cells. Anatomically, FRB has been found in intermediate layers of the neocortex (2/3 layer) (Gray & McCormick, 1996), but also in layer 5 or layer 6 (Steriade et al., 1998). Layer 6 receives strong input connections from the thalamus. In addition, FRB cells present profuse axonal projections to different types of neurons, including other FRB cells (Cunningham et al., 2004; Gray & McCormick, 1996; Steriade et al., 1998). Both findings strongly suggest that FRB cells may contact many postsynaptic targets, thus amplifying and distributing inputs from thalamic regions (Cardin, Palmer, & Contreras, 2005). On the basis of these findings, it has been suggested that FRB cells, rather than functioning as pacemakers of gamma-frequency oscillations, could act as a part of a corticothalamic network that on one hand facilitates the generation of corticothalamocortical oscillatory loops (Cardin et al., 2005; Castelo-Branco, Neuenschwander, & Singer, 1998), and also distributes rhythmic gamma-band activity (possibly generated by inhibitory interneuron networks) to the other types of neurons in the neocortex (Cunningham et al., 2004; Takekawa, Aoyagi, & Fukai, 2007).

In addition to FRB-cell-mediated gamma-band oscillation, inhibitory interneurons have been implicated in generating and sustaining robust neuronal oscillatory activity. Inhibitory interneurons are highly heterogeneous, differing in their morphology, expression of molecular markers, anatomical location, and physiological properties (Bartos et al., 2007; Freund & Buzsaki, 1996; McBain & Fisahn, 2001). Interneurons can be broadly classified according to their pattern of spike discharge into fast-spiking (FS) and non-fast-spiking neurons. Basket cells are FS parvalbumin neurons owing to the expression of the calcium binding protein parvalbumin. These cells play an important role in the generation of gamma-band oscillations (Bartos et al., 2007; Galarreta & Hestrin, 2001a; Hajos et al., 2004; Hestrin & Galarreta, 2005; McBain & Fisahn, 2001). FS parvalbumin neurons are abundant in the cortex (approximately 20% of all inhibitory interneurons) (Freund & Buzsaki, 1996). Their particular morphology and extensive dendritic and axonic arborizations reflect and extensive network of mutually connected inhibitory neurons (Buzsaki, Kaila, & Raichle, 2007; McBain & Fisahn, 2001). Importantly, FS neurons target perisomatic region of principal neurons. This inhibition controls the timing of spike discharges of the targeted principal neurons with respect to the rhythm imposed by the inhibitory interneuron network (Buzsaki et al., 2007; Cobb, Buhl, Halasy, Paulsen, & Somogyi, 1995). Similar to what has been proposed for FRB cells, inhibitory interneuron networks could act as pacemakers, imposing a timing signal by successive windows of inhibition over groups of principal neurons which are thereby rhythmically entrained (Bartos et al., 2007; Buzsaki & Chrobak, 1995; Buzsaki et al., 2007; Fries et al., 2007).

Several studies have shown a strong preference of FS neurons to generate action potentials at the ascending phase of LFP oscillations in the gamma-frequency band in hippocampal slices (Hajos et al., 2004; Hasenstaub et al., 2005); a phenomenon also described during in vivo extracellular recordings in behaving animals (Bragin et al., 1995; Csicsvari et al., 2003; Hasenstaub et al., 2005). Inhibitory and excitatory neurons fire at different phases of a gamma cycle. For instance, spikes of inhibitory interneurons are found in the ascending part of a gamma cycle, a few milliseconds before the firing of pyramidal cells (Csicsvari et al., 2003; Hajos et al., 2004; Hasenstaub et al., 2005). This activation pattern during gamma oscillations could reflect the interaction between excitatory and inhibitory networks, to provide an effective framework of communication during information processing (Fries, 2005; Fries et al., 2007).

The mechanisms that contribute to the generation of gamma oscillations have been mainly examined in hippocampal slices in vitro. In this experimental approach, evoked responses could be obtained through electrical or chemical stimulation. In concomitance, different neuronal networks could be isolated using pharmacological antagonists and neuronal activity is obtained using intracellular or adjacent electrodes. Also, artificial networks and simulations are used to compare the results of the experimental setup and to evaluate potential mechanisms.

With use of these techniques, several models of inhibitory–excitatory network interactions have been proposed (Bartos et al., 2007; Borgers & Kopell, 2003, 2005; Chow, White, Ritt, & Kopell, 1998; Kopell & Ermentrout, 2004; Kopell, Ermentrout, Whittington, & Traub, 2000; Netoff et al., 2005; Tiesinga et al., 2004; Tiesinga & Sejnowski, 2004; Traub et al., 2001; Traub, Whittington, Stanford, & Jefferys, 1996; Vida, Bartos, & Jonas, 2006; Wang & Buzsaki, 1996; Whittington, Traub, & Jefferys, 1995; Whittington et al., 2000). Importantly, it has been found that under appropriate stimulations conditions (tetanic stimulation), gamma-band oscillations could be evoked in the absence of excitatory synaptic transmission (Whittington et al., 1995). Furthermore, several agonists of metabotropic and ionotropic receptors could chemically evoke gamma oscillatory activity. Cholinergic agonists such as carbachol could induce gamma oscillations in hippocampal slices (Fisahn, Pike, Buhl, & Paulsen, 1998), mediated by muscarinic receptors. These oscillations could be blocked in the presence of 2,3-dihydroxy-6-nitro-7-sulfamoylbenzo[f]quinoxaline-2,3-dione, an antagonist of glutamatergic α-amino-3-hydroxy-5-methyl-4-isoxazolepropionic acid (AMPA) receptors, and bicuculline, an antagonist of γ-aminobutyric acid (GABA) A receptors (Fisahn et al., 1998). On the other side, kainate, a chemical agonist of glutamate receptors, produces gamma oscillations when it is inoculated on hippocampal slices (Fisahn et al., 2004). In this case, gamma oscillations are preserved in the presence of the glutamatergic antagonist AMPA, but they are completely blocked with the inoculation with bicuculline (Fisahn et al., 2004). These differences could be explained in terms of synaptic connectivity and cellular localization of the stimulated receptors (Fig. 1.3).

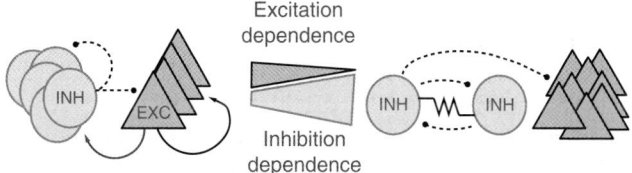

Fig. 1.3 Network architectures of inhibitory–excitatory neuronal population interaction. Several pharmacological experimental studies (see the main text for details) have described neuronal population in which the oscillatory behavior is affected by the blockade of excitatory or inhibitory receptors. On the *left* is a network which is partially dependent on an excitatory neuronal population. *Trapezoids* in the *center* describe the level of dependency. In this case, oscillations evoked by carbachol, an agonist of metabotropic receptors, could be blocked by excitatory and inhibitory receptor agonists. The scheme shows the putative neuronal architecture. In this case, carbachol-induced oscillations require a combination of phasic inhibition and phasic excitation. The *right side* shows a neural network in which oscillations are completely independent of excitatory blockade. In this case, inhibitory networks appear to be connected by electrical and chemical synapses that send forward projections to excitatory neurons. Excitatory neurons do not send their projections onto inhibitory cells. (Adapted from Bartos et al. 2007)

On the other hand, glutamatergic receptors are found mainly in the soma of the inhibitory interneurons (Fisahn et al., 2004; Fuchs et al., 2007). As a consequence, excitatory drive directly impinges on interneurons, triggering gamma-band oscillations. Additionally, muscarinic receptors are found in pyramidal neurons of the hippocampus (Fisahn et al., 2002). In this case, excitatory drive depends on the phasic excitation of pyramidal cells to the interneuron network. Thus, depending on the method of induction, hippocampal gamma oscillations could depend exclusively on inhibitory GABAergic activity, as in the case of stimulation with kainate, or in combination with phasic excitatory modulation, when excitatory drive is provided by the muscarinic agonist carbachol. However, despite the fact that both kinds of network architectures could generate gamma oscillations, the emergence of gamma oscillations critically depends on a highly homogeneous tonic excitatory drive onto that network (Traub et al., 1996; Wang & Buzsaki, 1996). This assumption is realistically under physiological conditions in which excitatory drive to gamma oscillating networks is highly heterogeneous and spikes could arrive at any phase of the gamma-band cycle. More recent models (Bartos et al., 2002; Bartos et al., 2007; Borgers & Kopell, 2008; Buzsaki, Geisler, Henze, & Wang, 2004; Kopell & Ermentrout, 2004; Vida et al., 2006; Vida & Frotscher, 2000) have explored membrane, synaptic, and action potential properties of FS inhibitory neurons that are advantageous for transmission of higher-frequency activity.

The presence of FS responses is based mainly on electrical synaptic transmission and shunting inhibition voltage membrane potentials; in combination with the network architectures previously described these have been considered as the basic mechanisms underlying neuronal oscillations (Bartos et al., 2007). These basic neuronal mechanisms have in common the facilitation of precise spike timing between different neurons of the ensemble, which impose a highly coherent activity even in conditions in which the excitatory drive is asynchronous.

Electrical synapses are prominently present in FS parvalbumin-containing neurons (Galarreta & Hestrin, 2001a). In visual cortex, a dense and widespread intercolumnar network of electrical synapses among dendrites of GABAergic interneurons has been found (Fukuda, Kosaka, Singer, & Galuske, 2006). Electrical coupling promotes fast synapse spiking between neurons (Deans, Gibson, Sellitto, Connors, & Paul, 2001; Galarreta & Hestrin, 2001a; Gibson, Beierlein, & Connors, 2005). Pairs of FS cells connected by both chemical and electrical synapses could act as coincidence detectors of synchronized inputs (Galarreta & Hestrin, 2001b), thus promoting synchronization through the network. Theoretical models of interneuron networks connected by both chemical and electrical synapses have shown that electrical synapses could decrease the effect of general suppression given by a heterogeneous excitatory drive during the initial states of the network, allowing less inhibitory conductance in the propagation of a synchronized gamma rhythm (Kopell & Ermentrout, 2004).

FS cells present an asymmetry between transportation of negative and positive ions through plasmatic membrane that prevents complete hyperpolarization during inhibitory GABAergic stimulation. This process has been called "shunting inhibition" (Bartos et al., 2007; Vida et al., 2006). Shunting inhibition shortens interspike intervals for low levels of drive but prolongs it for high levels, leading to homogenization of neuronal firing rates (Vida et al., 2006). Theoretical models have shown that shunting inhibition increases the gamma oscillation robustness in a network with fast and strong synapses, decreasing the dependence of tonic excitatory drive (Bartos et al., 2007).

In summary, networks of FS parvalbumin-positive interneurons are able to generate robust gamma oscillations with varying degrees of dependence on excitatory drive, depending of network architecture and intrinsic properties of the cell membrane. Gamma-band oscillations depend critically on the neuronal group properties imposing a highly specific timing of discharge to other neurons, to generate windows of effective connectivity across separate groups of neurons. Furthermore, neuron membrane properties of inhibitory interneurons, together with the network architectures described, allow precise excitability windows of pyramidal cells corresponding to a particular phase of the gamma cycle (Csicsvari et al., 2003; Fries et al., 2007; Hasenstaub et al., 2005). Thus, phase modulation of gamma-band oscillations from particular interneuron networks could modify the activity of pyramidal neurons. The consequences of such modulation will be discussed next.

Implementation of Attentional Gain in Attentional Networks

As we previously noted, the architecture and functioning of oscillatory inhibitory networks imposes high-frequency, synchronized synaptic input over excitatory neurons. The impact of such a regime has been studied in computational models that simulate integrate-and-fire excitatory neurons driven by hundreds of synaptic excitatory and inhibitory inputs (Fellous et al., 2003; Salinas & Sejnowski, 2000, 2001; Tiesinga & Sejnowski, 2004). These models have shown that, during balanced excitatory and inhibitory synaptic background, the response of the neurons depends on the fluctuation caused by their synaptic inputs. Thus, correlated inputs increase the variance of the voltage fluctuations on the membrane potential, producing an increase in the gain of balanced neurons, by increasing the output rate of the postsynaptic neurons (Chance, Abbott, & Reyes, 2002; Salinas & Sejnowski, 2000).

It has been postulated that in a regime of high-input correlations, it would be difficult for a neuron to distinguish truly synchronous inputs from spuriously synchronized inputs, i.e., those that occurred by chance (Shadlen & Newsome, 1998). As a consequence, the mechanisms of neuronal synchronization might not be functionally relevant, because they require temporal precision on the order of milliseconds to integrate the coincident firing of presynaptic neurons (Shadlen & Newsome, 1998). The studies of Azouz and Gray (1999, 2000, 2003) have provided convincing evidence that, even in a high-input regime, membrane properties of neurons allow an effective detection of coincident inputs, thus increasing the gain of the postsynaptic response (Azouz & Gray, 2003). In these studies, striate cortex neurons were recorded intracellulary in anesthetized cats, during spontaneous activity and visually evoked responses. The authors correlated the receiving input of a neuron, measuring the membrane potential fluctuation over time (a measure that depends on the presynaptic drive), with the output spike activity of the cell, expressed as the spike rate of the neuron. Their results demonstrated that the spike threshold, the membrane potential necessary to elicit a spike could be dynamically modified by presynaptic inputs. In particular, the spike threshold of a cell exposed to a high-input correlation regime, e.g., induced by an optimal stimulus, shows more variability than the spike threshold of neurons activated with suboptimal stimuli (Azouz & Gray, 2000). Moreover, fluctuations in the membrane potential determined the spiking variability of the neuron, with higher spikes rates being observed in a cell when the membrane potential fluctuates between wide ranges (Azouz & Gray, 1999). These results have been corroborated in theoretical models of high-input activity neurons (Chance et al., 2002; Fellous et al., 2003; Salinas & Sejnowski, 2001).

Thus, mechanisms that dynamically control synchronization of local neuronal populations, basically the interaction between excitatory and inhibitory neurons in local circuits, could also adjust the gain of selected neurons by means of intrinsic membrane properties of the cell, thereby allowing long-distance communication between brain regions. In this context, several models have been proposed to account for enhanced firing rates during selective attentional processing and enhanced neuronal gamma-band coherence (Borgers & Kopell, 2008; Buia & Tiesinga, 2006; Mishra et al., 2006; Salinas & Sejnowski, 2000; Tiesinga, 2005; Tiesinga et al., 2004; Tiesinga & Sejnowski, 2004). These mechanisms underlie the modulation of neuronal responses by voluntary attention.

Common to these models is the interplay of populations of excitatory neurons involved in the processing of different attributes of the stimuli and inhibitory neuronal populations that modulate the activity of the local circuitry. Tiesinga and colleagues (Buia & Tiesinga, 2006; Tiesinga, 2005; Tiesinga et al., 2004; Tiesinga & Sejnowski, 2004) have proposed a model in which attentional top-down effects modulate the synchronization on a local inhibitory neuronal population. When synchronization of an inhibitory network occurred in the gamma band, the presynaptic driving inputs produced an increment in the spike rate of neurons of the excitatory population (Tiesinga, 2005). In addition, neuronal synchronization is able to adjust the sensitivity of a neuron, enhancing the spike rate for weaker inputs, similar to what has been observed in physiological conditions when a stimulus is attended (see the discussion of contrast gain in "Attention Imposes a Gain Pattern on Sensory Neurons") (Reynolds & Chelazzi, 2004; Reynolds et al., 2000). Thus, according to these models, neuronal synchronization and firing rates are complementary processes that are modulated by voluntary attention to increase the gain of a neuron during stimulus processing.

In the particular case of selective attention, an additional question is related to the spatial and temporal selectivity of neuronal circuitry. Neurons in the models previously described require an increase of excitatory synaptic inputs to exhibit attentional modulation. However, this assumption implies a mechanism that allows recruitment of new excitatory synaptic inputs, or which can selectively increase the firing rate of bottom-up groups of neurons activated by the attended stimulus (Buia & Tiesinga, 2006). What are the mechanisms that increase the effective communication of attended information, while rendering less effective those connections that convey irrelevant information? As we previously noted, an increasing number of studies suggest that modulation of the phase of rhythmic synchronization between neurons allows an effective routing of the attended information with respect to the unattended information (Schoffelen et al., 2005; Womelsdorf et al., 2007). Phase modulation has the advantage of aligning temporal windows of excitability, promoting effective communication between neuronal populations (Fries, 2005; Womelsdorf et al., 2007). A recent model proposed by Mishra and colleagues (2006) implemented the modulation of the phase relationship between synchronous excitatory and inhibitory inputs as a critical key mechanism by which selective attention operates in visual cortex. In this model, the number of excitatory inputs (number or rate) remains unchanged, but top-down interactions could promote an attentional bias by means of phase shifts of the oscillatory activity of an inhibitory interneuron population (Fig. 1.4). As a consequence, a phase shift could produce a rapid change in the relative spike timing of these inputs such that spikes from different neurons arrive close together in time and have therefore a greater impact on postsynaptic neurons (Mishra et al., 2006).

The prediction obtained by these models is supported by several studies. As we noted previously, effective communication has been found when two neuronal populations oscillate with a particular phase relationship (Womelsdorf et al., 2007). The activity pattern of excitatory and inhibitory neuronal populations show a different phase preference in a gamma oscillatory cycle (Hasenstaub et al., 2005). Furthermore, a recent study by Mitchell, Sundberg, and Reynolds (2007) performed in awake monkeys found differences in the absolute strength of attentional modulation between inhibitory and excitatory neuronal populations during an attention-demanding task (Fig. 1.5). In the study, differences in the duration of the spike waveforms were used to identify putative interneurons and pyramidal cells. The

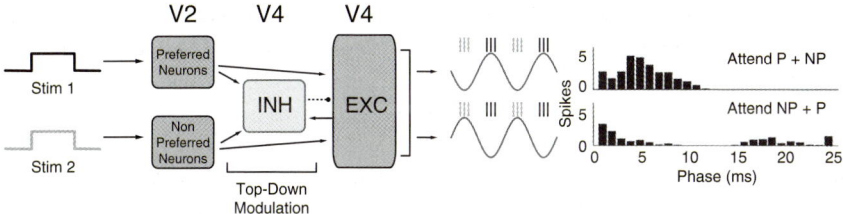

Fig. 1.4 The model proposed by Mishra et al. (2006) to explain phase shift in the inhibitory oscillations as a control mechanism of attention. Two competing stimuli are projecting over V2 neurons. Stimulus 1 represents an attended stimulus with preferred features (major number of synapses). Stimulus 2 represents a nonattended stimulus with nonpreferred features. V2 neurons project over a multicompartmental model in V4 composed of a pool of excitatory and inhibitory neurons with reciprocal connections. Top-down influences act over the inhibitory pool of V4 neurons. Oscillatory activity arising from the inhibitory network provides the windows of excitability necessary to process incoming stimuli. Phase shift produces changes in the excitability windows of the excitatory neurons. The histogram shows the number of spikes as a function of the phase of the oscillation. Stimulus 1 is efficiently processed in a limited range of phases, even in the presence of the competing stimulus. Unattended stimulus is unable to elicit a response. (Adapted from Mishra et al. 2006)

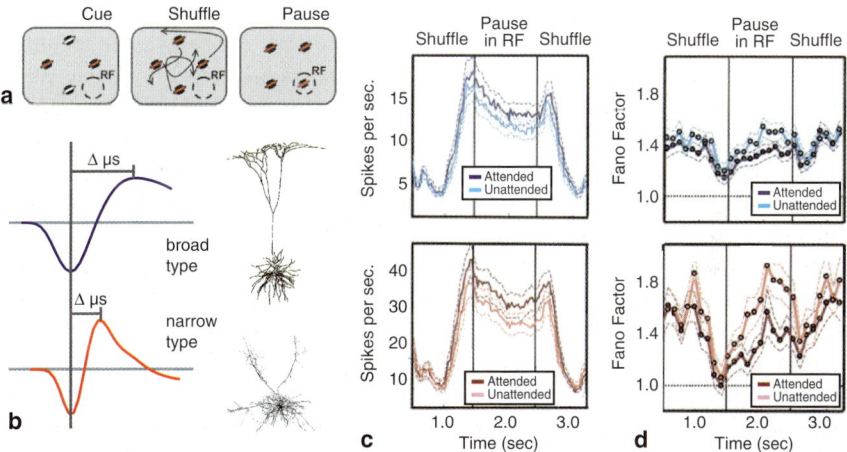

Fig. 1.5 Differential attentional modulation response through different cell classes in macaque visual area V4. (**a**) Stimulation paradigm. Monkeys were trained to maintain the fixation at a central point while covertly tracking two of four stimuli, previously cued by a flash, while they moved along independent randomized trajectories. After the period of motion, a pause period of 1,000 ms positioned one of the stimuli within the receptive field being studied. After that, a new period of random movement began, which ended when the fixation point disappearing and the monkeys had to make a target over attended stimuli. (**b**) Neuronal populations were sorted considering the waveform duration of the spike. The waveform duration was defined as the peak to trough distance measured in microseconds. Broad-type (*blue*) waveforms are putative pyramidal neurons. Narrow-type (*red*) waveforms are probably inhibitory interneurons. (**c**) Average firing rate of broad-spiking (*blue*) and narrow-spiking (*red*) neurons for attended (*dark red, dark blue*) and nonattended (*light red, light blue*) conditions. (**d**) Average Fano factor (variability index) for both populations and conditions (same color codes as for (**c**)). (Adapted from Mitchell et al. 2007)

duration of the extracellular spike waveform is directly related to the duration of the intracellular waveform (Bartho et al., 2004). Narrow-spiking neurons are considered as inhibitory, while broad-spiking waveforms are related to spikes originating from excitatory cells (Bartho et al., 2004; Mitchell et al., 2007). The effects of attention were compared between these two neuronal populations. Inhibitory neurons showed a higher firing rate and a stronger absolute enhancement of firing rate and reduction in variability compared with putatively excitatory neurons, although the proportional change of firing was similar in both types of neurons (Fig. 1.5).

These results support the notion that selective attention could be mediated by the activity of interneurons. Attention could modify the phase relationship between synchronized interneuron networks, adjusting the gain of excitatory postsynaptic neurons.

Summary

Studies from the past two decades have demonstrated that selective attention restructures cortical information processing at various. At the level of single neurons, selective attention can be conceptualized as a gain control. This gain control is likely implemented by local neuronal circuitry. We reviewed evidence that these local networks typically generate rhythmic beta- and gamma-band synchronization, which is highly specific in time and space. Attention is capable of determining which neurons synchronize to a local group processing the attended input and modulating oscillatory coupling across distant groups of neurons. Here, we hypothesized that the observed selective synchronization reflects the key mechanism recruited by attention to impose a selective neuronal communication structure promoting behaviorally relevant information.

We inferred the mechanistic relevance and computational feasibility of neuronal synchronization for selective neuronal communication from theoretical and physiological studies. These studies demonstrate that synchronization is a powerful gain-control mechanism acting at various levels of cortical circuitry: coincident input enhances spiking output, postsynaptic mechanisms actively enhance sensitivity to synchronized input, various cell types promote gamma-band rhythmicity, and perisomatic targeting inhibitory interneurons influences the timing and responsiveness of excitatory neurons. At a larger scale, inhibitory interneuron networks begin to integrate these insights and account for the empirically observed selective attentional effects based on the dynamic interplay of inhibitory and excitatory pools of neurons.

These theoretical models suggest functional architectures of attention, which have only begun to be explored in detail in physiological studies. However, the evidence surveyed suggests that acknowledgement of the basic physiological processes underlying the dynamic generation of selective synchronization will be pivotal to elucidate further the working principles of voluntary attention in the brain.

References

Aboitiz, F., Lopez, J., & Montiel, J. (2003). Long distance communication in the human brain: timing constraints for inter-hemispheric synchrony and the origin of brain lateralization. *Biological Research, 36*(1), 89–99.

Armstrong, K. M., & Moore, T. (2007). Rapid enhancement of visual cortical response discriminability by microstimulation of the frontal eye field. *Proceedings of the National Academy of Sciences of the United States of America, 104*(22), 9499–9504.

Azouz, R., & Gray, C. M. (1999). Cellular mechanisms contributing to response variability of cortical neurons in vivo. *Journal of Neuroscience, 19*(6), 2209–2223.

Azouz, R., & Gray, C. M. (2000). Dynamic spike threshold reveals a mechanism for synaptic coincidence detection in cortical neurons in vivo. *Proceedings of the National Academy of Sciences of the United States of America, 97*(14), 8110–8115.

Azouz, R., & Gray, C. M. (2003). Adaptive coincidence detection and dynamic gain control in visual cortical neurons in vivo. *Neuron, 37*(3), 513–523.

Bartho, P., Hirase, H., Monconduit, L., Zugaro, M., Harris, K. D., & Buzsaki, G. (2004). Characterization of neocortical principal cells and interneurons by network interactions and extracellular features. *Journal of Neurophysiology, 92*(1), 600–608.

Bartos, M., Vida, I., Frotscher, M., Meyer, A., Monyer, H., Geiger, J. R., et al. (2002). Fast synaptic inhibition promotes synchronized gamma oscillations in hippocampal interneuron networks. *Proceedings of the National Academy of Sciences of the United States of America, 99*(20), 13222–13227.

Bartos, M., Vida, I., & Jonas, P. (2007). Synaptic mechanisms of synchronized gamma oscillations in inhibitory interneuron networks. *Nature Reviews Neuroscience, 8*(1), 45–56.

Bichot, N. P., Rossi, A. F., & Desimone, R. (2005). Parallel and serial neural mechanisms for visual search in macaque area V4. *Science, 308*(5721), 529–534.

Borgers, C., Epstein, S., & Kopell, N. J. (2005). Background gamma rhythmicity and attention in cortical local circuits: A computational study. *Proceedings of the National Academy of Sciences of the United States of America, 102*(19), 7002–7007.

Borgers, C., & Kopell, N. (2003). Synchronization in networks of excitatory and inhibitory neurons with sparse, random connectivity. *Neural Computation, 15*(3), 509–538.

Borgers, C., & Kopell, N. (2005). Effects of noisy drive on rhythms in networks of excitatory and inhibitory neurons. *Neural Computation, 17*(3), 557–608.

Borgers, C., & Kopell, N. J. (2008). Gamma oscillations and stimulus selection. *Neural Computation, 20*(2), 383–414.

Bragin, A., Jando, G., Nadasdy, Z., Hetke, J., Wise, K., & Buzsaki, G. (1995). Gamma (40–100 Hz) oscillation in the hippocampus of the behaving rat. *Journal of Neuroscience, 15*(Pt 1), 47–60.

Buia, C., & Tiesinga, P. (2006). Attentional modulation of firing rate and synchrony in a model cortical network. *Journal of Computational Neuroscience, 20*(3), 247–264.

Buschman, T. J., & Miller, E. K. (2007). Top-down versus bottom-up control of attention in the prefrontal and posterior parietal cortices. *Science, 315*(5820), 1860–1862.

Buzsaki, G., & Chrobak, J. J. (1995). Temporal structure in spatially organized neuronal ensembles: A role for interneuronal networks. *Current Opinion in Neurobiology, 5*(4), 504–510.

Buzsaki, G., Geisler, C., Henze, D. A., & Wang, X. J. (2004). Interneuron Diversity series: Circuit complexity and axon wiring economy of cortical interneurons. *Trends in Neuroscience, 27*(4), 186–193.

Buzsaki, G., Kaila, K., & Raichle, M. (2007). Inhibition and brain work. *Neuron, 56*(5), 771–783.

Cardin, J. A., Palmer, L. A., & Contreras, D. (2005). Stimulus-dependent gamma (30–50 Hz) oscillations in simple and complex fast rhythmic bursting cells in primary visual cortex. *Journal of Neuroscience, 25*(22), 5339–5350.

Carrasco, M., Ling, S., & Read, S. (2004). Attention alters appearance. *Nature Neuroscience, 7*(3), 308–313.

Castelo-Branco, M., Neuenschwander, S., & Singer, W. (1998). Synchronization of visual responses between the cortex, lateral geniculate nucleus, and retina in the anesthetized cat. *Journal of Neuroscience, 18*(16), 6395–6410.

Chance, F. S., Abbott, L. F., & Reyes, A. D. (2002). Gain modulation from background synaptic input. *Neuron, 35*(4), 773–782.

Chow, C. C., White, J. A., Ritt, J., & Kopell, N. (1998). Frequency control in synchronized networks of inhibitory neurons. *Journal of Computational Neuroscience, 5*(4), 407–420.

Cobb, S. R., Buhl, E. H., Halasy, K., Paulsen, O., & Somogyi, P. (1995). Synchronization of neuronal activity in hippocampus by individual GABAergic interneurons. *Nature, 378*(6552), 75–78.

Connor, C. E., Preddie, D. C., Gallant, J. L., & Van Essen, D. C. (1997). Spatial attention effects in macaque area V4. *Journal of Neuroscience, 17*(9), 3201–3214.

Csicsvari, J., Henze, D. A., Jamieson, B., Harris, K. D., Sirota, A., Bartho, P., et al. (2003). Massively parallel recording of unit and local field potentials with silicon-based electrodes. *Journal of Neurophysiology, 90*(2), 1314–1323.

Cunningham, M. O., Whittington, M. A., Bibbig, A., Roopun, A., LeBeau, F. E., Vogt, A., et al. (2004). A role for fast rhythmic bursting neurons in cortical gamma oscillations in vitro. *Proceedings of the National Academy of Science of the United States of America, 101*(18), 7152–7157.

Deans, M. R., Gibson, J. R., Sellitto, C., Connors, B. W., & Paul, D. L. (2001). Synchronous activity of inhibitory networks in neocortex requires electrical synapses containing connexin36. *Neuron, 31*(3), 477–485.

Engel, A. K., Fries, P., & Singer, W. (2001). Dynamic predictions: Oscillations and synchrony in top-down processing. *Nature Review Neuroscience, 2*(10), 704–716.

Everling, S., Tinsley, C. J., Gaffan, D., & Duncan, J. (2002). Filtering of neural signals by focused attention in the monkey prefrontal cortex. *Nature Neuroscience, 5*(7), 671–676.

Fellous, J. M., Rudolph, M., Destexhe, A., & Sejnowski, T. J. (2003). Synaptic background noise controls the input/output characteristics of single cells in an in vitro model of in vivo activity. *Neuroscience, 122*(3), 811–829.

Fisahn, A., Contractor, A., Traub, R. D., Buhl, E. H., Heinemann, S. F., & McBain, C. J. (2004). Distinct roles for the kainate receptor subunits GluR5 and GluR6 in kainate-induced hippocampal gamma oscillations. *Journal of Neuroscience, 24*(43), 9658–9668.

Fisahn, A., Pike, F. G., Buhl, E. H., & Paulsen, O. (1998). Cholinergic induction of network oscillations at 40 Hz in the hippocampus in vitro. *Nature, 394*(6689), 186–189.

Fisahn, A., Yamada, M., Duttaroy, A., Gan, J. W., Deng, C. X., McBain, C. J., et al. (2002). Muscarinic induction of hippocampal gamma oscillations requires coupling of the M1 receptor to two mixed cation currents. *Neuron, 33*(4), 615–624.

Freund, T. F., & Buzsaki, G. (1996). Interneurons of the hippocampus. *Hippocampus, 6*(4), 347–470.

Frien, A., Eckhorn, R., Bauer, R., Woelbern, T., & Gabriel, A. (2000). Fast oscillations display sharper orientation tuning than slower components of the same recordings in striate cortex of the awake monkey. *European Journal of Neuroscience, 12*(4), 1453–1465.

Fries, P. (2005). A mechanism for cognitive dynamics: Neuronal communication through neuronal coherence. *Trends in Cognitive Science, 9*(10), 474–480.

Fries, P., Nikolic, D., & Singer, W. (2007). The gamma cycle. *Trends in Neuroscience, 30*(7), 309–316.

Fries, P., Reynolds, J. H., Rorie, A. E., & Desimone, R. (2001). Modulation of oscillatory neuronal synchronization by selective visual attention. *Science, 291*(5508), 1560–1563.

Fuchs, E. C., Zivkovic, A. R., Cunningham, M. O., Middleton, S., Lebeau, F. E., Bannerman, D. M., et al. (2007). Recruitment of parvalbumin-positive interneurons determines hippocampal function and associated behavior. *Neuron, 53*(4), 591–604.

Fukuda, T., Kosaka, T., Singer, W., & Galuske, R. A. (2006). Gap junctions among dendrites of cortical GABAergic neurons establish a dense and widespread intercolumnar network. *Journal of Neuroscience, 26*(13), 3434–3443.

Galarreta, M., & Hestrin, S. (2001a). Electrical synapses between GABA-releasing interneurons. *Nature Reviews Neuroscience, 2*(6), 425–433.

Galarreta, M., & Hestrin, S. (2001b). Spike transmission and synchrony detection in networks of GABAergic interneurons. *Science, 292*(5525), 2295–2299.

Gibson, J. R., Beierlein, M., & Connors, B. W. (2005). Functional properties of electrical synapses between inhibitory interneurons of neocortical layer 4. *Journal of Neurophysiology, 93*(1), 467–480.

Gilbert, C. D., & Sigman, M. (2007). Brain states: Top-down influences in sensory processing. *Neuron, 54*(5), 677–696.

Gottlieb, J. P., Kusunoki, M., & Goldberg, M. E. (1998). The representation of visual salience in monkey parietal cortex. *Nature, 391*(6666), 481–484.

Gray, C. M., & McCormick, D. A. (1996). Chattering cells: Superficial pyramidal neurons contributing to the generation of synchronous oscillations in the visual cortex. *Science, 274*(5284), 109–113.

Gray, C. M., & Singer, W. (1989). Stimulus-specific neuronal oscillations in orientation columns of cat visual cortex. *Proceedings of the National Academy of Science of the United States of America, 86*(5), 1698–1702.

Hajos, N., Palhalmi, J., Mann, E. O., Nemeth, B., Paulsen, O., & Freund, T. F. (2004). Spike timing of distinct types of GABAergic interneuron during hippocampal gamma oscillations in vitro. *Journal of Neuroscience, 24*(41), 9127–9137.

Hasenstaub, A., Shu, Y., Haider, B., Kraushaar, U., Duque, A., & McCormick, D. A. (2005). Inhibitory postsynaptic potentials carry synchronized frequency information in active cortical networks. *Neuron, 47*(3), 423–435.

Henrie, J. A., & Shapley, R. (2005). LFP power spectra in V1 cortex: The graded effect of stimulus contrast. *Journal of Neurophysiology, 94*(1), 479–490.

Hestrin, S., & Galarreta, M. (2005). Electrical synapses define networks of neocortical GABAergic neurons. *Trends in Neuroscience, 28*(6), 304–309.

Hopf, J. M., Luck, S. J., Boelmans, K., Schoenfeld, M. A., Boehler, C. N., Rieger, J., et al. (2006). The neural site of attention matches the spatial scale of perception. *Journal of Neuroscience, 26*(13), 3532–3540.

Ignashchenkova, A., Dicke, P. W., Haarmeier, T., & Thier, P. (2004). Neuron-specific contribution of the superior colliculus to overt and covert shifts of attention. *Nature Neuroscience, 7*(1), 56–64.

Inoue, M., & Mikami, A. (2007). Top-down signal of retrieved information from prefrontal to inferior temporal cortex. *Journal of Neurophysiology, 98*(3), 1965–1974.

Jefferys, J. G., Traub, R. D., & Whittington, M. A. (1996). Neuronal networks for induced '40 Hz' rhythms. *Trends in Neuroscience, 19*(5), 202–208.

Jensen, O., Kaiser, J., & Lachaux, J. P. (2007). Human gamma-frequency oscillations associated with attention and memory. *Trends in Neuroscience, 30*(7), 317–324.

Johnston, K., Levin, H. M., Koval, M. J., & Everling, S. (2007). Top-down control-signal dynamics in anterior cingulate and prefrontal cortex neurons following task switching. *Neuron, 53*(3), 453–462.

Kayser, C., & Konig, P. (2004). Stimulus locking and feature selectivity prevail in complementary frequency ranges of V1 local field potentials. *European Journal of Neuroscience, 19*(2), 485–489.

Khayat, P. S., Spekreijse, H., & Roelfsema, P. R. (2006). Attention lights up new object representations before the old ones fade away. *Journal of Neuroscience, 26*(1), 138–142.

Kim, Y. J., Grabowecky, M., Paller, K. A., Muthu, K., & Suzuki, S. (2007). Attention induces synchronization-based response gain in steady-state visual evoked potentials. *Nature Neuroscience, 10*(1), 117–125.

Kopell, N., & Ermentrout, B. (2004). Chemical and electrical synapses perform complementary roles in the synchronization of interneuronal networks. *Proceedings of the National Academy of Science of the United States of America, 101*(43), 15482–15487.

Kopell, N., Ermentrout, G. B., Whittington, M. A., & Traub, R. D. (2000). Gamma rhythms and beta rhythms have different synchronization properties. *Proceedings of the National Academy of Science of the United States of America, 97*(4), 1867–1872.

Kraskov, A., Quiroga, R. Q., Reddy, L., Fried, I., & Koch, C. (2007). Local field potentials and spikes in the human medial temporal lobe are selective to image category. *Journal of Cognitive Neuroscience, 19*(3), 479–492.

Kreiman, G., Hung, C. P., Kraskov, A., Quiroga, R. Q., Poggio, T., & DiCarlo, J. J. (2006). Object selectivity of local field potentials and spikes in the macaque inferior temporal cortex. *Neuron, 49*(3), 433–445.

Kreiter, A. K., & Singer, W. (1996). Stimulus-dependent synchronization of neuronal responses in the visual cortex of the awake macaque monkey. *Journal of Neuroscience, 16*(7), 2381–2396.

Lee, H., Simpson, G. V., Logothetis, N. K., & Rainer, G. (2005). Phase locking of single neuron activity to theta oscillations during working memory in monkey extrastriate visual cortex. *Neuron, 45*(1), 147–156.

Liu, J., & Newsome, W. T. (2006). Local field potential in cortical area MT: Stimulus tuning and behavioral correlations. *Journal of Neuroscience, 26*(30), 7779–7790.

Martinez Trujillo, J., & Treue, S. (2002). Attentional modulation strength in cortical area MT depends on stimulus contrast. *Neuron, 35*(2), 365–370.

Martinez Trujillo, J., & Treue, S. (2004). Feature-based attention increases the selectivity of population responses in primate visual cortex. *Current Biology, 14*(9), 744–751.

Maunsell, J. H., & Cook, E. P. (2002). The role of attention in visual processing. *Philosophical Transactions of the Royal Society of London – Series B: Biological Sciences, 357*(1424), 1063–1072.

Maunsell, J. H., & Treue, S. (2006). Feature-based attention in visual cortex. *Trends in Neuroscience, 29*(6), 317–322.

McAdams, C. J., & Maunsell, J. H. (1999). Effects of attention on orientation-tuning functions of single neurons in macaque cortical area V4. *Journal of Neuroscience, 19*(1), 431–441.

McBain, C. J., & Fisahn, A. (2001). Interneurons unbound. *Nature Reviews Neuroscience, 2*(1), 11–23.

McMains, S. A., Fehd, H. M., Emmanouil, T. A., & Kastner, S. (2007). Mechanisms of feature and space-based attention: Response modulation and baseline increases. *Journal of Neurophysiology, 98*(4), 2110–2121.

McPeek, R. M., & Keller, E. L. (2004). Deficits in saccade target selection after inactivation of superior colliculus. *Nature Neuroscience, 7*(7), 757–763.

Miller, B. T., & D'Esposito, M. (2005). Searching for "the top" in top-down control. *Neuron, 48*(4), 535–538.

Mishra, J., Fellous, J. M., & Sejnowski, T. J. (2006). Selective attention through phase relationship of excitatory and inhibitory input synchrony in a model cortical neuron. *Neural Networks, 19*(9), 1329–1346.

Mitchell, J. F., Sundberg, K. A., & Reynolds, J. H. (2007). Differential attention-dependent response modulation across cell classes in macaque visual area V4. *Neuron, 55*(1), 131–141.

Muller, J. R., Philiastides, M. G., & Newsome, W. T. (2005). Microstimulation of the superior colliculus focuses attention without moving the eyes. *Proceedings of the National Academy of Science of the United States of America, 102*(3), 524–529.

Netoff, T. I., Banks, M. I., Dorval, A. D., Acker, C. D., Haas, J. S., Kopell, N., et al. (2005). Synchronization in hybrid neuronal networks of the hippocampal formation. *Journal of Neurophysiology, 93*(3), 1197–1208.

Nowak, L. G., Azouz, R., Sanchez-Vives, M. V., Gray, C. M., & McCormick, D. A. (2003). Electrophysiological classes of cat primary visual cortical neurons in vivo as revealed by quantitative analyses. *Journal of Neurophysiology, 89*(3), 1541–1566.

Reynolds, J. H., & Chelazzi, L. (2004). Attentional modulation of visual processing. *Annual Review in Neuroscience, 27,* 611–647.

Reynolds, J. H., Pasternak, T., & Desimone, R. (2000). Attention increases sensitivity of V4 neurons. *Neuron, 26*(3), 703–714.

Roelfsema, P. R., Engel, A. K., König, P., & Singer, W. (1997). Visuomotor integration is associated with zero time-lag synchronization among cortical areas. *Nature, 385*(6612), 157–161.

Saalmann, Y. B., Pigarev, I. N., & Vidyasagar, T. R. (2007). Neural mechanisms of visual attention: How top-down feedback highlights relevant locations. *Science, 316*(5831), 1612–1615.

Salinas, E., & Sejnowski, T. J. (2000). Impact of correlated synaptic input on output firing rate and variability in simple neuronal models. *Journal of Neuroscience, 20*(16), 6193–6209.

Salinas, E., & Sejnowski, T. J. (2001). Correlated neuronal activity and the flow of neural information. *Nature Reviews Neuroscience, 2*(8), 539–550.

Scherberger, H., & Andersen, R. A. (2007). Target selection signals for arm reaching in the posterior parietal cortex. *Journal of Neuroscience, 27*(8), 2001–2012.

Scherberger, H., Jarvis, M. R., & Andersen, R. A. (2005). Cortical local field potential encodes movement intentions in the posterior parietal cortex. *Neuron, 46*(2), 347–354.

Schoenfeld, M., Hopf, J. M., Martinez, A., Mai, H., Sattler, C., Gasde, A., et al. (2007). Spatiotemporal Analysis of Feature-Based Attention. *Cerebral Cortex, 17*(10), 2468–2477.

Schoffelen, J. M., Oostenveld, R., & Fries, P. (2005). Neuronal coherence as a mechanism of effective corticospinal interaction. *Science, 308*(5718), 111–113.

Shadlen, M. N., & Newsome, W. T. (1998). The variable discharge of cortical neurons: Implications for connectivity, computation, and information coding. *Journal of Neuroscience, 18*(10), 3870–3896.

Siegel, M., Donner, T. H., Oostenveld, R., Fries, P., & Engel, A. K. (2007). High-frequency activity in human visual cortex is modulated by visual motion strength. *Cerebral Cortex, 17*(3), 732–741.

Siegel, M., & Konig, P. (2003). A functional gamma-band defined by stimulus-dependent synchronization in area 18 of awake behaving cats. *Journal of Neuroscience, 23*(10), 4251–4260.

Simons, D. J., & Rensink, R. A. (2005). Change blindness: Past, present, and future. *Trends in Cognitive Science, 9*(1), 16–20.

Steriade, M., Timofeev, I., Durmuller, N., & Grenier, F. (1998). Dynamic properties of corticothalamic neurons and local cortical interneurons generating fast rhythmic (30–40 Hz) spike bursts. *Journal of Neurophysiology, 79*(1), 483–490.

Takekawa, T., Aoyagi, T., & Fukai, T. (2007). Synchronous and asynchronous bursting states: Role of intrinsic neural dynamics. *Journal of Computational Neuroscience, 23*(2), 189–200.

Taylor, K., Mandon, S., Freiwald, W. A., & Kreiter, A. K. (2005). Coherent oscillatory activity in monkey area v4 predicts successful allocation of attention. *Cerebral Cortex, 15*(9), 1424–1437.

Thompson, K. G., Biscoe, K. L., & Sato, T. R. (2005). Neuronal basis of covert spatial attention in the frontal eye field. *Journal of Neuroscience, 25*(41), 9479–9487.

Tiesinga, P. H. (2005). Stimulus competition by inhibitory interference. *Neural Computation, 17*(11), 2421–2453.

Tiesinga, P. H., Fellous, J. M., Salinas, E., Jose, J. V., & Sejnowski, T. J. (2004). Inhibitory synchrony as a mechanism for attentional gain modulation. *Journal of Physiology (Paris), 98*(4–6), 296–314.

Tiesinga, P. H., & Sejnowski, T. J. (2004). Rapid temporal modulation of synchrony by competition in cortical interneuron networks. *Neural Computation, 16*(2), 251–275.

Tomita, H., Ohbayashi, M., Nakahara, K., Hasegawa, I., & Miyashita, Y. (1999). Top-down signal from prefrontal cortex in executive control of memory retrieval. *Nature, 401*(6754), 699–703.

Traub, R. D., Kopell, N., Bibbig, A., Buhl, E. H., LeBeau, F. E., & Whittington, M. A. (2001). Gap junctions between interneuron dendrites can enhance synchrony of gamma oscillations in distributed networks. *Journal of Neuroscience, 21*(23), 9478–9486.

Traub, R. D., Whittington, M. A., Stanford, I. M., & Jefferys, J. G. (1996). A mechanism for generation of long-range synchronous fast oscillations in the cortex. *Nature, 383*(6601), 621–624.

Treue, S. (2001). Neural correlates of attention in primate visual cortex. *Trends in Neuroscience, 24*(5), 295–300.

Treue, S., & Martinez Trujillo, J. C. (1999). Feature-based attention influences motion processing gain in macaque visual cortex. *Nature, 399*(6736), 575–579.

Tsodyks, M. V., & Markram, H. (1997). The neural code between neocortical pyramidal neurons depends on neurotransmitter release probability. *Proceedings of the National Academy of Science of the United States of America, 94*(2), 719–723.

Varela, F., Lachaux, J. P., Rodriguez, E., & Martinerie, J. (2001). The brainweb: Phase synchronization and large-scale integration. *Nature Reviews Neuroscience, 2*(4), 229–239.

Vida, I., Bartos, M., & Jonas, P. (2006). Shunting inhibition improves robustness of gamma oscillations in hippocampal interneuron networks by homogenizing firing rates. *Neuron, 49*(1), 107–117.

Vida, I., & Frotscher, M. (2000). A hippocampal interneuron associated with the mossy fiber system. *Proceedings of the National Academy of Science of the United States of America, 97*(3), 1275–1280.

von Stein, A., Chiang, C., & König, P. (2000). Top-down processing mediated by interareal synchronization. *Proceedings of the National Academy of Science of the United States of America, 97*(26), 14748–14753.

Wang, X. J., & Buzsaki, G. (1996). Gamma oscillation by synaptic inhibition in a hippocampal interneuronal network model. *Journal of Neuroscience, 16*(20), 6402–6413.

Wespatat, V., Tennigkeit, F., & Singer, W. (2004). Phase sensitivity of synaptic modifications in oscillating cells of rat visual cortex. *Journal of Neuroscience, 24*(41), 9067–9075.

Whittington, M. A., Traub, R. D., & Jefferys, J. G. (1995). Synchronized oscillations in interneuron networks driven by metabotropic glutamate receptor activation. *Nature, 373*(6515), 612–615.

Whittington, M. A., Traub, R. D., Kopell, N., Ermentrout, B., & Buhl, E. H. (2000). Inhibition-based rhythms: Experimental and mathematical observations on network dynamics. *International Journal of Psychophysiology, 38*(3), 315–336.

Williford, T., & Maunsell, J. H. (2006). Effects of spatial attention on contrast response functions in macaque area V4. *Journal of Neurophysiology, 96*(1), 40–54.

Womelsdorf, T., Anton-Erxleben, K., Pieper, F., & Treue, S. (2006). Dynamic shifts of visual receptive fields in cortical area MT by spatial attention. *Nature Neuroscience, 9*(9), 1156–1160.

Womelsdorf, T., & Fries, P. (2007). The role of neuronal synchronization in selective attention. *Current Opinion in Neurobiology, 17*(2), 154–160.

Womelsdorf, T., Fries, P., Mitra, P. P., & Desimone, R. (2006). Gamma-band synchronization in visual cortex predicts speed of change detection. *Nature, 439*(7077), 733–736.

Womelsdorf, T., Schoffelen, J. M., Oostenveld, R., Singer, W., Desimone, R., Engel, A. K., et al. (2007). Modulation of neuronal interactions through neuronal synchronization. *Science, 316*(5831), 1609–1612.

Wrobel, A., Ghazaryan, A., Bekisz, M., Bogdan, W., & Kaminski, J. (2007). Two streams of attention-dependent beta activity in the striate recipient zone of cat's lateral posterior-pulvinar complex. *Journal of Neuroscience, 27*(9), 2230–2240.

Chapter 2
Intracortical Recordings During Attentional Tasks

J.-P. Lachaux and T. Ossandón

Abstract In the last 10 years, a new field has emerged investigating in humans the functional role of high-frequency neural activity (so-called gamma-band activity, above 40 Hz) with intracerebral electroencephalography (ICE). Although restricted to patients, most of them suffering from intractable epilepsy, ICE has a combined spatiotemporal resolution unmatched by any other human brain imaging technique. In this chapter, we review the contribution of this field to the neuroscience of attention, with particular emphasis on studies investigating attentional modulation of high-level cognitive processes such as visual perception, memory, or language. Attention is shown to amplify gamma-band synchronization and desynchronization processes associated with cognition, both in early sensory areas and in temporal, parietal, and frontal associative areas. We argue that this effect of attention on gamma-band activity provides a sound basis for the development of quantitative indices of attention, and efficient attention-training systems based on noninvasive brain measures.

Abbreviations

DSI   Dynamical spectral imaging
EEG   Electroencephalography
fMRI   Functional magnetic resonance imaging
ICE   Intracerebral electroencephalography
MEG   Magnetoencephalography
PET   Positron emission tomography

J.-P. Lachaux
Centre Hospitalier Le Vinatier (Bât. 452), 95 Bd Pinel, 69500 Bron, France,
e-mail: jp.lachaux@inserm.fr

F. Aboitiz and D. Cosmelli (eds.) *From Attention to Goal-Directed Behavior.*
© Springer-Verlag Berlin Heidelberg 2009

Introduction

Attention has been arguably one of the main topics of interest of cognitive neuroscience for the last 30 years, especially since the recent boom of functional neuroimaging. The progress of positron emission tomography (PET) and functional magnetic resonance imaging (fMRI) has led to a precise delineation of the major neural networks underlying the various types of attention in the human brain (Corbetta & Shulman, 2002; Kanwisher & Wojciulik, 2000). Yet, our understanding of the precise neural mechanisms underlying attention within those networks is still fragmentary, mainly because the same imaging techniques used to describe the functional organization of the human brain, fMRI and PET, measure only metabolic changes induced by neural activity, with low temporal resolution, and not the actual, fast-changing, behavior of the neural networks themselves. Electroencephalography (EEG) and magnetoencephalography (MEG) have been used to record this behavior in humans; however, those techniques measure the average activity of hundreds of millions of neurons and do not possess the spatial precision required to access the fine neural mechanisms of attention. To date, the only data sufficiently precise to reveal those mechanisms have been obtained in animals, with invasive intracranial microelectrodes recording from groups of individual neurons. Recent animal studies have revealed a mechanism for attention based on neural synchronization (Fries, Reynolds, Rorie, & Desimone, 2001; Steinmetz, Roy, Fitzgerald, Hsiao, Johnson et al., 2000). The purpose of this chapter is to introduce this mechanism and discuss its validity in humans, based on results obtained in patients with intracerebral EEG electrodes, at a level of precision intermediate between that from animal microelectrode studies and traditional human noninvasive recordings. We will support the view that this neural mechanism of synchronization may indeed be widely at play during attentional processes in the human brain.

Mechanisms of Selective Visual Attention

Most of what we know about the neural mechanisms underlying attentional selection has been learned from experiments manipulating space-based visual attention in awake behaving monkeys, with microelectrode recordings from the visual cortex (Moran & Desimone, 1985). In such experimental set-ups, experimenters can record the activity of neurons stimulated by visual items presented in their receptive field, and compare this activity when the animal allocates or does not allocate attention to that spatial location. The results converge to show that attention acts as a contrast gain-control mechanism used to increase the firing rate of neurons whose receptive field is being attended, exactly as if the contrast of the attended object had been increased (Reynolds & Chelazzi, 2004). This mechanism is particularly effective when the animal is confronted with two stimuli shown at different locations, in which case the neural response to the attended stimulus is enhanced relative to that of the unattended stimulus, hence the attentional selection. This mechanism has now been well established and observed directly throughout the visual system (reviewed in Reynolds & Chelazzi, 2004). This effect is in fact so strong that it can be measured in humans at the scalp surface with EEG, as visual stimuli generate a stronger evoked potential over the visual cortex when they appear within the scope of visual attention (Mangun, Buonocore, Girelli, & Jha, 1998).

This relatively simple and elegant mechanism seemed perfectly adequate and somewhat sufficient to explain the mechanism of attention, at least in sensory cortices. This situation lasted until two puzzling studies found evidence for a complementary mechanism: Steinmetz et al. (2000) in the somatosensory cortex and Fries et al. (2001) in the visual cortex of awake behaving monkeys reported a phenomenon of synchronization linking together in a transient manner those neurons responding to attended stimuli. As we will see in the forthcoming sections, those results, however surprising at first, had in fact been anticipated. To understand why, we will now briefly review some important research on neural synchronization indicating that this phenomenon may be a cornerstone of neural processing.

The Functional Importance of Neural Synchronization

As in any network, performance in the nervous system relies on efficient coordination between its constitutive elements. The mechanisms underlying the coordination between neurons are certainly multiple and most of them remain unknown. However, evidence accumulated over the last 15 years points towards one strong candidate: a study by Gray and Singer (1989) in the primary visual cortex of anesthetized cats showed that neurons had a tendency to fire rhythmically and synchronously at a frequency around 40 Hz (called gamma band) when stimulated by a single coherent moving object; in contrast, they would fire independently of each other when stimulated by multiple, unrelated, visual stimuli. This mechanism allowed for the creation of transient "resonant" cell assemblies (Varela, 1995), characterized by synchronous gamma-band rhythmic activities, binding together the multiple neural responses to each portion of the contour of complex geometrical shapes.

Because of this property, gamma-band synchronization was first proposed as a solution to the visual "binding problem." The binding problem refers to the problem of how visual objects are "represented" within the visual system (Roskies, 1999). As we know, visual objects are essentially collections of visual features distributed in space (shapes, colors, etc.) which activate selective, specific, but large and widespread populations of neurons throughout the visual system. When, as in most visual scenes, the visual system is confronted with several, possibly overlapping, objects, a visual binding mechanism is necessary to mark neural activities triggered by the same object and enable scene segmentation, object identification, and proper behavior. Because it could be used to create labile neural assemblies, that is, to establish transient relationships between neurons, the synchronization mechanism observed by Gray and Singer was quickly proclaimed as a possible solution to the binding problem (Engel, Roelfsema, Fries, Brecht, & Singer, 1997).

Interestingly, this hypothesis established a strong link between gamma-band synchrony and visual attention, because of previous work by Treisman (1996, 1998), among others, showing the importance of attention for binding. Treisman referred to an effect called "illusory conjunction," which can be described as a failure to bind together the shape and color of an object presented outside the attentional focus: this situation occurs experimentally when participants are instructed to focus their attention on the periphery of a computer screen (while fixating on its center), before colored letters are flashed in the center (say, a red O and a blue X). When asked to report what they have seen, participants often report false associations (or "illusory conjunctions") between shape and color (a blue O and a red X). This phenomenon illustrated the existence of a visual binding mechanism, its occasional failure, and its dependence on attention. The necessity of attention for visual binding was also demonstrated by visual search experiments in which participants were asked to search for a specific item embedded in an array of distracters. For instance, the array would display 36 colored shapes, arranged in six rows and six columns, and the subject would have to find as quickly as possible a prespecified item, the target (say, a blue square). It is well known that if the target differs from the distracters along a simple dimension, such as color, it pops out, meaning that the search is fast and does not depend on the number of distracters. In contrast, if the target is defined as a unique conjunction of particular features (say, find the blue square among blue circles, red squares and red circles), the search time increases linearly with the number of distracters (Treisman, 1982). This indicates that the search requires a serial scanning of all the array, with attentional shifts from item to item. It is therefore commonly accepted that the search for particular conjunctions requires attention; or in other words, that attention is necessary to construct objects from elementary features.

In short, visual binding has been shown to require visual attention. Considering the stream of results initiated by the Singer group, this was enough to build a strong case for a tight connection between attention and gamma-band synchrony, one facilitating the other or vice versa. From a theoretical point of view, neural synchronization also provided an interesting property for attentional selection, in the context of the classic hierarchical model of the visual system. In this model, neurons in early visual areas, such as V1, V2, or V4, project to and activate neurons in higher-order visual areas such as the inferotemporal cortex. In this framework, attention facilitation is thought of as a mechanism which facilitates the propagation of neural activity, from early to late visual areas within neurons responding to the attended stimulus. Synchrony among neurons responding to the attended object at a given level of the hierarchy would produce the strongest possible impact on neurons responding to the same object at the next level, because the latter would receive simultaneous spikes whose effects add up constructively, as already proposed by others (Fries et al., 2001). Reciprocally, the activity of neurons responding to unattended objects would not propagate so effectively across the hierarchy because they are not synchronous.

For all the previous reasons, the notion that gamma-band synchrony and attention were somehow related was "in the air" even before the findings of two studies mentioned above (Fries et al., 2001; Steinmetz et al., 2000) were published. What was still missing was clear evidence at the neural level.

The first study (Steinmetz et al., 2000) measured the effect in the somatosensory cortex of a shift of attention between the somatosensory and visual modalities. Shifting attention towards somatosensory stimuli produced a clear synchrony increase in the gamma band among neurons of the corresponding sensory cortex, while shifting attention towards visual stimuli had, among those neurons, the opposite effect. In the second study, Fries et al. (2001) recorded neurons from V4 in monkeys attending to one of two visual stimuli. The effect of attention was also an increase in gamma-band synchrony among neurons responding to the attended stimulus. In contrast, the level of gamma-band synchrony among neurons selectively activated by the other, unattended stimulus was decreased. Importantly, the firing rate of either group of neurons was unaltered by attentional shifts.

In fact, two electrophysiological studies recording intracortical EEG signals in awake cats had already found much earlier a clear, positive, correlation between the cats' level of attention (as much as it could be interpreted from its behavior) and the level of gamma activity in EEG signals. An increase of EEG power in the gamma band was observed in the parietal and somatosensory cortex when cats observed a mouse (Bouyer, Montaron, & Rougeul, 1981), and even before that, Bauer and Jones (1976) had noticed that cats could be trained to increase the amplitude of gamma oscillations in their visual cortex via a biofeedback device; the result of this training was an seemingly increased attention to their visual environment. Later, studies by the Singer group (Herculano-Houzel, Munk, Neuenschwander, & Singer, 1999) showed that gamma-band synchronization between individual visual neurons was stronger when animals were more vigilant. Further, an experiment by the same group established a strong link between gamma synchronization and visual awareness (Fries, Roelfsema, Engel, Konig, & Singer, 1997). The study was done in amblyopic cats, stimulated with two simultaneous, contradictory, visual stimuli. Because the cats were amblyopic, and because of the experimental design, the cats would perceive at every moment only one of the two stimuli, with a spontaneous alternation between the two. Fries et al. (1997) were able to record from a group of neurons responding selectively to one of the two patterns possibly perceived and found an increase of synchrony within this group when the corresponding pattern was actually perceived by the animals.

Those two studies (Fries et al., 2001; Steinmetz et al., 2000) were in agreement with a very active research stream confirming the generality of the synchronization mechanism in a variety of animal species anesthetized or awake. Their particular merit has been to establish more specifically the link with selective attention, and the fact that gamma synchronization could bias competition between sensory stimuli. Since those two studies, several new experiments in awake behaving monkeys have further supported that same mechanism (Bichot & Desimone, 2006; Bichot, Rossi, & Desimone, 2005; Taylor, Mandon, Freiwald, & Kreiter, 2005), and it has even been shown that high levels of synchrony in the gamma band before the stimulus onset are predictive of faster reaction times when monkeys have to respond as fast as possible to the stimulus presentation (Womelsdorf, Fries, Mitra, & Desimone,

2006). This is especially interesting since a shorter reaction time is one of the psychophysical effects defining a high level of attention at the behavioral level (Fan, McCandliss, Fossella, Flombaum, & Posner, 2005). In summary, the manipulation of synchrony to implement attentional bias seems now like a plausible mechanism to complement the gain-control mechanism based on firing rate modulation (Jensen, Kaiser, & Lachaux, 2007).

Humans Studies on Neural Synchronization

The existence of large-scale oscillations in the human brain was discovered in the late 1920s, as soon as it became possible to record human EEG signals, by Hans Berger (1929). It became soon evident that the electrophysiological activity of the human brain, revealed by EEG, was the sum of slow and fast oscillations, ranging approximately, in the waking state, from the delta band (1–4 Hz), to the theta band (4–7 Hz), the alpha band (7–14 Hz), the beta band (15–30 Hz), and the gamma band (above 30 Hz). Early writers (Adrian, 1941; Berger, 1929) quickly formulated the hypothesis that each frequency band was associated with distinct functional and behavioral correlates, and was the result of synchronization mechanisms between neurons at different spatial scales. But more than 70 years after their discovery, the precise functional role of those oscillations remained elusive, mainly because their precise understanding required the conjunction of mathematical tools and computer power, which has become available only recently.

In the late 1990s, the progressive accumulation of evidence of neural synchronization in animals motivated the search for a similar phenomenon in humans, with the current technology to record electrophysiological activity in people, that is scalp EEG. Considering the large differences in spatial resolution between the recording techniques used in humans and animals, the objective was obviously not to observe neural synchronization directly, but to find modulations of the EEG signal, triggered by cognitive tasks, compatible with the effects likely to occur when synchronization emerges within the large populations of neurons producing EEG signals. As we know, with very few exceptions, human electrophysiological recordings do not record the activity of individual neurons. They rather record the average activity of hundreds of millions of them, usually from the scalp surface with scalp EEG electrodes. Still, there is a suspected relationship between neural synchronization in the gamma range and energy modulations in the same frequency range in EEG signals: when a group of neurons establishes synchronization around 40 Hz, an EEG electrode recording the average activity of that population should detect a transient oscillation at the same frequency. Such oscillations can be detected with the application of time–frequency analysis techniques to EEG signals. These analyses belong to a family of mathematical transforms which can measure the instantaneous variations of the power spectrum of a signal. Since the mid 1990s, a variety of techniques, more or less equivalent, have been used for this purpose, such as the Morlet wavelet decomposition, Wigner transform, short-term Fourier analysis, and matching pursuit.

At the end of the last century, fueled by the progress of computers and mathematical analysis, several groups started to look for gamma oscillations, and for their relation to cognition in humans, using scalp EEG (Gruber, Muller, Keil, & Elbert, 1999; Rodriguez et al., 1999; Tallon-Baudry, Bertrand, Delpuech, & Permier, 1997). Over the last 10 years, human and animal research developed mostly in parallel, to study the mechanisms and functional role of gamma synchronization at two very different spatial scales. Human studies have investigated complex cognitive functions sometimes unique to humans, while animal studies have used much simpler functions, but with incomparably higher spatial resolution.

Gamma-Band Responses in Humans

Since their beginning, the principle of EEG studies on gamma oscillations has remained the same: to evaluate statistically the probability that a cognitive, or motor, event is associated with an increase in gamma-band power in the electroencephalogram. This involves reproducing the same event many times, for instance, the perception of a face in a complex image, and performing statistical analysis on the EEG signals recorded after such events (Tallon-Baudry et al., 1997). Take the example of a study designed to test whether the perception of a face, embedded in a complex visual stimulus, is associated with gamma band energy increase in the electroencephalogram: in such a design, the experimenter shows the subject a series of 100 pictures (or more), and records the EEG signal in response to each image. The pictures are designed in such as way that the face is hard to detect, so that the experimenter can compare the EEG response when a face is actually detected and when it is not. Time–frequency analysis of each of the 100 EEG signals, for each picture presentation, provides a measure of the EEG spectral energy, at each frequency band, from, say, 1–200 Hz, as a function of the time elapsed since the stimulus was shown. For each time and frequency, it is then possible to compare statistically the 100 energy values measured for each stimulus presentation with the 100 energy values measured at the same frequency before stimulus presentation (during what is called the baseline level, that is, in the absence of the cognitive event of interest). It is also possible to compare the response to detected and to nondetected faces. Those statistical comparisons tell the experimenter (1) whether, on average, the cognitive processes which followed stimulus presentation were associated with an increased EEG gamma-band energy relative to the baseline and (2) to which degree such increase depends on the fact that the face is actually detected or not. It then becomes possible to associate specific aspects of the gamma-band response with face detection.

This exact procedure was used in a series of EEG studies starting in 1995 to associate such gamma band energy increases (called "gamma-band responses") with several stages of object perception and memorization (reviewed in Tallon-Baudry & Bertrand, 1999). In one of them (Tallon-Baudry et al., 1997), for instance, participants were presented with apparently random patterns of black and white dots which triggered neither meaningful perception nor significant gamma-band response. However, the pattern of the dots actually hid a Dalmatian dog, which could be clearly seen by the participant after the trick had been explained by the experimenter ("ah, ok, now I see it!"). After this explanation, in a second EEG session presenting the same stimuli, the same images, now interpreted by the subjects as meaningful images, triggered a gamma-band response over parietal and occipital electrodes. The difference between the results obtained in the two EEG sessions was explained by the fact that in the second session subjects had bound together some of the dots of the pictures to create the image of a Dalmatian dog, which is precisely what visual binding is about. This was an elegant demonstration, in humans, that visual binding is associated with increased EEG gamma-band activity.

The same procedure was then used to show EEG gamma band energy increases associated with very diverse cognitive or motor events, including language, memory, and their modulation by attention. Of particular interest for our present topic, several studies showed that EEG gamma-band responses were modulated by attention, and always more strongly for stimuli attended by the subjects (Gruber et al., 1999; Muller, Gruber, & Keil, 2000), in line with animal data on neural synchronization.

Intracerebral EEG studies in humans

In very recent years, starting around 2000, research on gamma-band synchronization in humans was considerably boosted by new studies using high-resolution intracerebral EEG (ICE) recordings in humans. Conventional—noninvasive—EEG, using electrodes placed on the scalp surface, suffers from some limitations owing to weak spatial resolution. In particular, the precise anatomical sources of the effects measured on the scalp surface are unknown, and several distant sources can combine to produce an effect visible at the surface. Also, the neural phenomena observed in scalp EEG in the gamma band are measured at a spatial scale so different from that for neuron-to-neuron synchronization, as observed in animal microrecordings, that it is not clear how, or even whether, the two are related. In addition, EEG signals are often contaminated by electromyographic activity, mainly from neck and face muscles, which produce a strong signal component above 20 Hz. For this reason, the ratio between true gamma-band activity of cerebral origin and this noise is often quite low, which complicated the detection of task-induced gamma-band synchrony in scalp EEG.

Invasive Recordings in Epileptic Patients

For this reason, several groups initiated collaborations with clinical groups specialized in the treatment of severe epilepsy to study brain oscillations in humans with invasive, ICE electrodes. Recordings were performed in patients suffering from drug-resistant epilepsy, where the only efficient therapy is the surgical resection of brain tissues critically involved in the generation of epileptic seizures. This complex procedure requires in most cases a careful identification of the epileptogenic cortical tissues, which can only be performed with ICE electrodes, positioned under the skull or directly inside the cortex. Compared with scalp EEG, ICE recordings are an order of magnitude more precise, with a spatial resolution on the order of 0.5 cm. Therefore, in this range, brain structures generating ICE signals can be identified readily and with no ambiguity. This resolution is in fact comparable to that of fMRI, with the additional millisecond temporal precision of EEG. Also, a very important feature for the study of gamma-band synchrony is that ICE signals are not contaminated by electromyographic activity.

The process of identifying the future site of cortical resection takes about 2 weeks of continuous ICE monitoring, in the patient's hospital room. During this time, patients are free to participate in cognitive tasks designed to understand the functional organization of their brain, and of the brain in general. The information that can be derived from such experiments is used to delineate functionally important brain structures which should be sparred by the surgery. This is also a rare opportunity to observe working human brains with a high spatiotemporal precision. Provided that certain precautions are taken to sort between healthy and pathological brain activity, conclusions can be drawn, across several patients, on neural mechanisms underlying human cognition in general (Lachaux, Rudrauf, & Kahane, 2003).

Dynamic Spectral Imaging

Initial ICE studies immediately produced significant improvement in our understanding of gamma-band synchrony in humans (Aoki, Fetz, Shupe, Lettich, & Ojemann, 1999; Crone, Miglioretti, Gordon, & Lesser, 1998; Lachaux et al., 2000). For instance, it was shown that the sources of the scalp EEG gamma-band responses to visual stimuli are multiple and complex. Their latencies, durations, and strength largely depend on their anatomical origin, even for very simple cognitive tasks, such as the perception of simple geometrical shapes (Lachaux et al., 2000).

The added-value of this technique was well illustrated by a study performed with ICE (Lachaux et al., 2005) after the same experiment had been done with scalp EEG (Rodriguez et al., 1999). The task was a face-detection task, in which participants had to respond whether briefly flashed stimuli contained a face picture or abstract shapes. The scalp EEG study had revealed a large gamma-band response around the 200-ms latency, with a widespread topography over parietocentral electrodes compatible with sources in the visual cortex. The ICE study brought a detailed description of the response networks that had been hypothesized from the scalp EEG analysis: it revealed that face stimuli elicited in fact a mosaic of gamma-band responses along the ventral visual pathway involved in object identification, propagating from early visual areas (around 100 ms) to higher-order, object-identification regions (around 200 ms); this was followed by responses in parietal regions involved in visual attention in the intraparietal sulcus. Moreover, ICE responses were stronger for pictures of faces in brain regions associated with face perception in fMRI, such as the fusiform gyrus, and in the parietal attentional network where the attention-grabbing effect of face stimuli coincided with gamma band responses in the intraparietal sulcus, another instance of the relationship between attention and gamma-band synchrony, in our interpretation.

Within a couple of years, ICE studies revealed task-specific gamma-band responses in a variety of cognitive processes, including not only visual perception, but also auditory (Crone, Boatman, Gordon, & Hao, 2001) and olfactory (Jung et al., 2006) perception, short- and long-term memory (Mainy, Kahane et al., 2007; Sederberg, Kahana, Howard, Donner, & Madsen, 2003; Sederberg et al., 2007), language perception and production (Crone, Hao et al., 2001; Jung et al., 2007; Mainy, Jung et al., 2007), and motor preparation and execution (Aoki et al., 1999; Brovelli, Lachaux, Kahane, & Boussaoud, 2005; Crone et al., 1998; Lachaux, Hoffmann, Minotti, Berthoz, & Kahane, 2006). The observation of gamma-band responses in cognitive tasks is now so common, that, at first glance, it may even seem that such responses are just general reactions of the active brain, with no functional specificity. This view is erroneous though, as the networks of gamma-band responses induced by a cognitive task have always been extremely sensitive to the task, in terms of both anatomical and temporal organization. Moreover, it turns out that the networks of gamma-band responses triggered by cognitive tasks are very similar to those found with fMRI (Lachaux, Fonlupt et al., 2007).

The results obtained so far support the models proposed by Singer (1999) and Fries (2005), which assert that gamma-band synchrony is a general mechanism for local communication between nearby neurons. As such, it would be indispensable for cognition: the recruitment of any local population of neurons, specialized in a precise cognitive process, would necessarily involve the establishment of transient but intense communication between those neurons, based on neural synchronization in the gamma band. According to this view, the detection of gamma-band synchrony in ICE signals would provide an efficient marker of the spatiotemporal dynamics of the neural networks underlying cognition, with a high spatial, frequency, and temporal resolution. For this reason, we called this technique "dynamical spectral imaging" (DSI).

To illustrate the advantages of DSI, let us consider two recent studies from our group. Recently, we asked whether reading was associated with gamma-band synchronization and, if it is, whether it would be possible to associate the various subcomponents of reading, such as graphophonological conversion or semantic analysis, with specific gamma-band responses. We recorded from ten patients in three experimental conditions (Mainy, Jung et al., 2007): a simple orthographic task in which they were shown consonant strings and had to decide whether those strings contained the same letter twice or not, a phonological task in which they had to mentally pronounce visually presented pseudowords, and a semantic task in which they had to decide whether visually presented words were living entities or not. Across the ten patients, we had the opportunity to record from most of the temporoparietofrontal network associated with reading as revealed by previous fMRI studies (Price, 2000). Our study showed that indeed, visual stimuli elicited a widely distributed network of gamma-band responses throughout the reading network, including early visual areas, the word form area in the fusiform gyrus, the superior temporal gyrus, and Broca's area. In addition, the degree to which these brain regions produced gamma-band responses varied across experimental conditions, matching the functional specificity known within this network. For instance, we found a functional dissociation within Broca's area: while the strongest gamma-band response in the most anterior part was associated with semantic processing, that elicited in the posterior part correlated with phonological processing.

In a second study (Mainy, Kahane et al., 2007), we investigated whether verbal working memory was also associated with gamma-band synchronization. Ten patients were recorded during a classic Sternberg task asking them to memorize a series of five letters, presented one by one in succession. We found that each letter in a series induced gamma-band responses in a network comprising both temporal and frontal structures. More interestingly, in some regions, such as the inferior frontal gyrus, the amplitude of those responses increased linearly with the number of letters stored in working memory, with a weaker response for the first letters in a series and stronger responses for the last ones. Those results provided a direct neural implementation of the classic model of phonological loop proposed by Baddeley (2003) to model verbal working memory.

Gamma-Band Synchrony and Attention – Insights from Human ICE Studies

The previous results laid the ground for complementary studies and analysis investigating the role of gamma-band synchrony in attentional processes in humans. What happens to the gamma-band responses observed in humans with ICE, when attention is manipulated? We now have elements to answer that ICE gamma-band responses are very sensitive to attention. Let us illustrate this point with three examples, drawn from the studies just reviewed on visual perception, reading, and verbal working memory:

1. Attention and visual perception. As we know now, the presentation of visual items triggers a chain of gamma-band responses in the visual system; but how does the duration of those responses vary with the time actually spent by the subjects paying attention to the scene and extracting visual information? If gamma-band synchrony is involved in visual attention, we might expect that a longer search would trigger a longer gamma-band response. This can be hypothesized considering a classic result from the psychophysiology literature (Treisman, 1998): when you search for a complex object among many distracters, such as a face in a crowd at a train station, there is no other strategy than to shift visual attention in space from face to face, until you eventually find your target. This is generally the case except if the face has a very distinctive feature, such as a large mustache, for example, in which case the target pops out, as

previously mentioned. In other words, the search for a specific item in an array of distracters is usually serial, with duration proportional to the number of distracters. Visual search is thus an ideal situation to study the effect of search duration on the timing of gamma-band responses. If visual attention requires gamma-band synchrony, we would expect longer gamma-band responses to arrays that require longer searches. This is precisely what we observed in a recent study (Ossandon, Kahane, & Lachaux, 2008): patients were shown arrays of letters containing one *T* and 35 *L*'s. The *T* was always gray, and was the target to be found in each display. *L*'s were either gray or black, so the duration of the search could be manipulated by varying the proportion of *L* with the same color as the target. Behaviorally, the longest search durations were obtained when all distracters were gray (long serial search), and the shortest search duration was when all distracters were black; the target being the only gray letter ("pop-out effect"). We found that the gamma-band responses induced by the arrays in the early visual areas (including V1) and in frontal attentional areas lasted longer for longer searches. Interestingly, this was not true for higher-level visual areas, along the ventral pathways: in the fusiform gyrus, for instance, arrays induced strong gamma-band responses whose duration was independent of the search duration. This clearly indicates that the correlation between the duration of attentional allocation and gamma-band response is anatomically specific: we argue that it is found only in regions actually involved in the extraction of information necessary for the task at hand, in the present case, the dissociation between a *T* and an *L* could be performed at a low level, based simply on the extraction of simple geometric configurations, and not necessarily requiring the participation of the fusiform gyrus, which would be specialized for processing more abstract features, such as faces (Lachaux et al., 2005).

2. Attention and reading. What happens when our attention fluctuates during reading, when we read certain portions of a text attentively, but shift to a forgetful mode for others? We had the opportunity to answer this question in the patients mentioned above, in whom networks of gamma-band responses had been identified for several subcomponents of reading (Jung et al., 2007). These patients volunteered to perform an additional reading task, in which they simply had to look at stories, shown word by word on a computer screen. The critical experimental manipulation of the task was that two stories were intermingled, with the words of the first story written in one color, and the words of the other story written in another color. Patients had to read only one of the two stories, say, the green story, and ignore the other. However, because words were flashed on screen for only 100 ms, and because patients had no way to guess the color of the upcoming word, each word was actually processed visually in the same way, irrespective of its color, at least at the early stages of the visual system. The only difference was that once a word had been identified as part of the attended story, most likely because of its color, or because of its congruence with that story, it would capture further attentional resources and be processed fully, while words of the nontarget story

would not. The analysis of the ICE signals in response to attended words showed that they triggered a chain of gamma-band responses throughout the reading network which was very similar to what had been observed in the semantic condition of the earlier reading task, (as discussed previously, see Mainy, Jung et al., 2007). This was expected in some sense, as comparably to that previous task, words also benefited from the patients' attention. But the most striking result was that gamma-band responses for unattended words were strongly attenuated compared with those for attended words. Furthermore, this attenuation, which was quantified by the ratio between the global energy of the "unattended" gamma-band response divided by the global energy of the attended response was stronger for frontal sites than for posterior sites, as if there was a gradient of attenuation from posterior to anterior brain structures. In some high-level structures such as Broca's area, the gamma-band response for unattended words was close to zero, as if the words had not been presented at all. In contrast, the attenuation was weak in visual regions such as the fusiform gyrus.

3. Attention and memory. Another effect of attention is that it facilitates memorization: it is a fact well known to every student that we memorize items and facts better when we are attentive. The study on verbal working memory summarized earlier, which asked subjects to memorize series of letters, investigated this effect. This was done with an experimental condition asking the subjects to pay attention to certain letters only (Mainy, Kahane et al., 2007). In that condition, each letter in a series was presented after a visual cue, a green or a red dot, informing the patient that the associated letter was important and had to be attended and remembered. We could then compare the gamma-band response to letters tagged as important or unimportant and found an effect similar to the one observed in the attentive reading study: in frontal regions, such as the inferior frontal gyrus, gamma-band responses followed only attended and later-remembered letters. This showed that the positive effect of attention on memorization manifested itself, at the neural level, as an increase in gamma-band synchronization, because only letters which received the full attention of the patient, for the clear purpose of memorizing them, triggered gamma-band responses.

The three studies discussed above converge to show amplification by attention of gamma-band responses associated with cognitive processing. In other words, when a process is done attentively, the gamma-band responses associated with this process are amplified. This was further demonstrated in premotor cortex during a spatial attention task (Brovelli et al., 2005), and in the ventral visual stream during object-based attention (Tallon-Baudry, Bertrand, Henaff, Isnard, & Fischer, 2005). Altogether, the results obtained on attention and gamma-band synchrony in ICE studies have extended into high-level cognitive tasks and associative cortical areas the validity of observations made in animals in simple tasks and lower-level sensory cortices. They fit the general view that attention and gamma-band synchrony go "hand in hand."

What we have learned too is that this relationship is always limited to very specific brain structures, completely determined by the task which is being attended. Our general experience is that an average of 10% of the cortical sites we record from generate gamma-band responses during a given task, a subset of which are modulated by attention. This was clearly illustrated by the visual search experiment in which there was correlation between the duration of the gamma-band responses and the time spent scanning the display attentively was in early visual areas but in which there was no correlation in higher-order areas along the ventral visual pathway.

The future will certainly add complexity to this apparently straightforward relationship. To launch the discussion, let us end this section with a thought-provoking result: in both the visual search experiment and the attentive experiment, we observed negative gamma-band responses to the stimulus presentation (Lachaux et al., 2008); they were found in the ventral lateral prefrontal cortex and had timing similar to the other, positive, gamma-band responses found in other cortical sites. The only difference was that they corresponded to decrease in gamma-band activity, possibly reflecting desynchronization, relative to prestimulus level. In fact, this decrease even brought the gamma-band energy below the level measured at rest, before the experiment started. Further, this phenomenon was strongly amplified by attention. The same effect of desynchronization was found in the lateral prefrontal cortex in the visual search experiment (*T* and *L*) described earlier: the duration of this desynchronization increased with the duration of the search. This suggests that in some brain structures, attention may act not by increasing gamma-band synchrony, but by decreasing it. This does not necessarily contradict the earlier conclusions, since our knowledge of the ventral lateral prefrontal cortex tells us that it could interfere with reading or visual search processes (Lachaux et al., 2008). Therefore, the negative gamma-band response may in fact correspond to the interruption of processes which could interfere with the task at hand, such as constant monitoring for occasional external events with possible general behavioral relevance, as the ventral lateral prefrontal cortex has been suggested to participate in this function (Corbetta & Shulman, 2002). One way to think about this effect is that the relationship between attention and gamma-band synchrony may have more degrees of freedom than previously thought, and that attentional bias may take the form of a balance between regional synchronization and desynchronization phenomena in the gamma band.

Perspectives

Given that the application of ICE to understand the functional role of gamma-band synchrony in attentional processes is relatively new, we will end up this chapter discussing its future, rather than its past. There are several major avenues for future ICE research which can be highlighted.

Real-Time DSI

In all the examples developed so far, the modulations of gamma-band activity triggered by cognitive events were inferred statistically, and presented as significant trends across multiple repetitions of the same cognitive situations. However, the energy of ICE signals in the gamma band can also be measured and displayed in real time, so that the experimenter, or the patient, can view its instantaneous modulations as the patient's brain reacts to its environment and generates imagination, emotion, attentional shifts, and so on. This system, called BrainTV because it enables patients to watch their brain activity in real time, as in live TV shows, has already been implemented successfully in a clinical environment in an effort to map major functional systems onto the patient's brain anatomy (Lachaux, Jerbi et al., 2007). With such a system, patients have helped the clinical staff to map out their brains, and have become actors in the research process. The reason for this success is that patients have direct access to aspects of their mental life, such as attentional or emotional components, which are hidden to external observers, but which strongly condition aspects of their brain activity measured by this system. In the context of attention studies, this provides an opportunity to discover neural activity patterns which strongly correlate with the attention level of an individual, as this individual modulates his/her attention to modify values displayed on the computer screen. We performed this experience in one patient with an electrode recording from the superior parietal lobule. The patient was able to modulate this activity at will by manipulating her visual attention. This kind of set-up will have two important applications in the future: first, to refine our understanding of attentional networks, obviously, but also to define novel ways to educate or reeducate attention, via biofeedback techniques based on precise electrophysiological indices reflecting the activity of nodes in such attentional networks.

Long-Distance Synchrony

In this chapter, we have been essentially dealing with one type of synchronization, called local, which binds together neurons belonging to a single, homogeneous, functional structure such as the primary visual cortex, or the posterior part of Broca's area. However, another type of synchronization, linking together widespread neural activities over the whole brain, has been evidenced in recent years as part of a mechanism of large-scale integration indispensable for cognition. In short, the need for integration mechanisms, which were first described in the visual system to solve the binding problem, is in fact general. Most cognitive acts involve the integration of processes carried out by distinct functional systems, sensory, motor, emotional, or cognitive, most of them anatomically widespread. All such aspects must be integrated to produce coherent experience of the world and adapted behavior.

This constraint has led to the idea that synchrony, as an integration mechanism, may well be used outside the context of "simple" sensory integration, as in visual binding, to achieve this large-scale integration across functional systems (Varela, Lachaux, Rodriguez, & Martinerie, 2001). Shortly after their first observations of synchrony in cat's area 18, Singer and colleagues showed that synchrony could occur between neurons in functionally segregated areas (areas 18 and 19) (Gray, Konig, Engel, & Singer, 1989) or between hemispheres (Konig, Engel, & Singer, 1995), providing the first examples of "nonlocal" synchrony. Since then, several studies have found instances of long-distance synchrony in relation to behavior, for instance, between parietal and motor cortices in cats involved in a sensory motor discrimination task, or in humans during face perception. It must be said, however, that in most instances long-distance synchrony was found at lower frequencies than local gamma-band synchrony—in the alpha and beta range, typically. It remains to be demonstrated, especially with the high anatomical precision of intracranial EEG, that attention facilitates the emergence of large-scale synchronization. We know of no convincing intracranial EEG evidence of large-scale synchrony in the gamma range facilitated by attention; this is the kind of research to watch for carefully in the future.

Other Frequency Bands

As said earlier, large-scale synchrony has often been found in frequency bands lower than the gamma range, between 8 and 30 Hz. For instance, Tallon-Baudry et al. (2001) have found in ICE recordings from the visual system that the maintenance of abstract shapes in working memory, over a couple of seconds, involves the synchronization of the lateral occipital sulcus with the fusiform gyrus around 20 Hz. Similar large-scale synchrony, in the same frequency range, was also found between the ventral and dorsal visual pathways in the ICE study on face perception described earlier (Lachaux et al., 2005). This clearly indicates that gamma-band synchronization is not the whole story. This is also true for local synchrony, as the theta, alpha, and beta bands are known to be modulated by cognitive activity: theta oscillations have been shown to be enhanced in the hippocampus mostly during spatial navigation, and across widespread cortical locations during memory tasks (Caplan, Madsen, Raghavachari, & Kahana, 2001; Kahana, Seelig, & Madsen, 2001). The reactivity of the alpha and beta bands has been repeatedly shown to be anticorrelated with that of the gamma band: in most cognitive situations studied so far, gamma-band energy increases have been simultaneous with alpha- and beta-band energy decreases, suggesting the hypothesis that the three phenomena are in fact "three sides" of the same coin (Lachaux et al., 2005). This possible claim should be qualified by the additional observation that the anatomical specificity of alpha and beta decreases is not as precise as for gamma increases. This was perfectly illustrated in a study by Crone et al. (1998), where they localized the alpha, beta and gamma band energy variations induced in the sensorimotor cortex by movements of various body parts: the homunculus drawn from movement-induced gamma band energy increases matched precisely the classic motor homunculus which characterizes the functional organization of the motor cortex. The distribution of alpha and beta decreases, in contrast, was by far less refined spatially, providing only a blurred version of the motor homunculus. Still, alpha-, beta-, and theta-band activities have the advantage over gamma-band activities of being much easier to detect at the scalp surface because they involve larger neural populations; for this reason, they provide valuable electrophysiological markers of a number of cognitive processes or states, including attentional states, which can be easily measured noninvasively with scalp EEG and MEG (Jokisch & Jensen, 2007).

Scalp Recordings

As precise as they can be, ICE recordings will never replace noninvasive EEG/MEG recordings. Because ICE is limited to patients and a handful of clinical centers, this type of recording is bound to remain a minor path of investigation of the human brain. Electrophysiological research in humans across large and healthy populations will rely on scalp EEG and MEG. Therefore, one can reasonably wonder how such noninvasive techniques can benefit from ICE research. There are several possible answers to this question. The first one is that ICE records neural activity at a level of resolution intermediate between that of microrecordings in animals and EEG/MEG, therefore constituting a missing link between the two extreme levels. Moreover, ICE studies provide the simplest way to associate EEG/MEG signal components, such as gamma band energy increases, with neural mechanisms observed directly at the neural level in animals. The second contribution of ICE is to reveal the sources of the signals recorded with EEG and MEG at the scalp surface. As we know, an entire field of research has developed to design mathematical techniques to enhance the spatial resolution of EEG and MEG (Sereno, 1998). Most of these techniques aim at reconstructing the cortical sources of EEG and MEG signals, and they suffer from the fact that any given set of EEG/MEG signals can be generated by a potentially infinite number of different source configurations. Solutions to the so-called source reconstruction problem require a priori knowledge about the actual sources, and this is exactly what ICE provides.

We believe that the future of attention research in humans, at least in electrophysiology, should involve studies combining ICE and EEG/MEG. Practically, this means EEG/MEG studies in normal subjects, using attentional tasks extensively studied in patients, so that the networks active during those tasks have been previously identified with ICE. EEG/MEG would then be used to validate those results in healthy subjects, refining our understanding of the parameters controlling those networks, with longer versions of the tasks and more control conditions than are acceptable in patients. The combination of ICE and EEG/MEG should be particularly fruitful for the development of attention-training paradigms based on real-time EEG analysis, such as BrainTV (Lachaux, Jerbi et al., 2007). With BrainTV, ICE is already providing electrophysiological indexes, with very well-defined anatomical origin and temporality, which correlate tightly with certain dimensions of attention. EEG/MEG research should focus on the specific traces of such indexes at the scalp surface to provide an ensemble of noninvasive attentional markers which could then be used for training aspects of attention in healthy individuals.

Acknowledgement T.O. acknowledges support from a President of the Republic Scholarship for graduate studies, Chilean Planning and Development Ministry.

References

Adrian, E. D. (1941). Afferent discharges to the cerebral cortex from peripheral sense organs. *Journal of Physiology, 100*(2), 159–191.

Aoki, F., Fetz, E. E., Shupe, L., Lettich, E., & Ojemann, G. A. (1999). Increased gamma-range activity in human sensorimotor cortex during performance of visuomotor tasks. *Clinical Neurophysiology, 110*(3), 524–537.

Baddeley, A. (2003). Working memory: Looking back and looking forward. *Nature Reviews Neuroscience, 4*(10), 829–839.

Bauer, R. H., & Jones, C. N. (1976). Feedback training of 36–44 HZ EEG activity in the visual cortex and hippocampus of cats: Evidence for sensory and motor involvement. *Physiology Behaviour, 17*(6), 885–890.

Berger, H. (1929). Uber das Elektrenkephalogramm des Menschen. *Archiv für Psychiatrie und Nervenkrankheiten,* (87), 527–570.

Bichot, N. P., & Desimone, R. (2006). Finding a face in the crowd: Parallel and serial neural mechanisms of visual selection. *Progress Brain Research, 155,* 147–156.

Bichot, N. P., Rossi, A. F., & Desimone, R. (2005). Parallel and serial neural mechanisms for visual search in macaque area V4. *Science, 308*(5721), 529–534.

Bouyer, J. J., Montaron, M. F., & Rougeul, A. (1981). Fast fronto-parietal rhythms during combined focused attentive behaviour and immobility in cat: Cortical and thalamic localizations. *Electroencephalography and Clinical Neurophysiology, 51*(3), 244–252.

Brovelli, A., Lachaux, J. P., Kahane, P., & Boussaoud, D. (2005). High gamma frequency oscillatory activity dissociates attention from intention in the human premotor cortex. *Neuroimage, 28*(1), 154–164.

Caplan, J. B., Madsen, J. R., Raghavachari, S., & Kahana, M. J. (2001). Distinct patterns of brain oscillations underlie two basic parameters of human maze learning. *Journal of Neurophysiology, 86*(1), 368–380.

Corbetta, M., & Shulman, G. L. (2002). Control of goal-directed and stimulus-driven attention in the brain. *Nature Reviews Neuroscience, 3*(3), 201–215.

Crone, N. E., Boatman, D., Gordon, B., & Hao, L. (2001). Induced electrocorticographic gamma activity during auditory perception. Brazier Award-winning article, 2001. *Clinical Neurophysiology, 112*(4), 565–582.

Crone, N. E., Hao, L., Hart, J., Jr., Boatman, D., Lesser, R. P., Irizarry, R., et al. (2001). Electrocorticographic gamma activity during word production in spoken and sign language. *Neurology, 57*(11), 2045–2053.

Crone, N. E., Miglioretti, D. L., Gordon, B., & Lesser, R. P. (1998). Functional mapping of human sensorimotor cortex with electrocorticographic spectral analysis. II. Event-related synchronization in the gamma band. *Brain, 121*(Pt 12), 2301–2315.

Engel, A. K., Roelfsema, P. R., Fries, P., Brecht, M., & Singer, W. (1997). Role of the temporal domain for response selection and perceptual binding. *Cerebral Cortex, 7*(6), 571–582.

Fan, J., McCandliss, B. D., Fossella, J., Flombaum, J. I., & Posner, M. I. (2005). The activation of attentional networks. *Neuroimage, 26*(2), 471–479.

Fries, P. (2005). A mechanism for cognitive dynamics: Neuronal communication through neuronal coherence. *Trends in Cognitive Science, 9*(10), 474–480.

Fries, P., Reynolds, J. H., Rorie, A. E., & Desimone, R. (2001). Modulation of oscillatory neuronal synchronization by selective visual attention. *Science, 291*(5508), 1560–1563.

Fries, P., Roelfsema, P. R., Engel, A. K., Konig, P., & Singer, W. (1997). Synchronization of oscillatory responses in visual cortex correlates with perception in interocular rivalry. *Proceedings of the National Academy of Sciences of the United States of America, 94*(23), 12699–12704.

Gray, C. M., Konig, P., Engel, A. K., & Singer, W. (1989). Oscillatory responses in cat visual-cortex exhibit inter-columnar synchronization which reflects global stimulus properties. *Nature, 338*(6213), 334–337.

Gray, C. M., & Singer, W. (1989). Stimulus-specific neuronal oscillations in orientation columns of cat visual cortex. *Proceedings of the National Academy of Sciences of the United States of America, 86*(5), 1698–1702.

Gruber, T., Muller, M. M., Keil, A., & Elbert, T. (1999). Selective visual-spatial attention alters induced gamma band responses in the human EEG. *Clinical Neurophysiology, 110*(12), 2074–2085.

Herculano-Houzel, S., Munk, M. H., Neuenschwander, S., & Singer, W. (1999). Precisely synchronized oscillatory firing patterns require electroencephalographic activation. *Journal of Neuroscience, 19*(10), 3992–4010.

Jensen, O., Kaiser, J., & Lachaux, J. P. (2007). Human gamma-frequency oscillations associated with attention and memory. *Trends in Neuroscience, 30*(7), 317–324.

Jokisch, D., & Jensen, O. (2007). Modulation of gamma and alpha activity during a working memory task engaging the dorsal or ventral stream. *Journal of Neuroscience, 27*(12), 3244–3251.

Jung, J., Hudry, J., Ryvlin, P., Royet, J. P., Bertrand, O., & Lachaux, J. P. (2006). Functional significance of olfactory-induced oscillations in the human amygdala. *Cerebral Cortex, 16*(1), 1–8.

Jung, J., Mainy, N., Kahane, P., Minotti, L., Hoffmann, D., Bertrand, O., et al. (2007). The neural basis of attentive reading. *Human Brain Mapping*, doi: 10.1002/hbm.20454.

Kahana, M. J., Seelig, D., & Madsen, J. R. (2001). Theta returns. *Current Opinion in Neurobiology, 11*(6), 739–744.

Kanwisher, N., & Wojciulik, E. (2000). Visual attention: Insights from brain imaging. *Nature Reviews Neuroscience, 1*(2), 91–100.

Konig, P., Engel, A. K., & Singer, W. (1995). Relation between Oscillatory Activity and Long-Range Synchronization in Cat Visual-Cortex. *Proceedings of the National Academy of Sciences of the United States of America, 92*(1), 290–294.

Lachaux, J. P., Fonlupt, P., Kahane, P., Minotti, L., Hoffmann, D., Bertrand, O., et al. (2007). Relationship between task-related gamma oscillations and BOLD signal: New insights from combined fMRI and intracranial EEG. *Human Brain Mapping, 28*(12), 1368–1375.

Lachaux, J. P., George, N., Tallon-Baudry, C., Martinerie, J., Hugueville, L., Minotti, L., et al. (2005). The many faces of the gamma band response to complex visual stimuli. *Neuroimage, 25*(2), 491–501.

Lachaux, J. P., Hoffmann, D., Minotti, L., Berthoz, A., & Kahane, P. (2006). Intracerebral dynamics of saccade generation in the human frontal eye field and supplementary eye field. *Neuroimage, 30*(4), 1302–1312.

Lachaux, J. P., Jerbi, K., Bertrand, O., Minotti, L., Hoffmann, D., Schoendorff, B., et al. (2007). A blueprint for real-time functional mapping via human intracranial recordings. *PLoS ONE, 2*(10), e1094.

Lachaux, J. P., Jung, J., Mainy, N., Dreher, J. C., Bertrand, O., Baciu, M., et al. (2008). Silence is golden: Transient neural deactivation in the prefrontal cortex during attentive reading. *Cerebral Cortex, 18*(2), 443–450.

Lachaux, J. P., Rodriguez, E., Martinerie, J., Adam, C., Hasboun, D., & Varela, F. J. (2000). A quantitative study of gamma-band activity in human intracranial recordings triggered by visual stimuli. *European Journal of Neuroscience, 12*(7), 2608–2622.

Lachaux, J. P., Rudrauf, D., & Kahane, P. (2003). Intracranial EEG and human brain mapping. *Journal of Physiology (Paris), 97*(4–6), 613–628.

Mainy, N., Jung, J., Baciu, M., Kahane, P., Schoendorff, B., Minotti, L., et al. (2007). Cortical dynamics of word recognition. *Human Brain Mapping*, doi: 10.1002/hbm.20457.

Mainy, N., Kahane, P., Minotti, L., Hoffmann, D., Bertrand, O., & Lachaux, J. P. (2007). Neural correlates of consolidation in working memory. *Human Brain Mapping, 28*(3), 183–193.

Mangun, G. R., Buonocore, M. H., Girelli, M., & Jha, A. P. (1998). ERP and fMRI measures of visual spatial selective attention. *Human Brain Mapping, 6*(5–6), 383–389.

Moran, J., & Desimone, R. (1985). Selective attention gates visual processing in the extrastriate cortex. *Science, 229*(4715), 782–784.

Muller, M. M., Gruber, T., & Keil, A. (2000). Modulation of induced gamma band activity in the human EEG by attention and visual information processing. *International Journal of Psychophysiology, 38*(3), 283–299.

Ossandon, T., Kahane, P., & Lachaux, J. P. (2008). Gamma-band activity during visual search in humans. *in preparation*.

Price, C. J. (2000). The anatomy of language: Contributions from functional neuroimaging. *Journal of Anatomy, 197(Pt 3)*, 335–359.

Reynolds, J. H., & Chelazzi, L. (2004). Attentional modulation of visual processing. *Annual Review of Neuroscience, 27*, 611–647.

Rodriguez, E., George, N., Lachaux, J. P., Martinerie, J., Renault, B., & Varela, F. J. (1999). Perception's shadow: Long-distance synchronization of human brain activity. *Nature, 397*(6718), 430–433.

Roskies, A. L. (1999). The binding problem. *Neuron, 24*(1), 7–9, 111–125.

Sederberg, P. B., Kahana, M. J., Howard, M. W., Donner, E. J., & Madsen, J. R. (2003). Theta and gamma oscillations during encoding predict subsequent recall. *Journal of Neuroscience, 23*(34), 10809–10814.

Sederberg, P. B., Schulze-Bonhage, A., Madsen, J. R., Bromfield, E. B., McCarthy, D. C., Brandt, A., et al. (2007). Hippocampal and neocortical gamma oscillations predict memory formation in humans. *Cerebral Cortex, 17*(5), 1190–1196.

Sereno, M. I. (1998). Brain mapping in animals and humans. *Current Opinion in Neurobiology, 8*(2), 188–194.

Singer, W. (1999). Neuronal synchrony: A versatile code for the definition of relations? *Neuron, 24*(1), 49–65, 111–125.

Steinmetz, P. N., Roy, A., Fitzgerald, P. J., Hsiao, S. S., Johnson, K. O., & Niebur, E. (2000). Attention modulates synchronized neuronal firing in primate somatosensory cortex. *Nature, 404*(6774), 187–190.

Tallon-Baudry, C., & Bertrand, O. (1999). Oscillatory gamma activity in humans and its role in object representation. *Trends in Cognitive Science, 3*(4), 151–162.

Tallon-Baudry, C., Bertrand, O., Delpuech, C., & Permier, J. (1997). Oscillatory gamma-band (30–70 Hz) activity induced by a visual search task in humans. *Journal of Neuroscience, 17*(2), 722–734.

Tallon-Baudry, C., Bertrand, O., & Fischer, C. (2001). Oscillatory synchrony between human extrastriate areas during visual short-term memory maintenance. *Journal of Neuroscience, 21*(20), RC177.

Tallon-Baudry, C., Bertrand, O., Henaff, M. A., Isnard, J., & Fischer, C. (2005). Attention modulates gamma-band oscillations differently in the human lateral occipital cortex and fusiform gyrus. *Cerebral Cortex, 15*(5), 654–662.

Taylor, K., Mandon, S., Freiwald, W. A., & Kreiter, A. K. (2005). Coherent oscillatory activity in monkey area v4 predicts successful allocation of attention. *Cerebral Cortex, 15*(9), 1424–1437.

Treisman, A. (1982). Perceptual grouping and attention in visual search for features and for objects. *Journal of Experimental Psychology: Human Perception and Performance, 8*(2), 194–214.

Treisman, A. (1996). The binding problem. *Current Opinion in Neurobiology, 6*(2), 171–178.

Treisman, A. (1998). Feature binding, attention and object perception. *Philosophical Transactions of the Royal Society of London – Series B: Biological Science, 353*(1373), 1295–1306.

Varela, F., Lachaux, J. P., Rodriguez, E., & Martinerie, J. (2001). The brainweb: Phase synchronization and large-scale integration. *Nature Reviews Neuroscience, 2*(4), 229–239.

Varela, F. J. (1995). Resonant cell assemblies: A new approach to cognitive functions and neuronal synchrony. *Biological Research, 28*(1), 81–95.

Womelsdorf, T., Fries, P., Mitra, P. P., & Desimone, R. (2006). Gamma-band synchronization in visual cortex predicts speed of change detection. *Nature, 439*(7077), 733–736.

Chapter 3
The Spatiotemporal Dynamics of Visual-Spatial Attention

S.J. Luck

Abstract The vast majority of research on visual-spatial attention has examined the allocation of attention to locations away from the point of fixation, tacitly assuming that we frequently detect, identify, and act upon objects that we have not fixated. In natural visually guided behavior, however, this does not appear to be how attention operates. Instead, shifts of attention away from fixation are generally used to find regions that are likely to contain useful information and are therefore good targets for eye movements. Once the eyes have been shifted to a location, it is necessary to adjust the size of the attended region around fixation to encompass the fixated object and filter out surrounding objects, a role of attention that has received little empirical investigation. This chapter reviews evidence for the role of attention in finding appropriate saccade targets and discusses the largely unstudied role of attention in shrinking or expanding around the currently fixated object. The ability to dynamically adjust these processes depending on the nature of the current visual input is emphasized.

Introduction

The human visual system is remarkably powerful, capable of rapidly perceiving countless different kinds of objects embedded within complex scenes (Thorpe, Fize, & Marlot, 1996; VanRullen, Reddy, & Fei-Fei, 2005). At the same time, however, the visual system shows striking limits in speed and accuracy, leading to phenomena such as the attentional blink (Shapiro, 1994), change blindness (Simons & Rensink, 2005), and inefficient visual search (Wolfe et al., 1990). One of the most fundamental limits to visual perception is the low visual acuity of the visual system

S.J. Luck
Center for Mind and Brain, University of California, Davis, 267 Cousteau Place, Davis, CA 95618, USA, e-mail: sjluck@ucdavis.edu

F. Aboitiz and D. Cosmelli (eds.) *From Attention to Goal-Directed Behavior.*
© Springer-Verlag Berlin Heidelberg 2009

outside the fovea. Detail vision, and hence object perception, is greatest for objects in the fovea, and a substantial quantity of neural machinery is devoted to the process of ensuring that gaze is directed toward important regions of the visual environment.

The importance of this fundamental limit on visual perception and the complexity of the processes of overcoming this limit are often underappreciated by attention researchers. Indeed, when I teach students about attention, I frequently begin by saying that although foveating an object is a powerful means of attending to the object, it is obvious how this means of attention works, and so there is little point in studying it. Most attention researchers apparently share this view, because the vast majority of research on visual-spatial attention (including most of my own research) either ignores eye movements (as in most visual search studies) or examines the consequences of attending to peripheral locations under conditions that disallow eye movements toward these locations (as in most spatial cuing studies). It is not clear that attention often (or ever) shifts to the periphery without a subsequent eye movement to the attended location; this issue has simply not been studied. Moreover, it is unlikely that people often sustain attention to the periphery for more than a few hundred milliseconds, as they must do in many studies of covert attention. In contrast, people spend a great deal of time attending to the object that they are also foveating. Thus, a great deal of research on visual attention examines a situation that is relatively rare during natural visually guided behavior, and very little research examines the attentional processes that occur during the most common situation, in which attention is directed to the fovea. I would like to make it clear that I have been as guilty of this as anyone. However, I have become increasingly uneasy with this approach to studying attention, and I believe it is time that research on attention seriously considers how attention is used "in the wild." This does not imply that attention researchers should abandon the use of well-designed artificial stimuli and tight control over eye position in favor of free viewing of natural scenes. It is simply a matter of thinking carefully about how vision operates in the natural environment so that we can have a more accurate view of the computational role of attention as we try to understand the algorithms and implementation of attention (to borrow the terms of Marr, 1982). The goal of this chapter, therefore, is to present a framework for understanding attention that takes the importance of foveation in natural vision very seriously.

In addition, because changes in gaze direction are driven by the current visual input but then produce a change in the input, the chapter will also emphasize the dynamics of attention in the context of foveating objects in scenes. Specifically, visual-spatial attention appears to have two main functions that are temporally linked with eye movements: (1) shifts of attention away from the currently foveated location are used to find an object that is likely to be task-relevant, triggering an eye movement to this object; and (2) once the eyes land on this object, attention expands or contracts around the foveated object to minimize interference from surrounding objects. The information available to the visual system after the second of these steps becomes the basis for the next eye movement, and the process continually unfolds over time.

Shifts of Attention Prior to Eye Movements

During normal scene perception, the eyes are largely motionless for periods of a few hundred milliseconds, and these periods of fixation are separated by sudden and brief saccadic eye movements that last tens of milliseconds (reviewed by Henderson, in press). The eyes do not move randomly, but instead tend to shift to regions of the scene that have a high information content for the current task (Torralba, Oliva, Castelhano, & Henderson, 2006). This implies that information from the to-be-fixated location – which may be tens of degrees from the currently fixated location – is used to plan the eye movement. Several studies have shown that attention necessarily shifts to the to-be-fixated location prior to the movement (Deubel & Schneider, 1996; Hoffman & Subramaniam, 1995), which further implies that attention plays a critically important role in using the parafoveal information to guide the eyes to important regions of the scene. But what, exactly, is the function of this shift of covert attention?

Most research on covert attention has assumed that attention is important for detecting, perceiving, or remembering nonfoveated objects. The best-known example of this perspective is Treisman's feature integration theory (Treisman, 1988; Treisman & Gelade, 1980), in which focusing attention onto an object binds together the features of the object and leads to the formation of an *object file*. However, it is not clear why the visual system would bother to form a detailed representation of an object without first fixating it. If the object is important enough to trigger the creation of an object file, then it is presumably important enough to foveate so that it can be perceived with high resolution. Attention may need to be focused on peripheral objects when multiple items must be tracked concurrently; however, this tracking appears to be achieved largely on the basis of the spatiotemporal properties of the objects rather than their identities (reviewed by Scholl, 2001).

Nonetheless, feature integration theory and its relatives may still be useful in understanding the role of attention in eye movements if we simply assume that covert attention is used to find potentially important objects, not so that they can be fully perceived at that moment, but so that they can be foveated and then fully perceived. Schneider (1995) has developed a model of attention – the visual attention model (VAM) – that combines several aspects of feature integration theory with the role of attention in eye movements (and other actions). Just like feature integration theory, VAM posits that attention is used both to localize features and to combine them into object tokens, but VAM further describes how these two functions of attention can be used both for recognizing objects and for controlling eye movements. Forming an object token is obviously a key part of recognizing an object, because the token brings together the object's features. However, VAM proposes that this is accomplished by using spatial information from the dorsal stream to link together the object features that are represented in the ventral stream. As a result, forming an object token also provides the visuomotor system with the location of the object, which is a necessary step in programming an eye movement to the object. Moreover, the representation of object identity in the ventral stream is fed into the frontal eye fields, which may determine whether or not an eye movement should be triggered to the object (Bichot & Schall, 1999; Schall & Hanes, 1993). Thus, by solving certain computational problems that arise in recognizing an object, attention may also solve computational problems that arise in locating and selecting objects for eye movements.

If, however, an object has been identified on the basis of a shift of attention to a peripheral location, why would the visual system bother making an eye movement to the object? There are two likely answers to this question. First, the peripheral shift of attention may provide enough information to indicate that an object is likely to be relevant to the current goals of the system, but it may not be sufficient to provide all of the information that the visual system needs about the object. Consider, for example, the task of finding the best apple in a fruit bowl. A peripheral shift of attention might allow the visual system to determine that a given object is an apple, but foveation of the apple may be necessary to determine whether it is optimally ripe, free from bruises, and lacking in worm holes.

A second factor is that a shift of attention to a peripheral object may allow the visual system to determine that this object is not, in fact, a good candidate for foveation. That is, preattentive information may indicate that relevant low-level features are present within a given region of space, but because feature localization is poor in the absence of attention, the relevant features may actually arise from two different objects, neither of which has all the desired target features. Consequently, peripheral shifts of attention may be useful for determining whether a given object has the correct combination of features. Thus, by shifting attention to a peripheral object, the visual system may be able to avoid making eye movements to objects that are not useful for the current task, and this may dramatically improve the overall efficiency of visually guided behavior. By this account, peripheral shifts of attention are just as useful for avoiding eye movements as for actually making them.

The overall picture implied by this view of attention is as follows. When the visual system is trying to find an object of a particular type (a target), information about the features of the target is used to guide covert attention to objects that contain these features. Wolfe's guided search theory has provided a detailed account of how this guidance of attention operates (Wolfe, 1994; Wolfe, Cave, & Franzel, 1989). According to this theory, low-level feature information is combined – weighted by the task-relevance of the different feature values – into a map of locations that specifies the likelihood that a target is present at each location. Covert attention is then shifted to the location most likely to contain the target. When attention is shifted to this object, an object token is formed that uses spatial information from the dorsal stream to collect the features of this object together in the ventral stream. A decision is then made about whether this object is likely to be a target object; because attention has now been focused on this object, this decision is much more accurate than the initial processing that identified this object as a potential target. If the new information provided by the shift of attention indicates that the object is still likely to be a target, this triggers an eye movement to the target, allowing the visual system to extract additional high-resolution information from the target. A negative decision, in contrast, will lead to a shift of covert attention to the next most likely target location, producing a new object token and yet another decision about whether to make an eye movement. This process iterates until an eye movement is generated or until there are no likely target locations in view.

When a shift of covert attention indicates that an object appears to be the target, the object token becomes available to awareness, is stored in visual short-term memory, and can trigger an arbitrary behavioral response (e.g., a button-press in a typical laboratory task). These events do not require that the observer first makes a saccade to the object; we can be aware of attended objects that we are not foveating (although we may almost always foveate such objects during natural visually guided behavior). But what is the fate of objects that have been covertly attended but have then been rejected as being nontargets before an eye movement has been triggered? Visual search experiments indicate that the object tokens that were formed when attention was directed to these objects fall back into unbound bundles of features when attention is removed from them (Wolfe, Klempen, & Dahlen, 2000) and that they are not stored in visual working memory (Woodman, Vogel, & Luck, 2001). More generally, Duncan (1980) has argued that very few cognitive resources are devoted to objects that are determined to be nontargets.

Although many attention researchers subscribe to the idea that covert attention may make multiple shifts prior to an eye movement, there is little or no direct evidence of such shifts under conditions that permit and encourage eye movements (although see later for evidence that multiple shifts occur when the eyes remain fixed). Indeed, some researchers have proposed that covert attention does not shift rapidly but instead evolves gradually over approximately 500 ms (Duncan, Ward, & Shapiro, 1994; Ward, Duncan, & Shapiro, 1996). It is now clear that this is not correct and that covert attention can shift in as little as 100 ms (reviewed by Woodman & Luck, 2003). However, the attention literature contains a significant gap demonstrating that multiple shifts of covert attention commonly occur prior to an eye movement.

Eye movement researchers have also provided evidence that some eye movements during visual search are not intended to bring attention to the target object, per se, but are instead intended to bring a group of objects closer to fixation so that the visual system can better determine which of them, if any, is a likely target (Najemnik & Geisler, 2005; Zelinksy, Rao, Hayhoe, & Ballard, 1997). Assuming that such eye movements are preceded by a shift of covert attention, these results indicate that covert shifts of attention may be directed to groups of objects rather than to individual objects. In this case, the computational goal of attention – once again – is not to directly identify the attended object, but instead to facilitate an eye movement that will eventually lead to object identification.

Shifts of covert attention prior to eye movements appear to play two additional roles that are directly related to eye movements. First, eye movements are fairly inaccurate, with errors occurring in 30–40% of saccades in typical laboratory tasks, and the errors are followed by fast, automatic, and unconscious corrective saccades (reviewed by Becker, 1991). When an eye movement does not land on an object, multiple objects may be near the currently fixated location, and the visual system needs a means of determining which of these objects was the intended target so that the correct object can be chosen for the corrective saccade. Attention appears to play a key role in this. Specifically, focusing attention onto a peripheral object causes that object to be stored in visual short-term memory (Irwin, 1991; Irwin & Andrews, 1996; Irwin, Zacks, & Brown, 1990), so shifting attention to a saccade target prior to making an eye movement to it will presumably create a short-term memory representation of the intended saccade target. This short-term memory representation could potentially be compared with the objects near the actual saccade landing position to find the original saccade target and make a corrective saccade. Hollingworth, Richard, and Luck (2008) demonstrated that this is exactly what happens when the eye movement fails to land on the saccade target. That is, a fast, automatic, and largely unconscious corrective saccade is made to an item that matches a short-term memory representation of the original saccade target. Given that eye movements occur approximately three times per second during normal scene viewing and that 30–40% of these eye movements fail to land on the intended target, Hollingworth et al. (2008) estimated that these memory-guided gaze corrections may occur as many as 17,000 times per day. This provides a compelling reason to shift attention to the saccade target prior to the saccade.

A second role for the shift of covert attention is that it may provide a reference point that can be used to link the presaccade visual input with the postsaccade input. Vision is suppressed during saccades, and the gradual accumulation of information from a scene therefore requires some means of integrating the memory of the presaccade input with the spatially shifted sensory input that is suddenly visible after the saccade (Irwin, 1993). Because attention shifts to the saccade target and causes the target to be stored in short-term memory, it is possible for the visual system to compare the location of this item before and after the saccade to determine which objects in the presaccade input should be integrated with which objects in the postsaccade input (Currie, McConkie, Carlson-Radvansky, & Irwin, 2000; Henderson, in press).

To summarize, shifts of covert visual-spatial attention to peripheral locations are not typically executed for the purpose of directly identifying objects at those peripheral locations. Instead, these shifts of attention play several roles that interact with eye movements, including (1) providing maximally precise information about the location of the upcoming saccade target; (2) determining whether a given object is likely to contain useful information and is thus worthy of being foveated; (3) facilitating shifts of gaze toward clusters of objects, which will in turn make it possible to determine whether a given cluster contains any objects of interest; (4) creating a short-term memory representation of the saccade target that can be used to make a corrective saccade if the eye movement fails to land on the target object; and (5) creating a short-term memory of the saccade target that can be used in aligning and integrating the presaccade input with the postsaccade input.

The Time Course of Attention Shifts

One common assumption about shifts of covert attention is that they are much more rapid than shifts of fixation. If this were true, it would provide a compelling reason to execute multiple shifts of covert attention before making an eye movement, thus maximizing the probability that a saccade will be directed to a useful object.

However, the evidence for this is not as straightforward as is usually thought. Behavioral studies of covert attention using the spatial cuing paradigm have been used to argue that attention can shift within approximately 50 ms of the onset of a peripheral cue (Lyon, 1990). This is based on the finding that behavioral performance for a target object can be facilitated by the presence of a cue at the target's location that appears 50 ms prior to the onset of the target. However, it is impossible to draw conclusions about the amount of time required to shift attention on the basis of manipulations of cue–target timing. There are several factors that could influence the effects of these manipulations. For example, most peripheral cues are readily detectable by the magnocellular processing pathway, which can very rapidly deliver the cue information to dorsal stream cortical regions. In contrast, the information used to discriminate the target is presumably provided by the parvocellular pathway, which is much slower than the magnocellular pathway. Thus, even though the cue is presented only 50 ms prior to the target, information about the cue may arrive in cortex well over 50 ms prior to the arrival of information about the target. Thus, it is impossible to use cue–target intervals to estimate the time required to shift attention.

The time course of attention can be estimated much more directly from electrophysiological data, which provide a continuous, real-time measure of processing. In highly trained monkeys, visually responsive cells in the frontal eye fields begin to respond differentially depending on the presence or absence of a visual search target approximately 125 ms after the onset of the search array (Schall & Thompson, 1999). In area V4 and the inferotemporal cortex, similar effects have been observed beginning around 150 ms after search array onset (Chelazzi, Duncan, Miller, & Desimone, 1998; Chelazzi, Miller, Duncan, & Desimone, 2001). Saccades to the target are typically generated approximately 100 ms after the onset of these attention effects, although this depends on the nature of the search task (Buschman & Miller, 2007).

Event-related potential (ERP) recordings from human observers have identified an attention-related electrophysiological response called the *N2pc component*, which appears to be closely related to the single-unit attention effects observed in V4 and inferotemporal cortex (Luck, Girelli, McDermott, & Ford, 1997; Luck & Hillyard, 1994b) and appears to be generated in the human homologues of these areas (Hopf et al., 2006). As illustrated in Fig. 3.1, the N2pc component is isolated from the rest of the ERP waveform by computing the difference in voltage between scalp sites contralateral versus ipsilateral to a visual search target. For salient visual search targets, the onset of the N2pc component occurs 150–175 ms after the onset of the search array. When eye movements are made, they again follow N2pc component onset by approximately 100 ms (Luck et al., 1997).

These measurements of the time course of attention make it possible to estimate the amount of time required to shift attention, at least for the first shift following the onset of a stimulus array. An upper bound on the shift time is 125–175 ms, because this is the amount of time that must pass between stimulus onset and the onset of selective neural activity. However, this time period also includes the amount of time required for preattentive feature analysis, leading to an overestimate

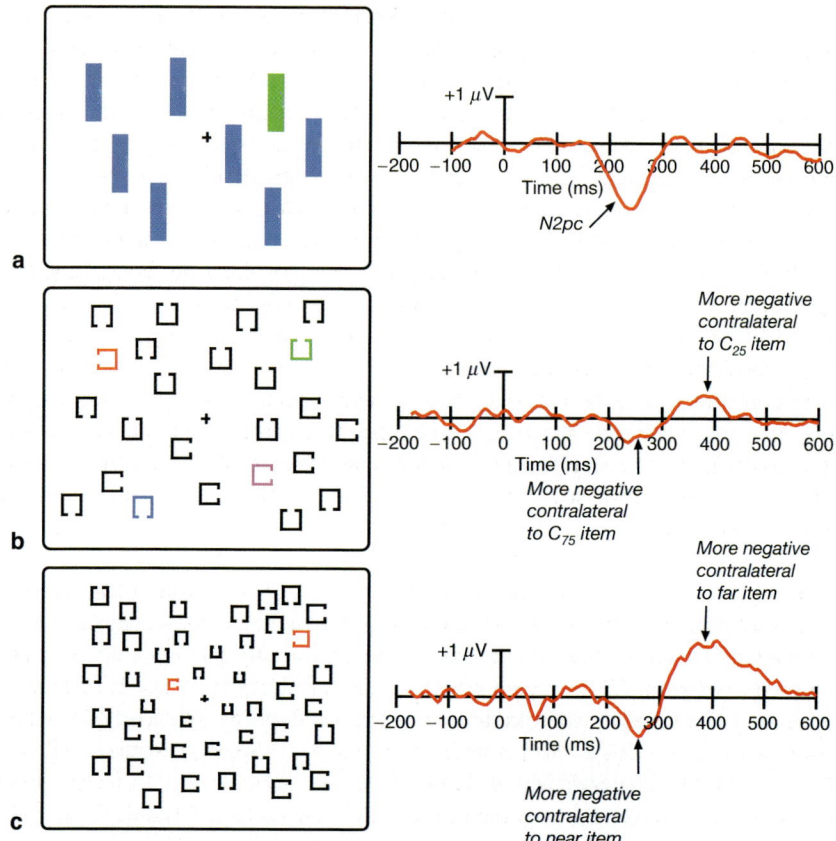

Fig. 3.1 (a) Stimuli and results of an experiment used to isolate the N2pc component (Luck & Hillyard, 1994a). Subjects indicated whether the stimulus array contained a target defined by a specific feature value (*green* in this example). The N2pc component was isolated by subtracting the waveform at scalp sites ipsilateral to the target from the waveform at scalp sites contralateral to the target. (**b**), (**c**) Stimuli and results from two experiments in which the N2pc component was recorded as subjects made multiple shifts of attention

of the time require to shift attention, per se. That is, information about the stimulus array does not reach primary visual cortex until 40–60 ms after the onset of the array (depending on a variety of factors). If we subtract this time period, then it appears that the time required to find the target and shift attention – once the preattentive information has reached the cortex – is no more than 75–125 ms. This is substantially faster than the amount of time required to generate an eye movement, which is on the order of 150–200 ms (again depending on a variety of factors).

All of the measurements of attention shift times described so far were made by examining the first shift of attention following the onset of a stimulus array. As described above, however, many attention researchers assume that covert attention may shift from one object to another until a likely target is found, at which point a saccade will be generated. How quickly can attention shift from item to item under these conditions? Traditionally, the rate of attention shifting has been measured on the basis of reaction time experiments in which the number of objects in the array (the *set size*) varies from trial to trial. Under many conditions, reaction time increases linearly as the set size increases (Wolfe, 1998), which is often interpreted as a result of a serial, self-terminating search process. That is, attention shifts randomly from object to object until the target is found or the array is exhausted. As more objects are added to the array, more objects must be searched to make an accurate target-present or target-absent response. Because the target may be found before the entire array has been searched in target-present trials, but all items must be searched to determine that the target is absent, reaction times in target-absent trials can be more directly related to the amount of time required to search each item. Specifically, the slope of the function relating reaction time to set size in target-absent trials can be used to estimate the amount of time required to search each item. Under conditions that are thought by many investigators to require a serial self-terminating search, the search rate is typically 25–100 ms per item (Wolfe, 1998). However, using these values to estimate the amount of time required to shift attention from one item to the next requires the assumption of a perfect serial, self-terminating search, in which (1) attention shifts to exactly one item at a time until the target is found, (2) all items are searched before a target-absent response is made, and (3) attention never returns to a previously searched item. All of these assumptions are likely to be incorrect under many or most conditions (Chun & Wolfe, 1996; McElree & Carrasco, 1999; Mordkoff, Yantis, & Egeth, 1990; Palmer, Verghese, & Pavel, 2000; Peterson, Kramer, Wang, Irwin, & McCarley, 2001). Thus, behavioral data do not provide an accurate estimate of the time required to make each shift of attention during visual search tasks.

The N2pc component has been used to assess the time course of the first two attention shifts in visual search tasks (Woodman & Luck, 1999, 2003). Because the N2pc component is observed over the hemisphere contralateral to the current focus of attention, it switches between the left and right hemispheres as attention shifts between the right and left hemifields, and vice versa. The time course of these shifts can be used to provide a relatively direct estimate of the time course of attention as it shifts from one object to the next. Unfortunately, ERP recordings are far too noisy to permit these shifts to be observed in individual trials, and averaging multiple trials together requires that the experimenter know which hemifield is attended first. This problem was solved by Woodman and Luck (1999; 2003) in two ways, as illustrated in Fig. 3.1b and c. In one set of experiments, each visual search array contained 20 black objects and four colored objects. One of the two colors was 75% likely to be the target (this was called the C_{75} item), and another was 25% likely to be the target (this was called the C_{25} item); the other two colors were used for counterbalancing purposes and were never targets. These probabilities motivated subjects to search the C_{75} item first, and then shift to the C_{25} item if the C_{75} item was not actually the target. In a second set of experiments, each array contained 40 black items and two red items, and the target was always one of the two red items. One of the red items was close to fixation and the other was far from fixation, but each was equally likely to be the target. Previous research demonstrated that observers have a strong tendency to search near-fixation items first, even when the items are scaled according to the cortical magnification factor (Carrasco, Evert, Change, & Katz, 1995; Wolfe, O'Neill, & Bennett, 1998). Thus, observers were expected to shift attention first to the hemifield of the near red item and then, if this item was not a target, they were expected to shift attention to the hemifield of the far red item.

For both of these sets of experiments, the N2pc component appeared contralateral to the first-attended item approximately 200 ms after the onset of the stimulus array, and then shifted to the hemisphere contralateral to the second-attended item after an additional 100 ms. These results indicate that covert attention does indeed shift serially from item to item during at least some visual search tasks. They also indicate that the time required for the second shift is approximately 100 ms. A fairly difficult discrimination task was used in these experiments, and it is possible that attention can shift more rapidly among objects that can be identified more rapidly. In contrast, eye movements occur every 200–250 ms in visual search tasks (Peterson, Beck, & Vomela, 2007). Thus, the visual system may gain substantial efficiency by using multiple shifts of covert attention prior to making an eye movement.

Adjusting the Scale of Attention After an Eye Movement

Although it is widely known that a shift of covert attention precedes an eye movement, spatial attention plays an important role after the completion of an eye movement that is less widely appreciated. Specifically, the fixated object may cast a large or small image on the retina depending on the size of the object and its distance from the observer, and potentially distracting information surrounding the fixated object will fall at an eccentricity that depends on the scale and density of the scene and the viewing distance. This is illustrated in Fig. 3.2a and b, which shows a bowl of fruit at two viewing distances. If the observer is trying to perceive the grape indicated by the yellow circle, then the surrounding "distractor" grapes are closer to fixation when at the greater viewing distance. To avoid interference from the surrounding objects, the size of the attended region around the fixated object must be adjusted to match the size of the fixated object and to reject the surrounding objects. Under most natural viewing conditions, in which the sizes and distances of objects vary across a given scene, the size of the attended region is presumably adjusted after virtually every eye movement. Thus, this postsaccade scaling of attention presumably occurs tens of thousands of times in a typical day, nearly as often as presaccade shifts of attention. However, this key role of attention has been largely neglected.

Some research has been conducted on two related topics. First, the size of the attended region has been explored in the context of shifts of attention to the periphery. The best known example of this is Eriksen's *zoom lens* hypothesis, which states

Fig. 3.2 (**a**) A fruit bowl viewed from an intermediate distance. (**b**) The same fruit bowl viewed from a shorter distance. (**c**) A hierarchical stimulus in which attention is focused on the central local item (*left*) or on the global form (*right*)

that the resolution of processing within the attended region will vary inversely with the size of the attended region (Eriksen & St. James, 1986; Eriksen & Webb, 1989; Eriksen & Yeh, 1985; McCormick & Jolicoeur, 1991; Müller, Bartelt, Donner, Villringer, & Brandt, 2003). However, this idea has been studied entirely within the context of shifts of attention away from the currently fixated location.

A second related topic is the perception of global versus local information within hierarchical stimulus displays, such as the large *H* formed from small *A*'s shown in Fig. 3.2c (Hopf et al., 2006; Lamb & Robertson, 1990; Navon, 1977; Shulman & Wilson, 1987). Under most natural conditions, an observer would foveate a local object when making a discrimination at the local level, and would foveate somewhere near the centroid of a global object when making a discrimination at the global level. In addition, it would be quite unusual for all of the local elements to be identical in a natural scene. Thus, the focus of attention is presumably narrowed around a single local object when observers make a local discrimination, whereas the focus of attention would be broadened when observers make a global discrimination so that the entire object falls within the attended region. However, the literature on local and global perception has focused on conditions that do not have these constraints of natural visual processing. Thus, rather than focusing on the narrowing and broadening of attention around the point of fixation, this literature has focused on hemispheric asymmetries for global and local information and on differences in sensitivity to high versus low spatial frequencies (reviewed by Ivry & Robertson, 1998).

It seems quite reasonable to suppose that differences in sensitivity to high versus low spatial frequencies accompanies differences in the size of the attended region, especially insofar as differences in the size of the attended region arise because of differences in viewing distance. That is, the spatial frequency composition of the attended region will vary in proportion to the viewing distance (all else being equal), so it makes sense to link the size of the attended region with sensitivity for a given range of spatial frequencies. Moreover, sensitivity to low spatial frequencies necessarily declines as the spatial window of integration becomes smaller, because a given frequency becomes ill-defined when the number of cycles becomes low. Thus, differential sensitivity to low and high spatial frequencies certainly plays a role in global versus local perception. Under the constraints imposed by natural vision, it is only a part of the story, and variations in the size of the attended region are a largely neglected but equally important part of the story.

Thus, although prior research has examined the spatial scale of attention in the periphery and the processing of global and local objects at fixation, very little research has directly examined how attention expands and contracts around the currently fixated object. Consequently, we know almost nothing about this key aspect of visual-spatial attention, and this would be a fertile ground for new research.

Summary and Conclusions

High-resolution vision in humans is limited to the central 2° of the visual field, and this places a highly significant constraint on vision. Humans, along with most other vertebrates, deal with this limitation by making frequent and rapid saccadic eye movements in the course of visually guided behavior. Although these eye movements are often treated as a confounding factor to be eliminated in studies of visual-spatial attention, the dynamic interplay between attention and eye movements may provide the key to understanding the nature of attention. That is, one cannot hope to fully understand how attention operates without first understanding the computational goals of attention, and these goals are closely tied to the operation of the oculomotor system.

Indeed, this chapter argues that the main computational goal of shifting attention toward extrafoveal objects is not to directly facilitate the perception of the attended object, as assumed by most theories of visual attention. Instead, attention makes it possible for sources of information that are likely to be relevant to current task goals to be foveated. That is, although the ultimate goal of shifting attention to the periphery may be to perceive useful information, the proximal goal of these shifts is to detect and localize regions of the input that are likely to contain task-relevant information so that they can be brought into the high-resolution foveal region of the retina and then perceived in detail.

However, the direct facilitation of perception is a direct computational goal of a second variety of visual-spatial attention. Specifically, once an object has been foveated, the spatial scale of attention must be adjusted to reflect the size of the object and the proximity of distracting objects. This presumably has a direct impact on perception. Imagine, for example, that the size of the attended region around fixation is constant. Reading this chapter would be quite difficult unless the page were held at a specific distance – neither too close nor too far – so that the words on the lines above and below the currently fixated line did not fall within the attended region and interfere with perception. The ability to flexibly adjust the size of the attended region to match the scale of the currently fixated object is clearly one of the most significant roles of attention in vision, but almost nothing is known about it. This is certainly a ripe topic for future research.

Acknowledgments Preparation of this chapter was supported by grant MH076226 from the US National Institutes of Health. The author wishes to thank Andrew Hollingworth for many interesting discussions of the oculomotor system and links between attention and eye movements.

References

Becker, W. (1991). Saccades. In R. H. S. Carpenter (Ed.), *Vision and visual dysfunction* (Vol. 8: Eye movements, pp. 93–137). London: MacMillan.

Bichot, N. P., & Schall, J. D. (1999). Saccade target seleciton in macaque during feature and conjunciton visual search. *Journal of Neuroscience, 16*, 81–89.

Buschman, T. J., & Miller, E. K. (2007). Top-down versus bottom-up control of attention in the prefrontal and posterior parietal cortices. *Science, 315*, 1860–1862.

Carrasco, M., Evert, D. L., Change, I., & Katz, S. M. (1995). The eccentricity effect: Target eccentricity affects performance on conjunction searches. *Perception and Psychophysics, 57*, 1241–1261.

Chelazzi, L., Duncan, J., Miller, E. K., & Desimone, R. (1998). Responses of neurons in inferior temporal cortex during memory-guided visual search. *Journal of Neurophysiology, 80*, 2918–2940.

Chelazzi, L., Miller, E. K., Duncan, J., & Desimone, R. (2001). Responses of neurons in macaque area V4 during memory-guided visual search. *Cerebral Cortex, 11*, 761–772.

Chun, M. M., & Wolfe, J. M. (1996). Just say no: How are visual searches terminated when there is no target present? *Cognitive Psychology, 30*, 39–78.

Currie, C., McConkie, G., Carlson-Radvansky, L. A., & Irwin, D. E. (2000). The role of the saccade target object in the perception of a visual stable world. *Perception and Psychophysics, 62*, 673–683.

Deubel, H., & Schneider, W. X. (1996). Saccade target selection and object recognition: Evidence for a common attentional mechanism. *Vision Research, 36*, 1827–1837.

Duncan, J. (1980). The locus of interference in the perception of simultaneous stimuli. *Psychological Review, 87*, 272–300.

Duncan, J., Ward, R., & Shapiro, K. (1994). Direct measurement of attentional dwell time in human vision. *Nature, 369*, 313–315.

Eriksen, C. W., & St. James, J. D. (1986). Visual attention within and around the field of focal attention: A zoom lens model. *Perception and Psychophysics, 40*, 225–240.

Eriksen, C. W., & Webb, J. M. (1989). Shifting of attentional focus within and about a visual display. *Perception and Psychophysics, 45*, 175–183.

Eriksen, C. W., & Yeh, Y. Y. (1985). Allocation of attention in the visual field. *Journal of Experimental Psychology: Human Perception and Performance, 11*, 583–597.

Henderson, J. M. (2008). Eye movements and visual memory. In S. J. Luck & A. Hollingworth (Eds.), *Visual memory* (pp. 87–121). New York: Oxford University Press.

Hoffman, J. E., & Subramaniam, B. (1995). The role of visual attention in saccadic eye movements. *Perception and Psychophysics, 57*, 787–795.

Hollingworth, A., Richard, A. M., & Luck, S. J. (2008). Understanding the function of visual short-term memory: Transsaccadic memory, object correspondence, and gaze correction. *Journal of Experimental Psychology: General, 137*, 163–181.

Hopf, J.-M., Luck, S. J., Boelmans, K., Schoenfeld, M. A., Boehler, N., Rieger, J., et al. (2006). The neural site of attention matches the spatial scale of perception. *Journal of Neuroscience, 26*, 3532–3540.

Irwin, D. E. (1991). Information integration across saccadic eye movements. *Cognitive Psychology, 23*(3), 420–456.

Irwin, D. E. (1993). Perceiving an integrated visual world. In D. E. Meyer & S. Kornblum (Eds.), *Attention and performance 14: Synergies in experimental psychology, artificial intelligence, and cognitive neuroscience* (pp. 121–142). Cambridge, MA: MIT.

Irwin, D. E., & Andrews, R. V. (1996). Integration and accumulation of information across saccadic eye movements. In T. Inui & J. L. McClelland (Eds.), *Attention and performance XVI* (pp. 125–155). Cambridge, MA: MIT.

Irwin, D. E., Zacks, J. L., & Brown, J. S. (1990). Visual memory and the perception of a stable visual environment. *Perception and Psychophysics, 47*, 35–46.

Ivry, R. B., & Robertson, L. C. (1998). *The two sides of perception*. Cambridge, MA: MIT.

Lamb, M. R., & Robertson, L. C. (1990). The effect of visual angle on global and local reaction times depends on the set of visual angles presented. *Perception and Psychophysics, 47*, 489–496.

Luck, S. J., Girelli, M., McDermott, M. T., & Ford, M. A. (1997). Bridging the gap between monkey neurophysiology and human perception: An ambiguity resolution theory of visual selective attention. *Cognitive Psychology, 33*, 64–87.

Luck, S. J., & Hillyard, S. A. (1994a). Electrophysiological correlates of feature analysis during visual search. *Psychophysiology, 31*, 291–308.

Luck, S. J., & Hillyard, S. A. (1994b). Spatial filtering during visual search: Evidence from human electrophysiology. *Journal of Experimental Psychology: Human Perception and Performance, 20*, 1000–1014.

Lyon, D. R. (1990). Large and rapid improvement in form discrimination accuracy following a location precue. *Acta Psychologica, 73*, 69–82.

Marr, D. (1982). Vision: A computational investigation into the human representation and processing of visual information. San Francisco: Freeman.

McCormick, P. A., & Jolicoeur, P. (1991). Predicting the shape of distance functions in curve tracing: Evidence for a zoom lens operator. *Memory and Cognition, 19*(5), 469–486.

McElree, B., & Carrasco, M. (1999). The temporal dynamics of visual search: Evidence for parallel processing in feature and conjunction searches. *Journal of Experimental Psychology: Human Perception and Performance, 25*, 1517–1539.

Mordkoff, J. T., Yantis, S., & Egeth, H. E. (1990). Detecting conjunctions of color and form in parallel. *Perception and Psychophysics, 48*, 157–568.

Müller, N. G., Bartelt, O. A., Donner, T. H., Villringer, A., & Brandt, S. A. (2003). A physiological correlate of the "zoom lens" of visual attention. *Journal of Neuroscience, 34*, 3561–3565.

Najemnik, J., & Geisler, W. S. (2005). Optimal eye movement strategies in visual search. *Nature, 434*, 387–390.

Navon, D. (1977). Forest before trees: The precedence of global features in visual perception. *Cognitive Psychology, 9*, 353–383.

Palmer, J., Verghese, P., & Pavel, M. (2000). The psychophysics of visual search. *Vision Research, 40*, 1227–1268.

Peterson, M. S., Beck, M. R., & Vomela, M. (2007). Visual search is guided by prospective and retrospective memory. *Perception and Psychophysics, 69*, 123–135.

Peterson, M. S., Kramer, A. F., Wang, R. F., Irwin, D. E., & McCarley, J. S. (2001). Visual search has memory. *Psychological Science, 12*, 287–292.

Schall, J. D., & Hanes, D. P. (1993). Neural basis of saccade target selection in frontal eye field during visual search. *Nature, 366*, 467–469.

Schall, J. D., & Thompson, K. G. (1999). Neural selection and control of visually guided eye movements. *Annual Review of Neuroscience, 22*, 241–259.

Schneider, W. X. (1995). VAM: A neuro-cognitive model for visual attention, control of segmentation, object recognition, and space-based motor action. *Visual Cognition, 2*, 331–375.

Scholl, B. J. (2001). Objects and attention: The state of the art. *Cognition, 80*, 1–46.

Shapiro, K. L. (1994). The attentional blink: The brain's "eyeblink." *Current Directions in Psychological Science, 3*(3), 86–89.

Shulman, G. L., & Wilson, J. (1987). Spatial frequency and selective attention to local and global information. *Perception, 16*, 89–101.

Simons, D. J., & Rensink, R. A. (2005). Change blindness: Past, present, and future. *Trends in Cognitive Sciences, 9*, 16–20.

Thorpe, S., Fize, D., & Marlot, C. (1996). Speed of processing in the human visual system. *Nature, 381*, 520–522.

Torralba, A., Oliva, A., Castelhano, M. S., & Henderson, J. M. (2006). Contextual guidance of eye movements and attention in real-world scenes: The role of global features in object search. *Psychological Review, 113*, 766–786.

Treisman, A. (1988). Features and objects: The fourteenth Bartlett memorial lecture. *Quarterly Journal of Experimental Psychology, 40*, 201–237.

Treisman, A. M., & Gelade, G. (1980). A feature-integration theory of attention. *Cognitive Psychology, 12*, 97–136.

VanRullen, R., Reddy, L., & Fei-Fei, L. (2005). Binding is a local problem for natural objects and scenes. *Vision Research, 45*, 3133–3144.

Ward, R., Duncan, J., & Shapiro, K. (1996). The slow time-course of visual attention. *Cognitive Psychology, 30*, 79–109.

Wolfe, J., Klempen, N. L., & Dahlen, K. (2000). Postattentive vision. *Journal of Experimental Psychology: Human Perception and Performance, 26*, 693–716.

Wolfe, J. M. (1994). Guided search 2.0: A revised model of visual search. *Psychonomic Bulletin and Review, 1*, 202–238.

Wolfe, J. M. (1998). What can 1 million trials tell us about visual search? *Psychological Science, 9*, 33–39.

Wolfe, J. M., Cave, K. R., & Franzel, S. L. (1989). Guided search: An alternative to the feature integration model for visual search. *Journal of Experimental Psychology: Human Perception and Performance, 15*, 419–433.

Wolfe, J. M., O'Neill, P., & Bennett, S. C. (1998). Why are there eccentricity effects in visual search? Visual and attentional hypotheses. *Perception and Psychophysics, 60*, 140–156.

Wolfe, J. M., Yu, K. P., Stewart, M. I., Shorter, A. D., Friedman-Hill, S. R., & Cave, K. R. (1990). Limitations on the parallel guidance of visual search: Color X Color and Orientation X Orientation Conjunctions. *Journal of Experimental Psychology: Human Perception and Performance, 16*, 879–892.

Woodman, G. F., & Luck, S. J. (1999). Electrophysiological measurement of rapid shifts of atten-
tion during visual search. *Nature, 400,* 867–869.
Woodman, G. F., & Luck, S. J. (2003). Serial deployment of attention during visual search.
Journal of Experimental Psychology: Human Perception and Performance, 29, 121–138.
Woodman, G. F., Vogel, E. K., & Luck, S. J. (2001). Visual search remains efficient when visual
working memory is full. *Psychological Science, 12,* 219–224.
Zelinksy, G. J., Rao, R. P. N., Hayhoe, M., & Ballard, D. (1997). Eye movements reveal the spatio-
temporal dynamics of visual search. *Psychological Science, 8,* 448–453.

Chapter 4
Attention and Neurodynamical Correlates of Natural Vision

P.E. Maldonado, J.P. Ossandón, and F.J. Flores

Abstract In the last decade, several lines of evidence have demonstrated that most sensory systems and particularly the visual system are intensely subject to dynamical top-down influences that depend on the current behavior of the organism. During natural vision, eye movements and neuronal activity in many visual areas are dependent on attention and goal-directed actions of the organism. Yet, current models of visual perception are mainly based on studies that have examined neuronal activity using very simple stimuli, or restrictive behavioral conditions. Moreover, current receptive field models based on these studies appear to fail when they are tested during experiments that used natural stimuli or complex visual behavior. In this chapter, we discuss new evidence showing that the classical receptive field is an incomplete description of the response of neurons in the visual system, largely because we have overlooked top-down influences in neuronal activity and behavior. We argue that the use of natural stimuli and natural behaviors such as free viewing, by including attention and other top-down mechanisms, can provide new insights into the neurodynamical correlates of visual perception.

Introduction

Let us examine what would happen if for a moment you stopped reading this text and visually explored your surroundings. You would likely move your eyes two to four times a second exploring most of your visual field. During each visual fixation that lasts from 100 to 350 ms, complex but structured images are projected on our retinas, producing changes in the receptor neurons as well as in all the rest of the

P.E. Maldonado
Programa de Fisiología y Biofísica, Facultad de Medicina, Universidad de Chile, Casilla 70005, Santiago, Chile, e-mail: pedro@neuro.med.uchile.cl

F. Aboitiz and D. Cosmelli (eds.) *From Attention to Goal-Directed Behavior.* 67
© Springer-Verlag Berlin Heidelberg 2009

network of neurons that participate in visual perception. When our eyes move, the images also move at great velocity over the cells in the retina, in marked contrast with our visual perception, which appears to be constituted by a uniform, clear, and stable world. Now, let us imagine that you are looking for something such as your pen. Again you will repetitively explore your visual surroundings, but will likely move your eyes more frequently, ignoring many features of the visual scene. What is different between freely exploring a scene and actively searching for an object? What neuronal mechanisms account for the different activity occurring in different visual areas of the brain? Vision is an active process in which both bottom-up and top-down mechanisms such as goal-directed behavior, attention, and perceptual learning are permanently involved. Moreover, as complex as they are, natural images were the actual type of visual stimuli under which visual systems evolved and from which animals extract relevant aspects or events from their environment. While much is know about bottom-up mechanisms of visual function, little is known about aspects of vision related to the visual stimuli to which animals are actually exposed and act upon when engaged in goal-oriented behaviors. In this chapter we briefly review our current understanding of visual perception, which has been primary based on the characterization of the receptive field of neurons in different areas of the visual system. We discuss several lines of evidence suggesting that it is constructive to depart from this classical paradigm because classical receptive fields cannot fully account for the response of neurons in the visual system to complex or natural stimuli. We propose that necessary complementary approaches may be incorporated in our models, mainly methods that include top-down influences as occur during goal-oriented behavior. We advocate that the use of natural stimuli, natural tasks, and the dynamical characterization of neuronal activity may be instrumental in explaining visual perception.

Functional Organization of the Visual System

The human retina possess about six million to seven million cones (concentrated on the central area), and 110 million to 130 million rods. These photoreceptors and several other cell types form a complex circuit in the retina. While we will not explore the complexity of this circuit here, it is worthwhile to note that the retina does not act only as a sensitive surface where the images simply become electric signals based on the brightness of stimuli. In fact, it is in the retina where three different systems, or visual pathways, originate, and which have different functional and anatomical properties. The magnocellular pathway originates in large ganglion cells, which in turn connect to large cells in the lateral geniculate nucleus (LGN), distributed in two ventral layers. From the LGN this pathway reaches the primary visual cortex (V1), where it synapses preferentially with a specific population of cells. V1 components of the magnocellular pathways connect preferably to specific regions of V2, which in turn project especially to area V5 as well as the lateral intraparietal area, the medial superotemporal area, the ventral intraparietal area, and area 7A. All these areas are located toward the dorsoparietal regions of the brain. This magnocellular visual pathway is concerned primarily with transitory stimuli and movement, exhibiting high contrast sensitivity and is achromatic. The parvocellular visual pathway, on the other hand, originates in smaller retinal ganglion cells and they project to small cells in the four dorsal layers of the LGN. From the primary visual cortex, parvocellular afferents project preferably to the interstripes and thin stripe regions of V2, and from there preferably to V4, posterior inferotemporal, central inferotemporal, and anterior inferotemporal areas of the inferotemporal cortex (Ungerleider & Haxby, 1994). The parvocellular pathway is preferably activated by chromatic and stationary stimuli. In the optic nerve of primates, 80% of the fibers are of parvocellular origin, while 10% of the axons contribute to the magnocellular pathway. This large pathway segregation in the visual system has corresponding functional consequences where different aspects of visual perception appear to be primarily linked to this distinctive anatomical arrangement. Thus, the magnocellular system, being characteristically activated by achromatic moving stimuli, is termed the "where" pathway, and the parvocellular system, responsive to chromatic and stationary stimuli, is termed the "what" pathway (Ungerleider & Haxby, 1994). A third system, the koniocellular pathway is a distinct but lesser-known system that account for up to 10% of the cells in the LGN. Is now known that these cells connect primarily with the cytochrome oxidase rich regions of the upper layer of V1, and thus are likely involved in color vision (Hendry & Reid, 2000).

The Classical Receptive Field

Common mechanistic concepts of visual perception depict an image projected on the retina that in turn is transmitted electrically by the visual pathways, to then be displayed on the primary visual cortex, where the image is reconstructed. This image is afterward interpreted by some higher areas in the brain. In this view, the neuronal activity of the retina is functionally transmitted from the eyes to the LGN and then conveyed toward the primary visual cortex. From there, the visual activity is transmitted successively toward higher visual cortices, where eventually they produce the final representation of the objects that are being observed. If we examine the functional characterization described above or a current diagram of the visual system in a neuroscience textbook, we will likely find that the neuronal activity is depicted like a flow chart with one-way arrows originating in the eye, pointing towards the "inside" of the brain. To provide explanations about visual perception, this scheme requires to determine how each of the neurons that participates in this chained neuronal processing responds to different visual stimuli. For this purpose, visual stimuli used in classical experiments usually have simple properties and are presented for a relatively long period of time in very constrained behavioral situations to produce the maximum possible activity and an adequate amount of data. The idea behind this strategy is to identify linear responses that can be summed to infer responses to more complex stimuli. Subsequently, the activity of many neurons is grouped to understand different aspects of the functional organization, such as the receptive field properties, perceptual correlates, or the mechanisms for codification of the stimulus. Thus, the central focus has been to determine the physical properties of the stimulus that provoke changes in the discharge rate of the recorded neurons, modifying physical parameters of the stimulus such as its brightness, contrast, orientation, form, color, etc. However, this approach implicitly includes two strong assumptions. First, it considers as meaningful neuronal events only those with significant increases of firing rates in arbitrarily defined time windows. Second, it assumes that the complexity of natural visual conditions and the neural activity that they generate rely only on the mixing of several physical features. This paradigm has been quite successful because we have been able to determine in which way the cells of the retina, the thalamus, and the cortex respond to a collection of simple stimuli with known properties. Yet, somehow our predictions of how neurons respond to visual stimuli are better in the retina that in the thalamus, and are similarly better in the latter nucleus than in the cerebral cortex. This occurs because as we move away from the initial sensory surface (the retina) the responses of neurons are increasingly variable. Besides the inconstancy of the responses, the stimulus needed to achieve a significant change in the discharge rate of the neurons becomes increasingly complex. In the retina, circles of light and darkness are good stimuli, in the primary visual cortex bars and gratings are better stimuli, but as higher cortical areas are examined, the visual stimulus needs to incorporate other features such as movement or complex geometrical forms to recognize complicated objects such as faces. This occurs because the neurons of upper cerebral areas begin to combine the simplest responses of earlier neurons, increasing the complexity necessary to change their activity, thus intensifying the specificity to different classes of stimuli. This specificity is translated in the proposal that perception is constituted (or codified) by the activity of one or a few neurons. Overall, we have worked with the supposition that perception is a function primarily based on the physical properties of the stimulus, and therefore if we adequately characterize the way neurons respond to these different properties, we will be able to explain how every perception occurs.

Nevertheless, in the last decade new studies have uncovered evidence that challenges this paradigm. The notion of the receptive field as the functional unit with which the brain represents the world seems to be much less appropriate or representative of brain function that it was thought. The receptive field is defined as the area of the sensory surface that when stimulated by a physical stimulus evokes changes in that neuron. A series of recent works (Bringuier, Chavane, Glaeser, & Fregnac, 1999; Jones, Grieve, Wang, & Sillito, 2001; Vinje & Gallant, 2000) have shown that each neuron is susceptible to changing its activity and to being influenced by stimuli located beyond its traditional receptive field (now called classical). Classical receptive fields could also be modulated by motor behaviors (Trotter & Celebrini, 1999) or EEG state (Worgotter et al., 1998). Moreover, this influence is neither symmetrical nor linear. In recent years, other studies have demonstrated that even that our best predictions on the responses of visual neurons to known stimuli such as bars and gratings fail systematically in situations of natural stimulation (Yen,

Baker, & Gray, 2007). All these new properties of the responses of neurons and the expanded properties of their receptive fields are now referred to as the nonclassical receptive field. Consequently, while the concept of receptive field has been useful for the description of the activity of neurons, now it seems too narrow to explain how the activities of neurons configure perception. The types of causes that produce a significant neuronal event reflected in an increase or decrease in firing rate include a large number of factors in addition to physical sensory stimulation. External factors such as local or global stimulation or internal ongoing activity related to top-down processes can strongly influence neuronal function. For a deeper understanding of neuronal function, it appears increasingly necessary to examine the kind of response of neurons when the experimental design involves such factors.

Classical Receptive Field and Attention

Although the classical receptive field model in visual perception research has long relied upon the use of simple, parametric stimuli, such as dots, bars, and gratings (Olshausen & Field, 2005), it can be argued that these simple, synthetic stimuli can be suboptimal in revealing not only a bottom-up mechanism that relies on more complex physical attributes of natural scenes but also top-down mechanisms that can influence all the stages of visual processing. One of these mechanisms, visual attention, has been one of the most studied. Attention is thought to give priority to behaviorally relevant stimuli at the expense of irrelevant distracters. Yet, simple or synthetic stimuli have proved useful to reveal some effects of visual attention in higher areas of the visual system, such as V4. These effects go from attentional topographic activation to potential target stimulus features (Motter, 1994), to increase in gamma-band power in response to attended stimuli (Fries, Reynolds, Rorie, & Desimone, 2001), and prediction of visually triggered behavior by means of response latency and gamma-band synchronization (Womelsdorf, Fries, Mitra, & Desimone, 2006). These effects are stronger in more upstream visual areas such as V4 and V5, but they have also been shown in early cortices such as V1 or even visual thalamus. Other studies have shown that the main differences between attentional modulation of neural activity of V1 and V4 are related to orientation selectivity, which is much broader in V4 than in V1 neurons (Motter, 1998). Owing to the small size of receptive fields in V1, some of the classical paradigms for testing attentional effects has been impossible to test in the same conditions that have been examined in V4 (Luck, Chelazzi, Hillyard, & Desimone, 1997). At the same time, attention studies in V1 compared attending to a stimulus (a properly oriented line) inside the receptive field and attending to a stimulus outside the receptive field, and showed that both in V1 and in V4 there is an increment in neural activity in response to attended stimuli, but there is a need for more distracters in V4 to drive neural response (Motter, 1993). Also, it has been shown that distributed attention increases the response to an oriented line, and this response can be increased even further if one or more flanking, collinear lines are placed near the target but outside the classical receptive field, suggesting a contextual facilitation effect (Gilbert, Ito, Kapadia, & Westheimer, 2000). Taken together, these results demonstrate that attentional effects can be seen both in V1 and in V4 using simple stimuli but at the same time they are strongly modulated by contextual cues, both outside and inside the receptive field. The question now is how the use of natural images could help to understand top-down processes such as attention by adding contextual cues and ecological, behaviorally relevant stimuli.

Neuronal Responses to Natural Stimuli

While synthetic stimuli have been very successful in revealing linear behavior in several regions of the early visual system, from the retina to V1, nonlinearities and nonclassical effects appear to play an important role in these structures, so natural images could help to expand our understanding of neuronal responses (Felsen & Dan, 2005) and to reveal top-down mechanisms that are not fully explained by current models. In fact, natural images should not only be useful for revealing neuronal mechanisms of visual perception, but are clearly the type of visual experience we want to explain. While we may learn a great deal about how the visual system responds to bars and gratings, ultimately the goal of visual neuroscience is to explain how vision occurs in natural conditions. The use of natural images for studying the visual system was proposed 50 years ago by H. Barlow and J. Gibson. Barlow (1961) and Gibson (1979) hypothesized that the goal of the visual system consists precisely of "represent the natural scenes (redundant, with nth-order statistics) by means of the activity of a dispersed selection of elements not redundant (independent, with leading alone statistics)."

Nevertheless, many researchers of the visual system have been at the least indifferent to this proposal, if not antagonistic (Rust & Movshon, 2005). There are several reasons for this. Usually, natural images have been considered to exist in such a large number that it will be very difficult to obtain a simple and general description of them, as is the case with parametric stimuli. As an example, a random image of 8 × 8 pixels with 64 levels of gray (6 bits) has 2^{384} (more than 10^{100}) combinations of pixel amplitudes (Olshausen & Field, 2005). However, natural images are not random combinations of pixels, but they have structure. In statistical terms, this structure accounts for redundancy, meaning that there is a high degree of correlation between adjacent zones of the scene. Redundancy can be illustrated as the possibility to predict hidden portions of the scene by an experienced observer (Kersten, 1987), and is the consequence of high-order statistics being present in natural images.[1] One of the most popular hypotheses about representation of natural images in visual system

---

[1] For every image, first-order statistics refers to the probability of a pixel having some value. Second-order statistics refers to the probability of a pixel having some value given the value of one neighboring pixel. Third-order statistics extends this definition for two neighboring pixels, and so on.

is Barlow hypothesis of redundancy reduction. This states that the visual system should represent highly redundant natural images by means of a small but sufficient number of independent, nonredundant elements such as neurons (Barlow, 1961). Redundancy reduction should be achieved after the photoreceptor stage, because these cells are still representing the highly redundant stimuli, and it was early proposed to be feasible through the antagonistic center-surround classical receptive fields of retinal ganglion cells by means of a simple weighted linear summation between the responsive neuron and surrounding interneurons (Srinivasan, Laughlin, & Dubs, 1982). Lately, it was proposed that second-order statistics, like the autocorrelation function and power spectra, provide a useful characterization of the structure of natural images. In contrast to signals with uniform or "white" spectra, natural scenes show a typical $1/f^2$ fall in power spectra, with high power at low frequencies and low power at high frequencies. If classical receptive fields of simple cells in V1 can effectively be modeled by means of space–time-centered functions like Gabor functions (Marcelja, 1980), then the structure of the classical receptive field is optimally coupled to the statistics of natural images, allowing one to represent them with a small, sparse, and independent number of neurons (sparse coding), each carrying the same amount of "information" (Field, 1987). A simulation study showed that an artificial neural network that attempts to find sparse, nonredundant codes for natural scenes develops a complete family of receptive fields with properties very similar to those found in V1 and to Gabor functions (Olshausen & Field, 1996). Explicitly, it has been proposed that the classical receptive fields of simple cells in V1 constitute an overcomplete basis set: the number of coding elements is greater than the number of effective dimensions in the stimulus, so just a few active elements are necessary to encode it (Olshausen & Field, 1997), thus satisfying the redundancy-reduction hypothesis. So far, all these ideas were based on hypothesis and simulations. Sparseness and redundancy reduction in neural representation of natural images gained experimental support when it was shown that cat's LGN X-cells present a uniform, white spectrum in response to stimulation with sequences of natural images (Dan, Atick, & Reid, 1996), demonstrating that LGN decorrelates the characteristic $1/f^2$ natural image spectrum, enhancing higher spatial frequencies. Psychophysical studies in humans have also shown that the visual system seems to be optimized for higher-order statistics of natural images. The performance in tests of visual discrimination deteriorates when second-order statistics of the images are altered, making them less "natural" (Parraga, Troscianko, & Tolhurst, 2000). Sparseness and decorrelation in monkey's V1 neurons increase with the stimulation of the nonclassical receptive field using a sequence of patches of natural scenes, which simulated the temporal course of eye movements on the image. This protocol also adds reliability to trial-to-trial neural responses (Vinje & Gallant, 2000). Similar results for sparseness and redundancy reduction have been found in ferret's and cat's primary visual cortex, using natural images and movies (Weliky, Fiser, Hunt, & Wagner, 2003; Yen et al., 2007). Evidence of sparse coding in other sensory systems suggests that this may be a widely employed strategy in sensory systems for representing just a few out of the enormous number of sensory inputs received at any given time (Olshausen & Field, 2004). In recent years, techniques have been developed to precisely map classical receptive fields in primary visual cortex using natural images (Theunissen et al., 2001; Smyth, Willmore, Baker, Thompson, & Tolhurst, 2003; Tolhurst, Tadmor, & Chao, 1992) and incorporating nonlinear and eye-movement influences (David, Vinje, & Gallant, 2004), fully accounting for the difference in predictive power with receptive field models obtained using simple stimuli. Using natural images has also made clear the influence of contextual effects on neural responses, at least in primary visual cortex, by shifting the preferred orientation of neurons toward those orientations more commonly found in natural images (vertical and horizontal), and suggesting an optimally coupled, sparse, and nonredundant mechanism for representing natural stimuli in the visual system (Felsen, Touryan, & Dan, 2005).

Altogether these studies show that a natural stimulus is not just a collection of simple stimuli displayed together in a single frame. The way the visual system is perturbed when exposed to natural visual stimuli cannot be explained by adding up the response of cells to simple stimuli. This situation reveals not only the complexity of the mutual influence of neurons, but also the influence of the entire network on each visual area. This bring us to the next point, which relates to the fact that not only does our visual system have to deal with complex visual stimuli, but often, if not always, our visual behavior is motivated by attention, a motor goal, or another internal state. In these cases, top-down influence modulation needs to be brought into our explanatory models.

Ecological Vision

As mentioned earlier, the use of natural images for studying the visual system was proposed long ago by Barlow and Gibson. Both authors proposed that the visual system must be "optimized" for the kind of stimuli that the system is most likely to encounter in the environment, i.e., natural scenes. Gibson stressed the fact that visual perception always occurs in ecological conditions, where the observer it is in constant movement. Therefore, an important issue relates to examining if our current experimental paradigms utilized by us so far to understand visual function can indeed give us a complete description of the nature of the organization of a system that almost always is engaged in complex behaviors. However, these behaviors are seldom tested in such experiments. Is notorious that the dissection of the cognitive domains and their functions is a useful approach for understanding how different behaviors can be produced. Moreover, it usually expands the dimensions in which a behavior can be explained and understood. However, when testing a function of a specific cognitive domain, from a low-level operation such as sensitivity to contrast, to high-level ones such as object binding, it is usually tested in isolation from other functions of the same domain. Furthermore, rarely is an attempt made to examine how different domains make up a unified agent that evokes our cognition as a whole. One of the characteristics that makes a complex response or behavior is that parameters and variables exist only in the interaction of multiple functions or systems, and this transforms the set of responses of the single function to a set that does not necessarily include or cannot necessarily be deducted from those obtained in conditions other than cooperative or unified behavior. In other words, it is not clear how knowledge of the responses of a set of neurons in constrained stimuli situations could explain the responses obtained in more natural conditions, if they are actually never tested in such natural conditions. Parsimoniously, testing the neural responses in more natural conditions with complex visual stimuli that are important for the subject's goals, navigation, or attentional states is far more complicated. For instance, electrophysiological techniques such as single unit recordings, which enable very fine temporal resolution of the neuronal activity, face complications to identify and register a set of neurons involved in a unitary function because of severe technical constraints such as the number of units that can be recorded, biased sampling, immobilization, anesthesia, limited time of recording, reduction of noise, and ethics. Electroencephalographic recordings – the other extreme – have similar time resolution and can be performed with far less restrictions, but the spatial resolution is orders of magnitude coarser that that of the single unit. Neuroimaging, on the other hand, can provide an intermediate spatial resolution (that reaches down to macrocolumnar structures but cannot resolve circuitry) but also has technical caveats such as a largely indirect relation between the electrical activity of neurons and the signal, and the limits in temporal resolution. This means that many cognitive phenomena that occur in the milliseconds range are filtered our by the seconds resolution of this technique. In addition there are also constraints with immobilization of the subjects, the architecture of the machines, or, simply, cost. So it does not come as a surprise that the bulk of our current knowledge of the mechanism related to natural vision comes from neuropsychological, behavioral, and psychophysical studies, while far less is known about the change in neural responses related to natural behavior. In the next section we discuss some of the evidence that demonstrates that vision is a highly integrative process where the visual system *acts vision*, as opposed to being a passive processing system of a collection of images falling upon the retina. We argue that the use of ecological conditions to study visual perception is a useful approach, mainly because it brings out the actual mechanisms involved in the types of behavior that we want ultimately to explain.

Active Vision

The structure of the retina determines the characteristics of visual acuity, possible wave-length discrimination, and spatial continuity in different locations of the visual field. Thus, detailed and colorful vision is limited to a few visual grades centered in the line of gaze. However, this limitation is commonly underrated because animals have the possibility to explore the visual world by changing the line of gaze through ocular movements, head movements, or spatial displacement. For instance, humans are permanently doing two to four ocular movements per second, exploring and changing the image focused on the retina. So, although we cannot see in a single fixation every detail of the visual field, we can have constant and rapid access to the whole. Furthermore, since ocular movements are generally effortless and nonconsciously directed, natural vision appears detailed, colorful, and continuous in the entire field, owing to the real capacity of access to those attributes (Cohen, 2002; Noe, 2002). In this continuous exploration of the world it is common that the focus of visual attention – the selection of which location and attributes of visual stimuli will undergo quicker and deeper processing – goes together with the line of sight (Motter, 1998; Motter & Belky, 1998). This capacity of continuously changing the focus of sight and attention allows optimization of the limited capacity of visual processing without losing rapid access to the entire visual field. Yet, this also has disadvantages. As shown in a series of studies of change and inattentional blindness (Simons & Rensink, 2005), subjects commonly fail to detect large alterations in explicit or implicit change tasks using static visual scenes (O'Regan, Deubel, Clark, & Rensink, 2000), motion pictures (Simons & Chabris, 1999; Levin & Simons, 1997), and even real-world interactions (Simons & Levin, 1998). Moreover, this perceptual blindness is strongly dependent on task-relevance dynamics. For example, when subjects have to manipulate simple objects characterized by a few attributes, the ability to detect changes in these objects not only depends on the task relevance of the changing attributes (even when seeing the change directly), but specifically on the relevance of the attribute for the task at the precise moment of the changes (Triesch, Ballard, Hayhoe, & Sullivan, 2003), suggesting that subjects fail to detect visual changes that shift with dynamics not coupled with the dynamics of the task, gaze, and attention. Consistent with this interpretation, different neurophysiological and functional neuroimaging studies have shown that attentional selection and oculomotor control are tightly related (Corbetta, 1998; Awh, Armstrong, & Moore, 2006). Successful goal-directed attentional selection and target detection is highly correlated with activity in a network including superior frontal and posterior parietal cortex (Corbetta & Shulman, 2002). For example, in functional MRI studies successful change detection has been correlated with enhanced activity in parietal and prefrontal cortex (Beck, Rees, Frith, & Lavie, 2001; Huettel, Guzeldere, & McCarthy, 2001). Since incorrect change identification generates similar patterns of activation, alternative explanations have correlated this activity only with reporting a change and redeployment of attention (Pessoa & Ungerleider, 2004), but leaving unchanged the involvement of these areas with dynamic visual attention processes. Similarly, visual search tasks produce widespread brain activation, especially in the frontal eye field and intraparietal sulcus in relation to search and sustained attention signals and in the temporoparietal junction related to target detection (Shulman et al., 2003). In the same way, oculomotor selection is supported by a distributed network of cortical areas that include frontal premotor areas, posterior parietal cortex, anterior cingulated cortex (Pierrot-Deseilligny, Milea, & Muri, 2004), and subcortical structures such as basal ganglia and superior colliculus (Carello & Krauzlis, 2004). This relationship between attention and oculomotor control is not only anatomical but also functional. For example, when the frontal eye field is stimulated, a saccade is directed to a location retinotopically specified by the stimulated neurons (Tehovnik, Sommer, Chou, Slocum, & Schiller, 2000). If microstimulation alone is given, no saccade is elicited, but there is an enhancement of the detection of changes in the corresponding locations of the visual field (Moore & Fallah, 2001) and an increase in the neuronal activity of extrastriate visual areas in response to the presentation of the object in their corresponding receptive field (Moore & Armstrong, 2003). Furthermore, attention and oculomotor control can enhance or suppress the neuronal activity from visual thalamus to almost all visual areas (Corbetta & Shulman, 2002; Kastner & Ungerleider, 2000), suggesting that the visual system is permanently working purposefully and based on previous experiences and expectancies. In this sense, what the visual system does in natural conditions is much more an on-time coupling between relevant visual world features and ongoing behavior than a representation of an outside world.

When the ocular movements are tested with tracking methods it is clear that the locations of fixations are nonhomogeneously distributed in the visual field (Yarbus, 1967). However, it has been difficult to determine the mechanism involved in the choice of the destination of the next eye fixation owing to the large variability of patterns produced within and between subjects, especially if tested against natural images in conditions of free viewing. On the other hand, what normally is considered free viewing in experimental designs (presentation of static images on a monitor) has little resemblance to natural vision, in which most of the time vision is engaged in goal-directed behavior. When ocular movements are tested during simple, common behaviors the patterns and fixation locations are much more predictable because of the dependence on the task. Semistationary tasks such as hitting a ball (Land & McLeod, 2000) and preparing a sandwich or a tea (Hayhoe, Shrivastava, Mruczek, & Pelz, 2003; Land & Hayhoe, 2001) and mobile tasks such as walking to a target (Turano, Geruschat, & Baker, 2003) or hand washing (Pelz & Canosa, 2001) show ocular movements that are tightly correlated with the dynamics of the task, strictly directed to objects or locations relevant to the goals, monitoring action, frequently in advance or "looking ahead," and leaning upon schematic knowledge (Hayhoe & Ballard, 2005). Land (2001) summarized the patterns of movement described in these experiments as four types: *locating* in advance objects or places that will be utilized next, *directing* different actions such as hand movements with a fixation of the target until it is contacted, *guiding* multiple manipulation tasks with ocular movement alternating between the objects involved, and *checking* the accomplishment of particular conditions such as waiting for the water to reach some level when a glass is filled with it. Although is has been difficult to identify directly the neural processes underlying these kinds of behaviors, there is much evidence pointing to a large and distributed network supporting goal-oriented visual behaviors in natural conditions that compromises almost all the brain. The same areas involved in attentional and oculomotor processes are also involved with others in the control and coordination of eye and body.

Contextual Vision

In addition to modifications of the response of visual systems due to the low-level structure of natural images, to goal-directed eye movements, and to attentional mechanisms, there are early modifications in the visual systems produced for "high-level" properties of natural scenes. The history of interactions between the brain and its environment heightens different regularities that identify categorical and specific spatial layouts and objects, making vision an efficient guide for behavior. So, visual knowledge compromises not only object knowledge but also scene memories and schemes useful for guiding exploration behavior. Different studies have aimed to examine the apparent fast access of the visual system to contents of the entire scene or "gist," which includes low-level features but also scene identity and local context characteristics. In a series of studies Schyns and Oliva showed that scene category discrimination can be achieved with presentation times of visual stimuli shorter than 50 ms. This ability seems to be based principally on low frequency spatial and color content (Oliva & Schyns, 1997; Oliva & Schyns, 2000). Other studies have confirmed that this discrimination can be done with very short reaction times (Rousselet, Joubert, & Fabre-Thorpe, 2005), similar to the time necessary for object recognition (Thorpe, Fize, & Marlot, 1996). Parallel to psychophysics of scene recognition, neuropsychological and more recently neuroimaging techniques have identified a brain region, the parahippocampal cortex, which activity correlates with seeing and identifying natural scenes in a viewpoint-specific manner (Epstein & Kanwisher, 1998; Epstein, Graham, & Downing, 2003; Epstein, Harris, Stanley, & Kanwisher, 1999; Aminoff, Gronau, & Bar, 2007). Other authors propose that the activity of the parahippocampal cortex is additionally correlated with spatial and nonspatial contextual associations (Aminoff, Gronau, & Bar, 2007). Together, the ability to rapid access the "gist" of a scene and the existence of specialized subsystems in context and scene analysis suggests that the visual system uses cues at different spatial scales simultaneously. In this sense, coarser visual contents in the form of low-frequency spatial features can be processed faster and then used for scene identification, oculomotor selection, and object identification (Bar, 2004). In a recent study, Bar et al. (2006) showed how low-frequency spatial images activate V1 and left orbitofrontal cortex earlier than other visual cortices in the ventral stream. In the same way, a study on functional anatomy based on the timing and pattern of laminar activation in V4 and inferotemporal cortex points to activation by inputs from the dorsal visual stream and directly from the thalamus more than traditional "feedforward" inputs by the ventral pathway (Chen et al., 2007). Together this evidence challenges traditional top-down versus bottom-up dichotomies, suggesting that "higher" functions such as semantic or object identity knowledge can rely on coarser visual features and be processed in parallel and even bias the proper identification of a object.

Coda

Most of our understanding of the neuronal mechanisms of visual perception comes from experiments that have used simple stimuli. Although these studies have significantly advanced our knowledge on how visual perception occurs, several lines of evidence suggest that we need to also depart from this classical paradigm. First, the classical receptive fields cannot account for the response of neurons in the visual system to complex or natural stimuli. Second, most of our visual perception occurs in combination with motor acts, more noticeably eye movements, that quickly change visual stimuli falling on the retina. Third, many aspects of vision, including goal-oriented behaviors and attentional mechanisms, heavily modify neuronal responses to all types of stimuli, thus influencing the entire neuronal network involved in vision. Finally, perceptual learning or the history of interactions between the brain and its environment heighten different regularities that identify categorical and specific spatial layouts and objects, making vision an efficient guide for behavior. We have argued that the inclusion and emphasis of reexamining neuronal activity during ecological and natural conditions may reveal those elusive mechanisms that can fully explain our visual perceptions.

References

Aminoff, E., Gronau, N., & Bar, M. (2007b). The parahippocampal cortex mediates spatial and nonspatial associations. *CerebralCortex, 17*, 1493–1503.

Awh, E., Armstrong, K. M., & Moore, T. (2006). Visual and oculomotor selection: Links, causes and implications for spatial attention. *Trends in Cognitive Science, 10*, 124–130.

Bar, M. (2004). Visual objects in context. *Nature Reviews Neuroscience, 5*, 617–629.

Bar, M., Kassam, K. S., Ghuman, A. S., Boshyan, J., Schmid, A. M., Dale, A. M. et al. (2006). Top-down facilitation of visual recognition. *Proceedings of the National Academy of Sciences of the United States of America, 103*, 449–454.

Barlow, H. B. (1961). Possible principles underlying the transformations of sensory images. In W. A. Rosenblith (Ed.), *Sensory communication* (pp. 217–234). Cambridge, MA: MIT.

Beck, D. M., Rees, G., Frith, C. D., & Lavie, N. (2001). Neural correlates of change detection and change blindness. *Nature Neuroscience, 4*, 645–650.

Bringuier, V., Chavane, F., Glaeser, L., & Fregnac, Y. (1999). Horizontal propagation of visual activity in the synaptic integration field of area 17 neurons. *Science, 283*, 695–699.

Carello, C. D. & Krauzlis, R. J. (2004). Manipulating intent: Evidence for a causal role of the superior colliculus in target selection. *Neuron, 43*, 575–583.

Chen, C. M., Lakatos, P., Shah, A. S., Mehta, A. D., Givre, S. J., Javitt, D. C. et al. (2007). Functional anatomy and interaction of fast and slow visual pathways in macaque monkeys. *Cerebral Cortex, 17*, 1561–1569.

Cohen, J. (2002). The grand grand illusion illusion. *Journal of Consciousness Studies, 9*, 141–157.

Corbetta, M. (1998). Frontoparietal cortical networks for directing attention and the eye to visual locations: Identical, independent, or overlapping neural systems? *Proceedings of the National Academy of Sciences of the United States of America, 95*, 831–838.

Corbetta, M. & Shulman, G. L. (2002). Control of goal-directed and stimulus-driven attention in the brain. *Nature Review Neuroscience, 3*, 201–215.

Dan, Y., Atick, J. J., & Reid, R. C. (1996). Efficient coding of natural scenes in the lateral geniculate nucleus: Experimental test of a computational theory. *Journal of Neuroscience, 16*, 3351–3362.

David, S. V., Vinje, W. E., & Gallant, J. L. (2004). Natural stimulus statistics alter the receptive field structure of v1 neurons. *Journal of Neuroscience, 24*, 6991–7006.

Epstein, R., Graham, K. S., & Downing, P. E. (2003). Viewpoint-specific scene representations in human parahippocampal cortex. *Neuron, 37*, 865–876.

Epstein, R., Harris, A., Stanley, D., & Kanwisher, N. (1999). The parahippocampal place area: Recognition, navigation, or encoding? *Neuron, 23*, 115–125.

Epstein, R. & Kanwisher, N. (1998). A cortical representation of the local visual environment. *Nature, 392*, 598–601.

Felsen, G. & Dan, Y. (2005). A natural approach to studying vision. *Nature Neuroscience, 8*, 1643–1646.

Felsen, G., Touryan, J., & Dan, Y. (2005). Contextual modulation of orientation tuning contributes to efficient processing of natural stimuli. *Network, 16*, 139–149.

Field, D. J. (1987). Relations between the statistics of natural images and the response properties of cortical cells. *Journal of the Optical Society of America A, 4*, 2379–2394.

Fries, P., Reynolds, J. H., Rorie, A. E., & Desimone, R. (2001). Modulation of oscillatory neuronal synchronization by selective visual attention. *Science, 291*, 1560–1563.

Gilbert, C., Ito, M., Kapadia, M., & Westheimer, G. (2000). Interactions between attention, context and learning in primary visual cortex. *Vision Research, 40*, 1217–1226.

Hayhoe, M. & Ballard, D. (2005). Eye movements in natural behavior. *Trends in Cognitive Science, 9*, 188–194.

Hayhoe, M. M., Shrivastava, A., Mruczek, R., & Pelz, J. B. (2003). Visual memory and motor planning in a natural task. *Journal of Vision, 3*, 49–63.

Hendry, S. H. & Reid, R. C. (2000). The koniocellular pathway in primate vision. *Annual Review in Neuroscience, 23*, 127–153.

Huettel, S. A., Guzeldere, G., & McCarthy, G. (2001). Dissociating the neural mechanisms of visual attention in change detection using functional MRI. *Journal of Cognitive Neuroscience, 13*, 1006–1018.

Jones, H. E., Grieve, K. L., Wang, W., & Sillito, A. M. (2001). Surround suppression in primate V1. *Journal of Neurophysiology, 86*, 2011–2028.

Kastner, S. & Ungerleider, L. G. (2000). Mechanisms of visual attention in the human cortex. *Annual Review in Neuroscience, 23*, 315–341.

Kersten, D. (1987). Predictability and redundancy of natural images. *Journal of Optical Society of America A, 4*, 2395–2400.

Land, M. F. & Hayhoe, M. (2001). In what ways do eye movements contribute to everyday activities? *Vision Research, 41*, 3559–3565.

Land, M. F. & McLeod, P. (2000). From eye movements to actions: How batsmen hit the ball. *Nature Neuroscience, 3*, 1340–1345.

Levin, D. T. & Simons, D. J. (1997). Failure to detect changes to attended objects in motion pictures. *Psychonomic Bulletin and Review, 4*, 501–506.

Luck, S. J., Chelazzi, L., Hillyard, S. A., & Desimone, R. (1997). Neural mechanisms of spatial selective attention in areas V1, V2, and V4 of macaque visual cortex. *Journal of Neurophysiology, 77*, 24–42.

Marcelja, S. (1980). Mathematical description of the responses of simple cortical cells. *Journal of Optical Society of America, 70*, 1297–1300.

Moore, T. & Armstrong, K. M. (2003). Selective gating of visual signals by microstimulation of frontal cortex. *Nature, 421*, 370–373.

Moore, T. & Fallah, M. (2001). Control of eye movements and spatial attention. *Proceedings of the National Academy of Sciences of the United States of America, 98*, 1273–1276.

Motter, B. C. (1993). Focal attention produces spatially selective processing in visual cortical areas V1, V2, and V4 in the presence of competing stimuli. *Journal of Neurophysiol, 70*, 909–919.

Motter, B. C. (1994). Neural correlates of attentive selection for color or luminance in extrastriate area V4. *Journal of Neuroscience, 14*, 2178–2189.

Motter, B. C. (1998). Inside and outside the focus of attention. *Neuron, 21*, 951–953.

Motter, B. C. & Belky, E. J. (1998). The guidance of eye movements during active visual search. *Vision Research, 38*, 1805–1815.

Noe, A. (2002). Is the visual world a grand illusion? *Journal of Consciousness Studies, 9*, 1–12.

O'Regan, J. K., Deubel, H., Clark, J. J., & Rensink, R. A. (2000). Picture changes during blinks: Looking without seeing and seeing without looking. *Visual Cognition, 7*, 191–211.

Oliva, A. & Schyns, P. G. (1997). Coarse blobs or fine edges? Evidence that information diagnosticity changes the perception of complex visual stimuli. *Cognitive Psychology, 34*, 72–107.

Oliva, A. & Schyns, P. G. (2000). Diagnostic colors mediate scene recognition. *Cognitive Psychology, 41*, 176–210.

Olshausen, B. A. & Field, D. J. (1996). Emergence of simple-cell receptive field properties by learning a sparse code for natural images. *Nature, 381*, 607–609.

Olshausen, B. A. & Field, D. J. (1997). Sparse coding with an overcomplete basis set: A strategy employed by V1? *Vision Research, 37*, 3311–3325.

Olshausen, B. A. & Field, D. J. (2004). Sparse coding of sensory inputs. *Current Opinion in Neurobiology, 14*, 481–487.

Olshausen, B. A. & Field, D. J. (2005). How close are we to understanding v1? *Neural Computation, 17*, 1665–1699.

Parraga, C. A., Troscianko, T., & Tolhurst, D. J. (2000). The human visual system is optimised for processing the spatial information in natural visual images. *Current Biology, 10*, 35–38.

Pelz, J. B. & Canosa, R. (2001). Oculomotor behavior and perceptual strategies in complex tasks. *Vision Research, 41*, 3587–3596.

Pessoa, L. & Ungerleider, L. G. (2004). Neural correlates of change detection and change blindness in a working memory task. *Cerebral Cortex, 14*, 511–520.

Pierrot-Deseilligny, C., Milea, D., & Muri, R. M. (2004). Eye movement control by the cerebral cortex. *Current Opinion in Neurology, 17*, 17–25.

Rousselet, G. A., Joubert, O. R., & Fabre-Thorpe, M. (2005). How long to get to the "gist" of real-world natural scenes? *Visual Cognition, 12*, 852–877.

Rust, N. C. & Movshon, J. A. (2005). In praise of artifice. *Nature Neuroscience, 8*, 1647–1650.

Shulman, G. L., McAvoy, M. P., Cowan, M. C., Astafiev, S. V., Tansy, A. P., d'Avossa, G. et al. (2003). Quantitative analysis of attention and detection signals during visual search. *Journal of Neurophysiology, 90*, 3384–3397.

Simons, D. J. & Chabris, C. F. (1999). Gorillas in our midst: Sustained inattentional blindness for dynamic events. *Perception, 28*, 1059–1074.

Simons, D. J. & Levin, D. T. (1998). Failure to detect changes to people during a real-world interaction. *Psychonomic Bulletin and Review, 5*, 644–649.

Simons, D. J. & Rensink, R. A. (2005). Change blindness: Past, present, and future. *Trends in Cognitive Science, 9*, 16–20.

Smyth, D., Willmore, B., Baker, G. E., Thompson, I. D., & Tolhurst, D. J. (2003). The receptive-field organization of simple cells in primary visual cortex of ferrets under natural scene stimulation. *Journal of Neuroscience, 23*, 4746–4759.

Srinivasan, M. V., Laughlin, S. B., & Dubs, A. (1982). Predictive coding: A fresh view of inhibition in the retina. *Proceedings of Royal Society of London – Series B: Biological Science, 216*, 427–459.

Tehovnik, E. J., Sommer, M. A., Chou, I. H., Slocum, W. M., & Schiller, P. H. (2000). Eye fields in the frontal lobes of primates. *Brain Research Brain Research Review, 32*, 413–448.

Theunissen, F. E., David, S. V., Singh, N. C., Hsu, A., Vinje, W. E., & Gallant, J. L. (2001). Estimating spatio-temporal receptive fields of auditory and visual neurons from their responses to natural stimuli. *Network, 12*, 289–316.

Thorpe, S., Fize, D., & Marlot, C. (1996). Speed of processing in the human visual system. *Nature, 381*, 520–522.

Tolhurst, D. J., Tadmor, Y., & Chao, T. (1992). Amplitude Spectra of Natural Images. *Ophthalmic and Physiological Optics, 12*, 229–232.

Triesch, J., Ballard, D. H., Hayhoe, M. M., & Sullivan, B. T. (2003). What you see is what you need. *Journal of Vision, 3*, 86–94.

Trotter, Y. & Celebrini, S. (1999). Gaze direction controls response gain in primary visual-cortex neurons. *Nature, 398*, 239–242.

Turano, K. A., Geruschat, D. R., & Baker, F. H. (2003). Oculomotor strategies for the direction of gaze tested with a real-world activity. *Vision Research, 43*, 333–346.

Ungerleider, L. G. & Haxby, J. V. (1994). 'What' and 'where' in the human brain. *Current Opinion in. Neurobiology, 4*, 157–165.

Vinje, W. E. & Gallant, J. L. (2000). Sparse coding and decorrelation in primary visual cortex during natural vision. *Science, 287*, 1273–1276.

Weliky, M., Fiser, J., Hunt, R. H., & Wagner, D. N. (2003). Coding of natural scenes in primary visual cortex. *Neuron, 37*, 703–718.

Womelsdorf, T., Fries, P., Mitra, P. P., & Desimone, R. (2006). Gamma-band synchronization in visual cortex predicts speed of change detection. *Nature, 439*, 733–736.

Worgotter, F., Suder, K., Zhao, Y., Kerscher, N., Eysel, U. T., & Funke, K. (1998). State-dependent receptive-field restructuring in the visual cortex. *Nature, 396*, 165–168.

Yarbus, A. L. (1967). *Eye movements and vision*. New York: Plenum.

Yen, S. C., Baker, J., & Gray, C. M. (2007). Heterogeneity in the responses of adjacent neurons to natural stimuli in cat striate cortex. *Journal of Neurophysiology, 97*, 1326–1341.

Chapter 5
Attending to the Stream of Consciousness – A Methodological Challenge

D. Cosmelli

Abstract Attention is usually conceptualized, and empirically approached, as a matter of selection, information reduction, and performance enhancement. In this context, a wealth of experimental approaches have been developed to study sustained attention, selective attention, orienting, divided attention, conflict resolution, and so on. However, much less importance has been traditionally accorded to a more intimate yet pervasive aspect of attention: how it continuously shifts and moves within the stream of consciousness – the ongoing flow of perceptions, thoughts, images, and feelings we all experience during any normal day. In this chapter we survey some of the traditional ways in which attention is experimentally studied while pointing out some limitations and potential interests these approaches have for the study of attention in the stream of consciousness. We highlight, based on a phenomenological approach to its dynamics, one crucial aspect of attention that has been systematically neglected, and that could have important consequences for its study. Taking into account the spontaneous nature of attentional shifts during the stream of consciousness leads us to consider recent developments in brain imaging, experimental psychology, and signal analysis that are beginning to establish a framework for the scientific study of this elusive phenomenon.

Introduction

The great majority of experimental approaches to the study of attention have focused on how it affects the way we perceive the world and how it enables us to navigate adaptively the myriad of alternatives we encounter (Posner & Rothbart, 1998). The

D. Cosmelli
Laboratorio de Neurociencias Cognitivas, Centro de Investigaciones Médicas, Escuela de Medicina, Pontificia Universidad Católica de Chile, Marcoleta 391, 2° Piso, Santiago, Chile, and Escuela de Psicología, Pontificia Universidad Católica de Chile, Vicuña Mackenna 4860, Macul, Santiago, Chile, e-mail: dcosmelli@uc.cl

F. Aboitiz and D. Cosmelli (eds.) *From Attention to Goal-Directed Behavior.*
© Springer-Verlag Berlin Heidelberg 2009

main metaphor to describe this is that of a process that limits the unmanageable amounts of information we are offered by the environment, selecting only what is relevant for further action (Itti, Rees, & Tsotsos, 2005). Extensive work has been done on how we select, choose, or enhance objects of our interest while at the same time attenuate, dim, or filter out those we are not concerned with (Corbetta & Shulman, 1998; Hillyard & Mangun, 1987). Attending facilitates detection of an object and enhances our capacity to discriminate its sensual characteristics; it shortens reaction times if we are asked to detect some change in it; it allows us to hold it in our mind's "gaze" without distraction; it facilitates recall and can ultimately determine if we perceive the object at all. These are well-known and rather uncontroversial findings regarding the nature and effects of attention (Parasuraman, 1998; Posner, 2004).

Now do the following thought experiment. Imagine yourself in a completely dark room, no sounds, no smells, nothing to see, hear or taste, just you, standing. Imagine there are no objects "out there" for you to perceive, no perturbations on your sensory surfaces to catch your interest, nothing to discriminate, nothing to attend to. It is reasonable to think that in this unusual situation your attention would still be on the move: from one thought to the next, from a feeling to an image, from your tired feet to the unease in your back and then to a new idea of what you will do when you get out of this weird experiment, continuously shifting between a multiplicity of objects that need not be external in any straightforward way. Our daily life is not only a matter of attending to perceptual objects, to choose and discriminate among them, to recall them or avoid them, etc. Indeed, most of the time we are immersed in an internal discourse (Ericsson & Simon, 1993), or fluctuating between mental images and what is in front of us, what we are listening to, what we think we should do later, or should have done a moment ago, what we feel, and so on (Pope & Singer, 1978). The stream of consciousness[1] permeates the totality of our experience, and its flow seems to be made of an almost frantic shifting of our focus of interest. Why is it that this very intimate side of attention, *its restless dynamics regardless of the object of interest*, has been the subject of so little experimental inquiry in the cognitive neurosciences?

Our main contention is that while traditional experimental approaches can probe specific components of the attentional process in highly controlled situations, they present important limitations for the study of attention in the most personal and probably significant of human experiences, namely, the stream of consciousness. We believe that the main reason for this is methodological difficulty, but also a limited notion of attention that does not take into account the spontaneous nature of the fluctuations and shifts that take place during everyday experience. Importantly, new strategies are emerging in terms both of experimental design and signal analysis to

---

[1] We understand the stream of consciousness as the flow of experience, encompassing the daily shifts of interest among sensory, perceptual, and ideal (mental) objects.

deal with this complicated issue; the objective of this chapter is therefore to provide a survey of these developments and how they might be relevant for the study of attention, in light of a critical appraisal of attentional dynamics from a phenomenological perspective.

In the following we will briefly mention some the main approaches to the study of attention under controlled situations. Numerous excellent reviews exist that discuss different aspects of the overwhelming amount of data that have been produced in the study of attention (Corbetta & Shulman, 2002; Fan & Posner, 2004; Johnston & Dark, 1986; Kastner & Pinsk, 2004; Kastner & Ungerleider, 2000; Kinchla, 1992; Raz & Buhle, 2006; Shipp, 2004), so we will not go into actual results. Our emphasis will be on the experimental strategies used to study the orienting and maintaining of attention, as these are the most relevant to the problem at hand. We will then discuss both limitations and potential interests of the standard experimental approaches for studying attention in the flow of consciousness. After a short phenomenological detour, we will review recent developments in the study of mind-wandering, stimulus-independent cognition, and adapted signal analysis strategies. These approaches are part of an emerging trend that has begun the bold enterprise of tackling experimentally this elusive but deeply significant phenomenon.

The Standard Approach to Attention – A Selective Survey

In current cognitive neuroscience literature attention is far from being a monolithic construct.[2] Whether it is selective attention, sustained attention, orienting attention, vigilance, divided attention, spatial or object-based attention, attention to language, inattention, atypical attention, etc., the general trend is to consider attention as something that chooses among alternatives for the benefit of a limited-capacity perceptual system. Accordingly, it is conceptualized in terms of how it affects *performance* when one is confronted with a demanding situation.[3] In this context, two attentional types are especially relevant when dealing with the stream of consciousness: sustained attention and orienting/selecting.

---

[2] An exhaustive review of the details and modifications in the myriad of approaches that have been developed to explore the function of attention is beyond the scope of the present chapter. Only some of the most prevailing strategies will be presented as a context.

[3] This is certainly due in great measure to the fact that attention has traditionally been defined behaviorally, where change in performance is the index of whether the subject is in the attending or nonattending state. Nevertheless, as we discuss in "Bent on Performance – The Informational Bias" this performance-based concetpualization carries the legacy of an informational view of the nervous system, which might not be the only possible approach.

Vigilance, Sustained Attention, and Alerting

The relation between vigilance, sustained attention, and alerting is all but uncontroversial (Oken, Salinsky, & Elsas, 2006; Sarter, Givens, & Bruno, 2001). At least in cognitive neuroscience, vigilance and sustained attention are used interchangeably to refer to the *continuous* monitoring of particular sources of information. Alerting, on the other hand, is commonly conceived as *phasic*, temporally restricted, task-related readiness to respond to an upcoming event (Raz & Buhle, 2006; Smid, de Witte, Homminga, & van den Bosch, 2006).[4] Thus, the main difference seems to be temporal, so what changes is mainly the duration of the stabilization of the attentional gesture.

Sustained attention is approached experimentally through lengthy (often tedious) paradigms such as the general class of continuous performance tasks. Here, stimuli (visual, auditory, tactile, etc.) are presented recursively on the timescale of minutes to hours. The subject is asked to detect an infrequent target by either responding to it or withholding a response when it appears. Increase in false alarms, decrease in hit rate, and lengthening of reaction times are indicative of the waning of sustained attention or vigilance. Alertness in its more phasic version can be probed on a trial-by-trial basis, using cues that indicate an upcoming event. A classic example for the latter is the central cue condition in the Attention Network Test of Posner and colleagues (Fan, McCandliss, Fossella, Flombaum, & Posner, 2005; Fan, McCandliss, Sommer, Raz, & Posner, 2002).

It is worth noting that, especially in the case of phasic alerting (Sturm & Willmes, 2001), it can be difficult to disentangle this variety of attention from another well-known psychological construct, namely, expectation (Bastiaansen, Bocker, Brunia, de Munck, & Spekreijse, 2001; Brunia & van Boxtel, 2001). Indeed, as in alerting, expectation is usually assessed through tasks where there is a given time interval between a cue and a target. In this sense both alerting and expectation share the gesture of attending to a (future) moment in time (Nobre, 2001).

Orienting and Selecting

Selective attention is the most studied aspect of attention in cognitive neuroscience (Hillyard & Anllo-Vento, 1998; Luck & Ford, 1998; Mangun, 1995). The aim is to

---

[4] In contrast to arousal, where changes are best defined by neurobiological parameters (e.g., variations in the EEG signal due to ascending brainstem modulation resulting in relatively nonspecific changes of cortical excitability), sustained attention and alertness are commonly studied on the basis of goal-directed tasks, using changes in reaction times, false alarms, and hit rates as the main measures, thus defined mainly in behavioral terms.

understand how singling out a relevant target among competing distracters is achieved. To do so, it appears that attention has to be directed to the location of the object, to some or all of its relevant features, or to both.[5]

Probably the most used paradigm in the study of orienting and selection is the attentional cueing paradigm (Posner, 1980).[6] In its general form, during this task subjects receive a cue indicating where an expected target is most likely to appear, or what feature of the object to attend. The subject therefore needs to orient his or her attentional focus to the cued location/feature to accurately select the relevant target among distracters. Importantly, this is not necessarily restricted to one sensory modality, so intersensory cueing experiments are common (Eimer, 2001; Eimer & Driver, 2000; Foxe & Simpson, 2005; Foxe, Simpson, Ahlfors, & Saron, 2005). An extensively used alternative in the study of attentional orienting that does not rely on directional cues is the visual search paradigm (Davis & Palmer, 2004). In this approach one has to find a relevant target in a set of simultaneously presented objects. The use of natural scenes (see Chap. 4 by Maldonado et al.) is a form of visual search paradigm. Importantly, one difference from Posner's classical paradigms is that in visual search selection is mostly endogenous, while in the former both exogenous and endogenous selection are probed.

One could argue that orienting attention and selecting are two different processes, orienting being the displacement of the focus of attention and selection the singling out of a target among distracters. But this is not such a clear-cut distinction. To orient you have to choose a location, object, feature, etc. among alternatives (see Chap. 3 by Luck). Conversely, to select you have to orient the focus of attention in space and/or among features of a potential target (see Chap. 3 by Luck). This is why several authors use selection and orienting as referring to the general process of finding a target among distracters. Nevertheless, especially when restricting oneself to spatial and temporal orientation, it seems possible to disentangle these two components. For example, attentional blink[7] and, in general, rapid serial visual presentation paradigms, minimize spatial orienting and can therefore probe selection more specifically (Arnell, Howe, Joanisse, & Klein, 2006; Hommel et al., 2006; Popple & Levi, 2007). On the other hand, studying what happens after the cue but prior to the appearance of the target in visual endogenous cueing paradigms has been used to isolate the orienting component of attention from the selective processes (Foxe & Simpson, 2005; Foxe, Simpson, & Ahlfors, 1998; Sauseng et al., 2005).

---

[5] Note that objects and features can be nonspatial as in searching for a deviant tone in a single auditory stream.

[6] Visual cueing paradigms can use symbolic central (endogenous) or peripheral (exogenous) cues, inducing either voluntary or reflexive orienting, respectively.

[7] The attentional blink consists in the failure to detect the second of two targets embedded in a rapid serial stream of sensory input, if both targets are separated by less than approximately 500 ms.

A Qualification of the Standard Approach

The aforementioned cases are examples of classic situations in which attention is highly controlled to disentangle specific aspects of the process. In all but the (free) visual search cases, the subject has to perform, following very specific instructions, some kind of simplified – often restricted – task on perceptual objects *which have been defined and chosen by the experimenter*. But is this what we do throughout a regular day? Do we usually sit in front of a stream of repetitive objects waiting for something to happen? Are we continuously being indicated where to attend? Is attention only about enhancing performance? These are pertinent questions if we are to study attention in its daily, intimate expression.

Bent on Performance – The Informational Bias

"Regardless of one's methodology, discipline, and intuitions, there is only one core issue that justifies attentional processes: *information reduction*" (Itti et al., 2005, p. *xxx*).

Standard approaches to attention are based on the information-reduction metaphor as the main characterization. The origins of this can probably be traced to the way cognition is conceptualized in the first place, and how organisms are understood (Varela, 1979). Apart from the more recent embodied cognition and dynamical systems approach (Clark, 1999; Thompson & Varela, 2001), the usual way of operationalizing the study of cognitive processes is by – allegedly – fixing the state of the subject and varying the environment. In the context of an information processing system view, this naturally leads to asking how the subject selects from the multiple sources of information that arrive from the world, and how doing so affects its performance.

Yet organisms are not only reactive, but are fundamentally motivated, moody, easily interested, and easily bored. They seek, explore, rest, actively ignore, and choose according to their internal state of affairs, and how external situations match or not these prior expectations. Why does a cat sit there, looking into nothing, almost as if meditating? Is it just a monitoring device ready to jump at any minimal change in the environment? Or could it be possible that it might be shifting its attention spontaneously between multiple internal and external aspects of its cat-experience? This however inscrutable question seems to suggest that attending is not only about selection of externally predefined objects, for the benefit of performance. Both selection and performance are ultimately relative to the organisms' motivational state, not just the world's state. If the only function of attention were to enhance performance on some task, we would have to answer that the cat is either in a pure monitoring state or in a full blank – mentally void – state. Neither seems to capture the situation.

The upshot is that while selection and changes in performance are undoubtedly fundamental functions of attention, they do not exhaust it as a phenomenon. As we will see further later, mind-wandering is a pervasive state in human beings, yet it has been demonstrated to impair performance as measured by recall and perceptual discrimination (Smallwood, Fishman, & Schooler, 2007; Smallwood & Schooler, 2006). Why would it therefore be evolutionarily selected if performance were the only driving force? The existence of ongoing attentional fluctuations, between a multiplicity of possible poles of interest – many of which are not immediately related to the environmental state of affairs – should be incorporated into our characterization of attention, regardless of whether we deem them useful in any a priori sense (Ballard et al., 2002; Laufs et al., 2003; Sonuga-Barke & Castellanos, 2007).[8]

I'll Tell You What's Good for You

Experimental manipulations of attention generally consider an endogenous versus exogenous distinction in relation to the cueing procedure that is used (Hopfinger, Buonocore, & Mangun, 2000; Hopfinger & West, 2006). Endogenous attention is called into action when symbolic cues are presented at or near fixation instructing the subject where and when to attend. The subject has to understand the cue and make a voluntary decision to shift attention, thus the "endo-" prefix. On the other hand, exogenous attention refers to orienting attention reflexively, driven by a salient event that is not the current focus of attention. Here cues are presented away from fixation, inducing an involuntary shift of attention towards its location, justifying the "exo-" denomination.

This distinction is extremely useful as numerous neuropsychological studies have demonstrated. It has been shown that certain brain lesions affect these two types of attention differentially (Corbetta & Shulman, 2002). Indeed, important differences exist at the level of cortical networks involved: whereas endogenous attention recruits rather symmetric frontoparietal networks, exogenous attention depends on the integrity of a strongly lateralized right-hemisphere network involving inferior parietal, temporoparietal junction, and lateral prefrontal cortex (Fox, Corbetta, Snyder, Vincent, & Raichle, 2006).

However, especially in the endogenous attention case, these approaches are based on the experimenter determining what is relevant (what counts as a target or a distracter). The notion of voluntarily shifting attention is therefore quite narrow: you shift where you are instructed to, to perform a task. But to what extent is this the way attention is shifted in the stream of consciousness? Certainly someone can

---

[8] One could argue that when studying the executive attentional system (which determines priorities and builds saliency maps according to short and long-term goals) we are already dealing with this exploratory aspect. But as will become evident in the following sections, this does not undermine the main point being made here. Rather, we stress that besides the endogenous–exogenous distinction (where the executive system is a major player), there is a spontaneous ongoing character to attentional shifts that exist regardless of a predefined performance criterion.

come and tell you "Think of what you had for lunch yesterday" or ask you "Do you hear the sound of the cars far away?" and you will turn your attention accordingly. Yet most of the time it is us who spontaneously – and sometimes despite ourselves – choose the focus of our current interest. Put another way, even in such highly controlled cases as these, it seems impossible to completely exclude the ongoing, variable, activity of the organism from the experimental situation. This reveals an important limitation in standard approaches to attentional dynamics when dealing with more ecological settings.

Singling Out the Shift

There is, however, an aspect in the standard approach that deals precisely with a structural feature of attention that is highly relevant for the study of the stream of consciousness – the *shifting* or *reorienting* of attention.[9]

Using classical cue–target paradigms, a growing number of studies are exploring what happens when attention changes focus and prior to it selecting a given object (Anllo-Vento, 1995; Belmonte, 1998; Gitelman et al., 1999; Hopf & Mangun, 2000; Kelley, Serences, Giesbrecht, & Yantis, 2008; Kiss, Velzen, & Eimer, 2007; Muller, Teder-Salejarvi, & Hillyard, 1998; Sauseng et al., 2005; Wager, Jonides, & Reading, 2004; Yamaguchi, Tsuchiya, & Kobayashi, 1994, 1995). In the visuospatial domain, it has been shown that activity in a prefrontal parietal network is specific for the process of shifting the focus of attention. Interestingly, studies using tactile and auditory stimuli suggest that at least a part of neural activity in this network is supramodal (Eimer & Van Velzen, 2002; Eimer, van Velzen, Forster, & Driver, 2003), possibly underlying a generic displacement of the attentional focus. Moreover, high-density EEG recordings have shown that a similar frontoparietal network participates in a more complex version of reorienting: task switching (Wylie, Javitt, & Foxe, 2003). Although the precise limits of such a potential supramodel network remain to be elucidated – especially in the case of task switching (Rushworth, Passingham, & Nobre, 2005; Wager, Jonides, Smith, & Nichols, 2005; Wylie et al., 2003) – its activation could serve as a neurodynamical signature for the shifts that punctuate the stream of consciousness (Corbetta M, Patel G & Shulman GL., 2008).

Two limitations are nevertheless evident in standard cue–target paradigms and related task-switching studies. We touched upon the first one in the previous section: shifts of attention are cued in time, therefore neglecting their spontaneity (see "The Spontaneity of Attention"). The second limitation has to do with the poor

---

[9] Here "shifting" or "reorienting" attention is used to refer to the dynamical change of the main focus of attention, and does not refer uniquely to the visuospatial domain (see also following paragraph).

signal-to-noise ratio that current brain imaging techniques have. Being able to track the evolution of brain activity as it unfolds in time, without resorting to averaging, is a crucial methodological problem to which we will return in "Some Considerations Regarding Signal Analysis Strategies."

A Phenomenological Detour

> *"The mind chooses to suit itself, and decides what particular sensation shall be held more real and valid than all the rest"*
> *(James, 1890, p. 286).*

Many authors in the philosophical phenomenology tradition have dealt with the problem of attention from Husserl onwards. We cannot expand here on this important issue so we will refer the interested reader to a number of excellent reviews and synthesis as a point of entry (Arvidson, 1996, 2003; Depraz, 2004; Steinbock, 2004; Vermersch, 2004). We will, however, draw from one author's account that is probably the best known in cognitive neuroscience: William James.[10]

It is notable that one of the fathers of contemporary empirical psychology and cognitive neuroscience considered the stream of thought as probably the most fundamental source for the study of attention. James delved into this complex matter in several chapters of his magnum opus of the end of the nineteenth century (James, 1890). He produced detailed descriptions that remain today as some of the most compelling first-person accounts of the dynamics of attention. Here we will focus on two aspects of his description.

The Spontaneity of Attention[11]

In discussing the different kinds of attention,[12] James states that "there is no clear line to be drawn between immediate and derived attention of an intellectual sort." We would contend that this also applies to the case of the daily wanderings of

---

[10] Although James is not strictly a phenomenologist (he advocated a "radical empiricism" basing much of his work on physiological studies of his time) we put him in this category because he did not fear going into the description of experience, and did so with impressive insight. Indeed Husserl himself – the founder of phenomenology – studied and admired his work (Depraz, 2004).

[11] I am greatly indebted to Evan Thompson for drawing my attention to the problem of spontaneity.

[12] James made in what he called "the varieties of attention," several distinctions regarding the different kinds of attention: either it is to objects of sense (sensory) or to ideal objects (intellectual); it is immediate (when the object is interesting in itself) or derived (when interest depends on some kind of relation to another object); it is passive, reflex, nonvoluntary, effortless or active and voluntary (James, 1890, p. 416). Importantly these three categories can be shared differentially among kinds, e.g., passive immediate sensory attention, where a relevant perceptual object reflexively draws our attention to it.

attention, as well as to the traditional endogenous/voluntary versus exogenous/ reflexive distinction (see "I'll Tell You What's Good for You"). Moreover, the traditional dichotomy that equates endogenous with voluntary and exogenous with passive-reflexive does not fully capture the dynamics of attention. This is because it leaves out the deep spontaneous, ongoing dynamics of attentional displacements, where shifts can occur towards personally relevant issues, while the shift itself is, in fact, hardly controllable (Smallwood & Schooler, 2006).

One could argue for a clear-cut distinction by saying that attention can be, at a given moment, completely in exogenous mode: if a car honks as it violently stops 20 cm away from your leg, there is no use in reflecting about the situation, you orient, you freeze, or you jump; you do not really have the choice. But, as any driver with some experience knows, a person who is deeply absorbed in some engaging daydream or preoccupation can keep walking across the street, only to realize a moment later that something important just happened.

During the ongoing stream of consciousness we are confronted with a complex coalescence of attentional types that poses a nontrivial methodological challenge. Even within a moment of concentration on a single object, which of the local displacements of attention are voluntary and effortful and which are passive and effortless is an open – and difficult – question. James cites Helmholtz: "The relation of attention to will is … less one of immediate than of mediate control" (James, 1890, p. 423). This seems to capture a crucial structural feature of attention that is almost paradoxical: it is spontaneous and *at the same time* it can be willfully "pushed" in one sense or another. This is probably the least explored aspect of attention, an aspect that the phenomenology of the flow of consciousness singles out as fundamental.

Rate of Change and the Substantive/Transitive Distinction

The second aspect of the Jamesian description that we wish to single out is not in the chapter on attention in his *Principles of Psychology*. In Chap. IX, "The Stream of Thought," he distinguishes two important aspects: substantive, stable moments, in which one is actually conscious of something, and transitive, fleeting moments, in which one passes from one content to another.

James remarks that substantive moments can be recognized as such, whereas transitive moments are quite difficult to pinpoint. They present themselves as tendencies and changes between states, and not as distinct contents immediately definable in themselves, save by some retrospective analysis. He then goes on to discuss that the main difference between these two aspects of the flow is the rate of change – slow during substantive parts and fast during transitive parts.

What is interesting for us here is that transitive parts between substantive moments are reminiscent of a basic attentional shift between more or less distinct contents. If in fact the main difference between substantive and transitive moments resides therein, following the rate of change of relevant neurophysiological parameters could serve as an objective marker of the ongoing deployment of attention during the stream of consciousness.

Emerging Trends

It seems that leaving out the study of attentional dynamics during the flow of coscious-ness owes much to the technical challenges involved. Experimental psychology and neuroimaging paradigms require very precise conditions to make them reproducible and to afford consistent interpretations of the results. But the flow of consciousness is all but precise and well delimited: it is structurally changing and thus fundamen-tally irreproducible, at least content-wise. Is it possible therefore to bring together phenomenological insights, traditional experimental psychology paradigms, and brain imaging approaches to bear on this question? We believe this line of inquiry is already at work in the cognitive neuroscience community, and worth considering in the general framework of the study of attention.

The study of no-task, no-input conditions goes back at least to the invention of the electroencephalograph (Berger, 1969), yet only recently has it become a relevant question in cognitive neuroscience (Lehmann, 1971; Lehmann, Henggeler, Koukkou, & Michel, 1993; Lehmann, Strik, Henggeler, Koenig, & Koukkou, 1998). We wish to highlight two major approaches to this question: mind-wandering or stimulus-independent thought, and studies on the default mode of brain function.

When the Mind Drifts

Mind-wandering and the related concept of stimulus-independent thought or stimu-lus-independent cognition has been studied since the late 1960s (Antrobus & Singer, 1964; Antrobus, Singer, Goldstein, & Fortgang, 1970; Singer, 1974; Singer & Antrobus, 1963; Wollman & Antrobus, 1986). Smallwood and Schooler (2006) have suggested the general term of "mind-wandering" to refer to task-unrelated thought, task-unrelated images, stimulus-independent thought, mind pops, zone outs, day-dreaming, etc. All these appeal to the fact that while doing something, say reading a novel, it is not uncommon to drift into thoughts that are not relevant to the action at hand, only to realize after a while that one has been distracted. Nevertheless, in the specific case of stimulus-independent thought, this can be elicited by specific instructions to perform the task (see below). In such cases, the nature of the shift of attention that takes place is different (Burgess, Dumontheil, & Gilbert, 2007).

There is a rich literature on this complex issue that is usually neglected in the more brain-function-inclined community (but see Smallwood, Beach, Schooler, & Handy, 2007 for a recent attempt). For the sake of conciseness we refer the reader to the work of Smallwood and Schooler (2006) that provides a thorough treatment of this question from the cognitive psychology perspective, as well as the pioneering book by Pope and Singer (1978). However, some general aspects are worth mentioning regarding experimental paradigms. Depending on the emphasis on mind-wandering or stimulus-independent thought, the experimental strategy used will vary: in the context of cognitive neuroscience, mind-wandering paradigms usually rely on the fluctuations of sustained attention and vigilance; continuous performance tasks such as the sustained attention to response task (Robertson, Manly, Andrade, Baddeley, & Yiend, 1997) are a common choice. By random thought-sampling (Pope and Singer, 1978) alone or by combining this with an electro-physiological analysis of brain activity evoked by stimuli just prior to errors, it is possible to determine when, approximately, the subject's focus of attention has drifted and therefore analyze retrospectively the corresponding brain states (Smallwood, Beach, Schooler, & Handy, 2007). On the other hand, in order to focus on stimulus-independent thought, a number of experimental tasks have been developed (Burgess et al., 2007; Gilbert, Frith, & Burgess, 2005). Generally, two conditions are contrasted: when the subject has to perform a given task fully dependent on the perceptually available information, and when he or she has to perform a task that demands some kind of intervening mental operation (for example, sum successive 13's starting from the number which is presented on the screen in red). In these cases the task either does not rely on perceptual information at all or, as in the previous example, only does so as a way to trigger the mental process (Burgess et al., 2007; Griffin & Nobre, 2003; Nobre et al., 2004).

Especially in the case of mind-wandering, these studies begin to take into account the spontaneous character of the shift of attention, therefore complementing the standard approach of endogenous/exogenous cueing. On the other hand, the stimulus-independent thought experiments sacrifice the spontaneous aspects but are able to capture more precisely another very relevant aspect of the stream of consciousness, i.e., attentional shifts from an external object towards a mental image.

Nothing to Do, Yet So Busy

A related question is the discovery of what has become known as the brain's "default-mode" network (Raichle et al., 2001). This group of brain regions – comprising typically medial prefrontal areas, posterior cingulate/precuneus, and lateral parietal cortex – is important from both theoretical and empirical points of view, because it shows higher activation during rest periods than when actively engaged in a task.

An important consequence of this discovery has been the study of spontaneous correlated fluctuations among distributed brain regions – as measured by functional magnetic resonance imaging (fMRI) blood oxygenation level dependent (BOLD) signals – during rest states (Biswal, Yetkin, Haughton, & Hyde, 1995). Correlation between spontaneous fluctuations of the default-mode network and alpha activity (8–12 Hz) and a subband in the beta range (17–23 Hz) have been claimed to reveal attentional fluctuations of the subject's state during a no-task condition (Laufs et al., 2003; Sonuga-Barke & Castellanos, 2007). Changes in this network have also been related to changes in behavioral performance (Fox, Snyder, Vincent, & Raichle, 2007; Fox, Snyder, Zacks, & Raichle, 2006; Kelly, Uddin, Biswal, Castellanos, & Milham, 2008). Moreover, activity in this set of brain areas has been related to stimulus-independent thought (McGuire et al., 1996) and self-reflection (Gusnard, Akbudak, Shulman, & Raichle, 2001). Nevertheless, results showing that correlated fluctuations in the BOLD signal persist even in sleep and anesthesia (Horovitz et al., 2007; Peltier et al., 2005; Vincent et al., 2007) support the idea that, in addition, a more structural level of brain activity is at play continuously (Fox & Raichle, 2007).

Whatever the case, it seems that at least part of the activity of this network is related to ongoing attentional processes, as the work by Fox et al. (2006) suggests. On the basis of the analysis of correlated fluctuations in the BOLD signal, the authors where able to distinguish the bilateral dorsal and right ventral networks that have been claimed to underlie the two attentional systems – reflexive and voluntary – of the brain (see also "I'll Tell You What's Good for You"). This could be a quite relevant finding in the context of the stream of consciousness, as it might serve as a "brain network template" that can be followed in a more constrained situation. Indeed, recent work has put together the default-mode network (albeit a restricted version of it) and the mind-wandering phenomena discussed previously, showing that activity in this network correlates with mind-wandering and stimulus-independent thought (Gilbert, Dumontheil, Simons, Frith, & Burgess, 2007; Mason et al., 2007a, 2007b). It remains to be seen how such a network would be engaged during a more complex situation where multiple alternatives are available to spontaneous selection by the subject, including "external" objects, mental imagery, and bodily sensations (see Fig. 1 in Mason et al., 2007b). Moreover, given the temporal limits of the fMRI approach, complementing it with electrophysiological measures such as EEG or MEG seems necessary to follow the rapid ongoing shifts that punctuate the stream of consciousness (Cosmelli & Thompson, 2007).

Some Considerations Regarding Signal Analysis Strategies

The last issue we wish to draw the reader's attention to is the problem of signal analysis that is inherent to the study of a fluctuating, nonstationary, and unconstrained flow of events. Traditional approaches in brain imaging and human electrophysiology rely heavily on signal averaging as a way of enhancing the signal-to-noise ratio (Hillyard & Kutas, 1983). However, adopting this strategy when dealing with the stream of consciousness is not necessarily adequate because there are no clear temporal markers to indicate time zero in the averaging procedure. Moreover, as it is fundamentally a matter of shifting between contents that rarely repeat, seeking an average activity does not seem to capture what would be related to the subjects' experience.

Recent developments in signal analysis do, however, provide alternative means to deal with this problem (Makeig, Debener, Onton, & Delorme, 2004; Corbetta M.

et al., 2008). Related to classical evoked potentials research is the notion of single trial analysis. Here the aim is to extract significant structure from the signal obtained during individual events. Numerous groups have developed sophisticated algorithms to detect and classify such events from different analytical perspectives (Bénar, Clerc, & Papadopoulo, 2007; Delorme & Makeig, 2004; Duann et al., 2002; Ioannides, 2001; Iyer & Zouridakis, 2007; Jung et al., 2001; Lachaux et al., 2002; Röschke et al., 1996). This general approach seems well suited to analyzing what is going on during the stream of consciousness as long as some kind of marker can be obtained. In the specific case of freely exploring a scene, eye movements would be a prima facie choice (see Chap. 3 by Maldonado et al. and Chap. 4 by Luck). Although this restricts in principle the analysis to events that deal with external exploratory actions (Leopold & Logothetis, 1999), it could serve as a complementary strategy to other approaches that do not rely on well-defined temporal points (see later). Finally the use of steady-state evoked potentials (SSEP) – both in visual (Burkitt, Silberstein, Cadusch, & Wood, 2000) and in auditory (Kaiser & Bertrand, 2003; Lins & Picton, 1995) modalities – can reveal the relative amount of attention directed upon an object. By following both the waxing and the waning of the SSEP, one should be able to track spontaneous shifts of attention to and from a given (frequency-tagged) object (Cosmelli et al., 2004; Cosmelli & Thompson, 2007). Because it has been shown that mind-wandering reduces performance in perceptual tasks (Smallwood, Beach, Schooler, & Handy, 2007; Smallwood, Fishman, & Schooler, 2007), using an SSEP approach could indicate the moment when attention is shifted (either spontaneously or instructed) from the external object to an internal mental image.

An alternative approach is to consider the entire sequence of data as the target for analysis. For example, independent component analysis has been shown to be able to blindly extract independent sources of variation in the signal (Delorme & Makeig, 2004; Makeig et al., 2004) and is currently used in studying variations in the default-mode network, mainly in fMRI recordings (McKeown et al., 1998; McKeown & Sejnowski, 1998). Another related multivariate strategy is to use classification algorithms (Besserve et al., 2007; Byvatov & Schneider, 2003). In these approaches the aim is to discriminate ongoing periods in the signal as belonging to one of a set of classes based on a multidimensional representation of the original data. A notorious potential of these methods lies in the capacity to distinguish with high accuracy the relative state of the subject using data windows as short as 1 s (Besserve et al., 2007) as long as an adequate training subset is available. In a similar vein, dynamical system-inspired approaches profit from the fact that dynamical systems exhibit recurrent but temporally variable trajectories in phase space (Cosmelli, Lachaux, & Thompson, 2007; Faure & Korn, 2001). Accordingly, relaxing the constraint on temporal regularity and focusing on recurrent patterns throughout an extended period is a promising avenue (David, Cosmelli, Hasboun, & Garnero, 2003). In the context of an MEG experiment on multistable perception we have been able to distinguish dominant and transition periods without any averaging procedures (Cosmelli et al., 2004). What is interesting in this case is that transitions during multistable perception are spontaneous and occur between substantive (i.e., dominant) states (see "Rate of Change and the Substantive/Transitive Distinction"), in a similar way to how attention shifts during the stream of consciousness. A complementary approach to recurrence-based methods would therefore be to consider a measure of the (dynamical) rate of change of a neurophysiological parameter to pinpoint transitions (Le Van Quyen, Chavez, Rudrauf, & Martinerie, 2003; Rudrauf et al., 2006). Numerous other approaches exist that deal directly with the dynamical analysis of ongoing electrophysiological signals based either on fluctuation analysis (Buchanan, Kelso, & Fuchs, 1996; Fox et al., 2005; Linkenkaer-Hansen, Nikouline, Palva, & Ilmoniemi, 2001; Stam & de Bruin, 2004) or on the discrimination of stationary local states embedded in the EEG signal (Lehmann, 1984, 1989; Lehmann et al., 1998).

Conclusions and Challenges – Towards an Ecology of Attention

The study of the dynamics of attention during the ongoing stream of consciousness, regardless of its specific object, is crucial if we are to come up with a complete theory of attention. Focusing on attention in restricted conditions, where it is either reflexively driven or symbolically cued, does not exhaust its varieties (James, 1890). In this context, recent studies in mind-wandering and stimulus-independent thought offer a promising avenue for research that considers a more ecological setting. In these situations two basic distinctions are brought together, the study of sustained attention, which focuses on the background state of the subject, and the moment this is lost, resulting in a reorienting – a shift – of the attentional focus. This can then be expanded by taking into account a more varied setting where several tasks are simultaneously available for the subject to choose by himself or herself (Alderman, Burgess, Knight, & Henman, 2003; Smilek, Birmingham, Cameron, Bischof, & Kingstone, 2006). This approach would be a general example of how it might be possible to tackle this elusive phenomenon while still profiting from the valuable know-how of experimental psychology and doing justice to phenomenological insight.

Many challenges raised by these questions are of a methodological nature. But they also point to a rather fundamental aspect of the way we conceptualize brain function and its relation to everyday experience. In this regard, the main conclusion we would like to advance is the following: attention in particular, and the stream of consciousness in general, cannot be fully understood if its spontaneous nature is not taken into account. As a way of operationalizing this claim, we propose that the reflexive/voluntary distinction in the study of the orienting of attention has to be complemented with the notion of *spontaneous* shifts, which are deeply motivated but essentially out of immediate control. Recent developments in brain imaging, experimental paradigms, and signal analysis are beginning to tap into this much neglected aspect of the human experience of attention.

Acknowledgment D.C. acknowledges support from the PBCT/CONICYT Initiative: Ring for Sensory Neuroscience, ACT-45.

References

Alderman, N., Burgess, P. W., Knight, C., & Henman, C. (2003). Ecological validity of a simplified version of the multiple errands shopping test. *Journal of International Neuropsychological Society, 9*(1), 31–44.

Anllo-Vento, L. (1995). Shifting attention in visual space: The effects of peripheral cueing on brain cortical potentials. *The International Journal of Neuroscience, 80*(1–4), 353–370.

Antrobus, J. S., & Singer, J. L. (1964). Eye movements accompanying daydreaming, visual imagery, and thought suppression. *Journal of Abnormal Psychology, 69*, 244–252.

Antrobus, J. S., Singer, J. L., Goldstein, S., & Fortgang, M. (1970). Mindwandering and cognitive structure. *Transactions of the New York Academy of Sciences, 32*(2), 242–252.

Arnell, K. M., Howe, A. E., Joanisse, M. F., & Klein, R. M. (2006). Relationships between attentional blink magnitude, RSVP target accuracy, and performance on other cognitive tasks. *Memory and Cognition, 34*(7), 1472–1483.

Arvidson, P. S. (1996). Towards a phenomenology of attention. *Human Studies, 19*, 71–84.

Arvidson, P. S. (2003). A lexicon of attention: From cognitive science to phenomenology. *Phenomenology and the Cognitive Sciences, 2*, 99–132.

Ballard, C. G., Aarsland, D., McKeith, I., O'Brien, J., Gray, A., Cormack, F., et al. (2002). Fluctuations in attention: PD dementia vs. DLB with parkinsonism. *Neurology, 59*(11), 1714–1720.

Bastiaansen, M. C., Bocker, K. B., Brunia, C. H., de Munck, J. C., & Spekreijse, H. (2001). Event-related desynchronization during anticipatory attention for an upcoming stimulus: A comparative EEG/MEG study. *Clinical Neurophysiology, 112*(2), 393–403.

Belmonte, M. (1998). Shifts of visual spatial attention modulate a steady-state visual evoked potential. *Brain Research. Cognitive Brain Research, 6*(4), 295–307.

Bénar, C., Clerc, M., & Papadopoulo, T. (2007). Adaptive time-frequency models for single-trial M/EEG analysis. *Information Process in Medical Imaging, 20*, 458–469.

Berger, H. (1969). On the electroencephalogram of man. *Electroencephalography and Clinical Neurophysiology,* (Suppl 28), 37–73.

Besserve, M., Jerbi, K., Laurent, F., Baillet, S., Martinerie, J., & Garnero, L. (2007). Classification methods for ongoing EEG and MEG signals. *Biological Research, 40*(4), 415–437.

Biswal, B., Yetkin, F. Z., Haughton, V. M., & Hyde, J. S. (1995). Functional connectivity in the motor cortex of resting human brain using echo-planar MRI. *Magnetic Resonance in Medicine, 34*(4), 537–541.

Brunia, C. H., & van Boxtel, G. J. (2001). Wait and see. *International Journal of Psychophysiology, 43*(1), 59–75.

Buchanan, J. J., Kelso, J. A., & Fuchs, A. (1996). Coordination dynamics of trajectory formation. *Biological Cybernetics, 74*(1), 41–54.

Burgess, P. W., Dumontheil, I., & Gilbert, S. J. (2007). The gateway hypothesis of rostral prefrontal cortex (area 10) function. *Trends in Cognitive Science, 11*(7), 290–298.

Burkitt, G. R., Silberstein, R. B., Cadusch, P. J., & Wood, A. W. (2000). Steady-state visual evoked potentials and travelling waves. *Clinical Neurophysiology, 111*(2), 246–258.

Byvatov, E., & Schneider, G. (2003). Support vector machine applications in bioinformatics. *Applied Bioinformatics, 2*(2), 67–77.

Clark, A. (1999). An embodied cognitive science? *Trends in Cognitive Science, 3*(9), 345–351.

Corbetta, M., & Shulman, G. L. (1998). Human cortical mechanisms of visual attention during orienting and search. *Philosophical Transactions of the Royal Society of London, 353*(1373), 1353–1362.

Corbetta, M., & Shulman, G. L. (2002). Control of goal-directed and stimulus-driven attention in the brain. *Nature Reviews Neuroscience, 3*(3), 201–215.

Corbetta, M., Patel, G., & Shulman, G. L. (2008) The reorienting system of the human brain: From environment to theory of mind. *Neuron, 58*(3):306–24.

Cosmelli, D., David, O., Lachaux, J. P., Martinerie, J., Garnero, L., Renault, B., et al. (2004). Waves of consciousness: Ongoing cortical patterns during binocular rivalry. *Neuroimage, 23*(1), 128–140.

Cosmelli, D., Lachaux, J.-P., & Thompson, E. (2007). Neurodynamical approaches to consciousness. In P. Zelazo, M. Moscovitch, & E. Thompson (Ed.), *The cambridge handbook of consciousness*. Cambridge: Cambridge University Press.

Cosmelli, D., & Thompson, E. (2007). Mountains and valleys: Binocular rivalry and the flow of experience. *Consciousness and Cognition, 16*(3), 623–641.

David, O., Cosmelli, D., Hasboun, D., & Garnero, L. (2003). A multitrial analysis for revealing significant corticocortical networks in magnetoencephalography and electroencephalography. *Neuroimage, 20*(1), 186–201.

Davis, E. T., & Palmer, J. (2004). Visual search and attention: An overview. *Spatial Vision, 17*(4–5), 249–255.

Delorme, A., & Makeig, S. (2004). EEGLAB: An open source toolbox for analysis of single-trial EEG dynamics including independent component analysis. *Journal of Neuroscience Methods, 134*(1), 9–21.

Depraz, N. (2004). Where is the phenomenology of attention that Husserl intended to perform? A transcendental pragmatic-oriented description of attention. *Continental Philosophy Review, 37*, 5–20.

Duann, J. R., Jung, T. P., Kuo, W. J., Yeh, T. C., Makeig, S., Hsieh, J. C., et al. (2002). Single-trial variability in event-related BOLD signals. *Neuroimage, 15*(4), 823–835.

Eimer, M. (2001). Crossmodal links in spatial attention between vision, audition, and touch: Evidence from event-related brain potentials. *Neuropsychologia, 39*(12), 1292–1303.

Eimer, M., & Driver, J. (2000). An event-related brain potential study of cross-modal links in spatial attention between vision and touch. *Psychophysiology, 37*(5), 697–705.

Eimer, M., & Van Velzen, J. (2002). Crossmodal links in spatial attention are mediated by supramodal control processes: Evidence from event-related potentials. *Psychophysiology, 39*(4), 437–449.

Eimer, M., van Velzen, J., Forster, B., & Driver, J. (2003). Shifts of attention in light and in darkness: An ERP study of supramodal attentional control and crossmodal links in spatial attention. *Brain Research, 15*(3), 308–323.

Ericsson, K. A., & Simon, H. A. (1993). *Protocol analysis: Verbal reports as data*. Cambridge, MA: e MIT.

Fan, J., McCandliss, B. D., Fossella, J., Flombaum, J., & Posner, M. (2005). The activation of attentional networks. *Neuroimage, 26*(2), 471–479.

Fan, J., McCandliss, B. D., Sommer, T., Raz, A., & Posner, M. I. (2002). Testing the efficiency and independence of attentional networks. *Journal of Cognitive Neuroscience, 14*(3), 340–347.

Fan, J., & Posner, M. (2004). Human attentional networks. *Psychiatrische Praxis, 31*(Suppl 2), S210–S214.

Faure, P., & Korn, H. (2001). Is there chaos in the brain? I. Concepts of nonlinear dynamics and methods of investigation. Comptes Rendus de l'Académie Des Sciences. Série III, *324*(9), 773–793.

Fox, M. D., Corbetta, M., Snyder, A. Z., Vincent, J. L., & Raichle, M. E. (2006). Spontaneous neuronal activity distinguishes human dorsal and ventral attention systems. *Proceedings of the National Academy of Sciences of the United States of America, 103*(26), 10046–10051.

Fox, M. D., & Raichle, M. E. (2007). Spontaneous fluctuations in brain activity observed with functional magnetic resonance imaging. *Nature Reviews Neuroscience, 8*(9), 700–711.

Fox, M. D., Snyder, A. Z., Vincent, J. L., Corbetta, M., Van Essen, D. C., & Raichle, M. E. (2005). The human brain is intrinsically organized into dynamic, anticorrelated functional networks. *Proceedings of the National Academy of Sciences of the United States of America, 102*(27), 9673–9678.

Fox, M. D., Snyder, A. Z., Vincent, J. L., & Raichle, M. E. (2007). Intrinsic fluctuations within cortical systems account for intertrial variability in human behavior. *Neuron, 56*(1), 171–184.

Fox, M. D., Snyder, A. Z., Zacks, J. M., & Raichle, M. E. (2006). Coherent spontaneous activity accounts for trial-to-trial variability in human evoked brain responses. *Nature Neuroscience, 9*(1), 23–25.

Foxe, J. J., & Simpson, G. V. (2005). Biasing the brain's attentional set: II. effects of selective intersensory attentional deployments on subsequent sensory processing. *Experimental Brain Research. Experimentelle Hirnforschung, 166*(3–4), 393–401.

Foxe, J. J., Simpson, G. V., & Ahlfors, S. P. (1998). Parieto-occipital approximately 10 Hz activity reflects anticipatory state of visual attention mechanisms. *Neuroreport, 9*(17), 3929–3933.

Foxe, J. J., Simpson, G. V., Ahlfors, S. P., & Saron, C. D. (2005). Biasing the brain's attentional set: I. cue driven deployments of intersensory selective attention. *Experimental Brain Research. Experimentelle Hirnforschung, 166*(3–4), 370–392.

Gilbert, S. J., Dumontheil, I., Simons, J. S., Frith, C. D., & Burgess, P. W. (2007). Comment on "Wandering minds: The default network and stimulus-independent thought". *Science, 317*(5834), 43; author reply 43.

Gilbert, S. J., Frith, C. D., & Burgess, P. W. (2005). Involvement of rostral prefrontal cortex in selection between stimulus-oriented and stimulus-independent thought. *European Journal of Neuroscience, 21*(5), 1423–1431.

Gitelman, D. R., Nobre, A. C., Parrish, T. B., LaBar, K. S., Kim, Y. H., Meyer, J. R., et al. (1999). A large-scale distributed network for covert spatial attention: Further anatomical delineation based on stringent behavioural and cognitive controls. *Brain, 122 (Pt 6)*, 1093–1106.

Griffin, I. C., & Nobre, A. C. (2003). Orienting attention to locations in internal representations. *Journal of Cognitive Neuroscience, 15*(8), 1176–1194.

Gusnard, D. A., Akbudak, E., Shulman, G. L., & Raichle, M. E. (2001). Medial prefrontal cortex and self-referential mental activity: Relation to a default mode of brain function. *Proceedings of the National Academy of Sciences of the United States of America, 98*(7), 4259–4264.

Hillyard, S. A., & Anllo-Vento, L. (1998). Event-related brain potentials in the study of visual selective attention. *Proceedings of the National Academy of Sciences of the United States of America, 95*(3), 781–787.

Hillyard, S. A., & Kutas, M. (1983). Electrophysiology of cognitive processing. *Annual Review of Psychology, 34*, 33–61.

Hillyard, S. A., & Mangun, G. R. (1987). Sensory gating as a physiological mechanism for visual selective attention. *Electroencephalography and Clinical Neurophysiology, 40*, 61–67.

Hommel, B., Kessler, K., Schmitz, F., Gross, J., Akyurek, E., Shapiro, K., et al. (2006). How the brain blinks: Towards a neurocognitive model of the attentional blink. *Psychological Research, 70*(6), 425–435.

Hopf, J. M., & Mangun, G. R. (2000). Shifting visual attention in space: An electrophysiological analysis using high spatial resolution mapping. *Clinical Neurophysiology, 111*(7), 1241–1257.

Hopfinger, J. B., Buonocore, M. H., & Mangun, G. R. (2000). The neural mechanisms of top-down attentional control. *Nature Neuroscience, 3*(3), 284–291.

Hopfinger, J. B., & West, V. M. (2006). Interactions between endogenous and exogenous attention on cortical visual processing. *Neuroimage, 31*(2), 774–789.

Horovitz, S. G., Fukunaga, M., de Zwart, J. A., van Gelderen, P., Fulton, S. C., Balkin, T. J. & Duyn J.H. (2008). Low frequency BOLD fluctuations during resting wakefulness and light sleep: A simultaneous EEG-fMRI study. *Human Brain Mapping, 29*(6), 671–682.

Ioannides, A. A. (2001). Real time human brain function: Observations and inferences from single-trial analysis of magnetoencephalographic signals. *Clinical Electroencephalography, 32*(3), 98–111.

Itti, L., Rees, G., & Tsotsos, J. K. (2005). Neurobiology of Attention. Elsevier Academic Press.

Iyer, D., & Zouridakis, G. (2007). Single-trial evoked potential estimation: Comparison between independent component analysis and wavelet denoising. *Clinical Neurophysiology, 118*(3), 495–504.

James, W. (1890). *The principles of psychology*. (Vol. 1) New York: Dover.

Johnston, W. A., & Dark, V. (1986). Selective attention. *Annual Review of Psychology, 37,* 43–75.

Jung, T. P., Makeig, S., Westerfield, M., Townsend, J., Courchesne, E., & Sejnowski, T. J. (2001). Analysis and visualization of single-trial event-related potentials. *Human Brain Mapping, 14*(3), 166–185.

Kaiser, J., & Bertrand, O. (2003). Dynamics of working memory for moving sounds: An event-related potential and scalp current density study. *Neuroimage, 19*(4), 1427–1438.

Kastner, S., & Pinsk, M. A. (2004). Visual attention as a multilevel selection process. *Cognitive, Affective, and Behavioral Neuroscience, 4*(4), 483–500.

Kastner, S., & Ungerleider, L. G. (2000). Mechanisms of visual attention in the human cortex. *Annual Review Neuroscience, 23,* 315–341.

Kelley, T. A., Serences, J. T., Giesbrecht, B., & Yantis, S. (2008). Cortical mechanisms for shifting and holding visuospatial attention. *Cerebral Cortex, 18*(1), 114–125.

Kelly, A. M. C., Uddin, L. Q., Biswal, B. B., Castellanos, F. X., & Milham, M. P. (2008). Competition between functional brain networks mediates behavioral variability. *Neuroimage, 39*(1), 527–537.

Kinchla, R. A. (1992). Attention. *Annual Review Psychology, 43,* 711–742.

Kiss, M., Van Velzen, J., & Eimer, M. (2008). The N2pc component and its links to attention shifts and spatially selective visual processing. *Psychophysiology, 45*(2), 240–249

Lachaux, J. P., Lutz, A., Rudrauf, D., Cosmelli, D., Le Van Quyen, M., Martinerie, J., et al. (2002). Estimating the time-course of coherence between single-trial brain signals: An introduction to wavelet coherence. Neurophysiologie Clinique-Clinical Neurophysiology, 32(3), 157–174.

Laufs, H., Krakow, K., Sterzer, P., Eger, E., Beyerle, A., Salek-Haddadi, A., et al. (2003). Electroencephalographic signatures of attentional and cognitive default modes in spontaneous brain activity fluctuations at rest. *Proceedings of the National Academy of Sciences of the United States of America, 100*(19), 11053–11058.

Le Van Quyen, M., Chavez, M., Rudrauf, D., & Martinerie, J. (2003). Exploring the nonlinear dynamics of the brain. *Journal of Physiology, (Paris), 97*(4–6), 629–639.

Lehmann, D. (1971). Topography of spontaneous alpha EEG fields in humans. *Electroencephalography and Clinical Neurophysiology, 30*(2), 161–162.

Lehmann, D. (1984). EEG assessment of brain activity: Spatial aspects, segmentation and imaging. *International Journal of Psychophysiology, 1*(3), 267–276.

Lehmann, D. (1989). Brain electrical mapping of cognitive functions for psychiatry: Functional micro-states. *Psychiatry Research, 29*(3), 385–386.

Lehmann, D., Henggeler, B., Koukkou, M., & Michel, C. M. (1993). Source localization of brain electric field frequency bands during conscious, spontaneous, visual imagery and abstract thought. *Brain Research. Cognitive Brain Research, 1*(4), 203–210.

Lehmann, D., Strik, W. K., Henggeler, B., Koenig, T., & Koukkou, M. (1998). Brain electric micro-states and momentary conscious mind states as building blocks of spontaneous thinking: I. Visual imagery and abstract thoughts. *International Journal of Psychophysiology, 29*(1), 1–11.

Leopold, D. A., & Logothetis, K. (1999). Multistable phenomena: Changing views in perception. *Trends in Cognitive Science 3*(7), 254–264.

Linkenkaer-Hansen, K., Nikouline, V. V., Palva, J. M., & Ilmoniemi, R. J. (2001). Long-range temporal correlations and scaling behavior in human brain oscillations. *Journal of Neuroscience, 21*(4), 1370–1377.

Lins, O. G., & Picton, T. W. (1995). Auditory steady-state responses to multiple simultaneous stimuli. *Electroencephalography and Clinical Neurophysiology, 96*(5), 420–432.

Luck, S. J., & Ford, M. A. (1998). On the role of selective attention in visual perception. *Proceedings of the National Academy of Sciences of the United States of America, 95*(3), 825–830.

Makeig, S., Debener, S., Onton, J., & Delorme, A. (2004). Mining event-related brain dynamics. *Trends in Cognitive Sciences, 8*(5), 204–210.

Mangun, G. R. (1995). Neural mechanisms of visual selective attention. *Psychophysiology, 32*(1), 4–18.

Mason, M. F., Norton, M. I., Horn, J. D. V., Wegner, D. M., Grafton, S. T., & Macrae, C. N. (2007a). Response to comment: "Wandering minds: The default network and stimulus-independent thought." *Science, 317,* 43c.

Mason, M. F., Norton, M. I., Horn, J. D. V., Wegner, D. M., Grafton, S. T., & Macrae, C. N. (2007b). Wandering minds: The default network and stimulus-independent thought. *Science, 315*(5810), 393–395.

McGuire, P. K., Silbersweig, D. A., Murray, R. M., David, A. S., Frackowiak, R. S., & Frith, C. D. (1996). Functional anatomy of inner speech and auditory verbal imagery. *Psychological Medicine, 26*(1), 29–38.

McKeown, M. J., Makeig, S., Brown, G. G., Jung, T. P., Kindermann, S. S., Bell, A. J., et al. (1998). Analysis of fMRI data by blind separation into independent spatial components. *Human Brain Mapping, 6*(3), 160–188.

McKeown, M. J., & Sejnowski, T. J. (1998). Independent component analysis of fMRI data: Examining the assumptions. *Human Brain Mapping, 6*(5–6), 368–372.

Muller, M. M., Teder-Salejarvi, W., & Hillyard, S. A. (1998). The time course of cortical facilitation during cued shifts of spatial attention. *Nature Neuroscience, 1*(7), 631–634.

Nobre, A. C. (2001). Orienting attention to instants in time. *Neuropsychologia, 39*(12), 1317–1328.

Nobre, A. C., Coull, J. T., Maquet, P., Frith, C. D., Vandenberghe, R., & Mesulam, M. M. (2004). Orienting attention to locations in perceptual versus mental representations. *Journal of Cognitive Neuroscience, 16*(3), 363–373.

Oken, B. S., Salinsky, M. C., & Elsas, S. M. (2006). Vigilance, alertness, or sustained attention: Physiological basis and measurement. *Clinical Neurophysiology, 117*(9), 1885–1901.

Parasuraman, R. (1998). *The attentive brain.* Cambridge, MA: MIT.

Peltier, S. J., Kerssens, C., Hamann, S. B., Sebel, P. S., Byas-Smith, M., & Hu, X. (2005). Functional connectivity changes with concentration of sevoflurane anesthesia. *Neuroreport, 16*(3), 285–288.

Pope, K. S., & Singer, J. L. (1978). *The stream of consciousness.* New York: Plenum Press.

Popple, A. V., & Levi, D. M. (2007). Attentional blinks as errors in temporal binding. *Vision Research, 47*(23), 2973–2981.

Posner, M. I. (1980). Orienting of attention. *Quarterly Journal of Experimental Psychology, 32*(1), 3–25.

Posner, M. I. (Ed.). (2004). *Cognitive neuroscience of attention.* New York: The Guilford Press.

Posner, M. I., & Rothbart, M. K. (1998). Attention, Self-regulation and Consciousness. *Philosophical Transactions of the Royal Society of London B, 353,* 1915–1927.

Raichle, M. E., MacLeod, A. M., Snyder, A. Z., Powers, W. J., Gusnard, D. A., & Shulman, G. L. (2001). A default mode of brain function. *Proceedings of the National Academy of Sciences of the United States of America, 98*(2), 676–682.

Raz, A., & Buhle, J. (2006). Typologies of attentional networks. *Nature Reviews, 7*(5), 367–379.

Robertson, I. H., Manly, T., Andrade, J., Baddeley, B. T., & Yiend, J. (1997). 'Oops!': Performance correlates of everyday attentional failures in traumatic brain injured and normal subjects. *Neuropsychologia, 35*(6), 747–758.

Röschke, J., Mann, K., Wagner, M., Grözinger, M., Fell, J., & Frank, C. (1996). An approach to single trial analysis of event-related potentials based on signal detection theory. *International Journal of Psychophysiology, 22*(3), 155–162.

Rudrauf, D., Douiri, A., Kovach, C., Lachaux, J. P., Cosmelli, D., Chavez, M., Adam C., Renault B., Martinerie J., Le Van Quyen M. (2006). Frequency flows and the time-frequency dynamics of multivariate phase synchronization in brain signals. *Neuroimage, 31*(1), 209–227.

Rushworth, M. F., Passingham, R. E., & Nobre, A. C. (2005). Components of attentional set-switching. *Experimental Psychology, 52*(2), 83–98.

Sarter, M., Givens, B., & Bruno, J. P. (2001). The cognitive neuroscience of sustained attention: Where top-down meets bottom-up. *Brain Research Brain Research Review, 35*(2), 146–160.

Sauseng, P., Klimesch, W., Stadler, W., Schabus, M., Doppelmayr, M., Hanslmayr, S., et al. (2005). A shift of visual spatial attention is selectively associated with human EEG alpha activity. *The European Journal of Neuroscience, 22*(11), 2917–2926.

Shipp, S. (2004). The brain circuitry of attention. *Trends in Cognitive Sciences, 8*(5), 223–230.

Singer, J. L. (1974). Daydreaming and the stream of thought. *American Scientist, 62*(4), 417–425.

Singer, J. L., & Antrobus, J. S. (1963). A factor-analytic study of daydreaming and conceptually-related cognitive and personality variables. *Perceptual and Motor Skills, 17,* 187–209.

Smallwood, J., Beach, E., Schooler, J. W., & Handy, T. C. (2008). Going AWOL in the brain: Mind wandering reduces cortical analysis of external events. *Journal of Cognitive Neuroscience, 20*(3), 458–469.

Smallwood, J., Fishman, D. J., & Schooler, J. W. (2007). Counting the cost of an absent mind: Mind wandering as an underrecognized influence on educational performance. *Psychonomic Bulletin and Review, 14*(2), 230–236.

Smallwood, J., & Schooler, J. W. (2006). The restless mind. *Psychological Bulletin, 132*(6), 946–958.

Smid, H. G. O. M., de Witte, M. R., Homminga, I., & van den Bosch, R. J. (2006). Sustained and transient attention in the continuous performance task. *Journal of Clinical Experimental Neuropsychology, 28*(6), 859–883.

Smilek, D., Birmingham, E., Cameron, D., Bischof, W., & Kingstone, A. (2006). Cognitive Ethology and exploring attention in real-world scenes. *Brain Research, 1080*(1), 101–119.

Sonuga-Barke, E. J. S., & Castellanos, F. X. (2007). Spontaneous attentional fluctuations in impaired states and pathological conditions: A neurobiological hypothesis. *Neuroscience and Biobehavioral Reviews, 31*(7), 977–986.

Stam, C. J., & de Bruin, E. A. (2004). Scale-free dynamics of global functional connectivity in the human brain. *Human Brain Mapping, 22*(2), 97–109.

Steinbock, A. (2004). Affection and attention: On the phenomenology of becoming aware. *Continental Philosophy Review, 37,* 21–43.

Sturm, W., & Willmes, K. (2001). On the functional neuroanatomy of intrinsic and phasic alertness. *Neuroimage, 14*(1 Pt 2), S76–S84.

Thompson, E., & Varela, F. J. (2001). Radical embodiment: Neural dynamics and consciousness. *Trends in Cognitive Sciences, 5,* 418–425.

Varela, F. J. (1979). *Principles of biological autonomy*. New York: Elsevier North Holland.

Vermersch, P. (2004). Attention between phenomenology and experimental psychology. *Continental Philosophy Review, 37,* 45–81.

Vincent, J. L., Patel, G. H., Fox, M. D., Snyder, A. Z., Baker, J. T., Essen, D. C. V., et al. (2007). Intrinsic functional architecture in the anaesthetized monkey brain. *Nature, 447*(7140), 83–86.

Wager, T. D., Jonides, J., & Reading, S. (2004). Neuroimaging studies of shifting attention: A meta-analysis. *NeuroImage, 22*(4), 1679–1693.

Wager, T. D., Jonides, J., Smith, E. E., & Nichols, T. E. (2005). Toward a taxonomy of attention shifting: Individual differences in fMRI during multiple shift types. *Cognitive, Affective & Behavioral Neuroscience, 5*(2), 127–143.

Wollman, M. C., & Antrobus, J. S. (1986). Sleeping and waking thought: Effects of external stimulation. *Sleep, 9*(3), 438–448.

Wylie, G. R., Javitt, D. C., & Foxe, J. J. (2003). Task switching: A high-density electrical mapping study. *NeuroImage, 20*(4), 2322–2342.

Yamaguchi, S., Tsuchiya, H., & Kobayashi, S. (1994). Electroencephalographic activity associated with shifts of visuospatial attention. *Brain, 117 (Pt 3)*, 553–562.

Yamaguchi, S., Tsuchiya, H., & Kobayashi, S. (1995). Electrophysiologic correlates of visuo-spatial attention shift. *Electroencephalography and Clinical Neurophysiology, 94*(6), 450–461.

Chapter 6
Crossmodal Attention – The Contribution of Event-Related-Potential Studies

R. Ortega and V. López

Abstract In everyday experience we are exposed to complex stimuli that impact sensorial receptors in more than one sensory modality. For example, human social interactions through language necessarily imply integration of auditory information with visual information referring to facial expression and gesture. In such a context, selective attention seems to operate favoring some of those stimuli on the basis of preexisting contextual information and physical or psychological salience, among other criteria. Although most studies of attentional mechanisms are still conducted using stimulation of a single sensory modality, which allows simpler hypothesis testing and minimizes confounding factors, in the last decade we have witnessed great developments in the study of so-called crossmodal attention. Evidence has accrued that both conflicting and congruent information presented to different sensory modalities interferes or influences the processing at each modality at different levels. Behavioral and electrophysiological studies support the existence of early crossmodal integration between almost all sensory modalities and stable crossmodal plasticity as a long-term effect. In this context, event-related potentials have been extensively used, especially owing to some advantages that this technique offers for the study of attentional mechanisms, such as its high temporal resolution. Attentional modulation of event-related potentials elicited by crossmodal stimuli partially resembles what has been described using monomodal stimulation. For example, an early negativity named "Nd" increases its amplitude when the eliciting stimulus is attended. In the present chapter, we attempt to revise the most recent developments in this cutting-edge research area and their implications for the theoretical framework in which we understand attention and attentional processes.

V. López
Escuela de Psicología, Pontificia Universidad Católica de Chile, Av. Vicuña Mackenna 4860, Macul, Santiago, Chile, e-mail: vlopezh@uc.cl

F. Aboitiz and D. Cosmelli (eds.) *From Attention to Goal-Directed Behavior.*
© Springer-Verlag Berlin Heidelberg 2009

Introduction

Ten years after the publication of the emblematic study "Crossmodal attention" (Driver & Spence, 1998), we face a complete different scenario regarding the task of summarizing the main lines of research and landmark results in this field. This decade has witnessed a growing interest in the study of the mechanisms that permit the integration of information from different sensory modalities and especially how this integration is affected by cognitive processes such as perception, attention, and memory. The perceptible boom in this research field is related to some relevant facts. First, the knowledge consolidated in this area emerges as a consequence of the integration of results from disciplines with very different levels of analysis, from single-cell electrophysiological recordings in animals to behavioral studies in humans. This exchange and complementation is quite new in the field. Usually neuroscience disciplines operating at different levels of analysis tend to work in a rather independent way despite the commonalities in the goals and the object of the study. Second, the crossmodal, or multimodal, approach is nowadays widely considered as a more ecological way of experimenting considering the fact that most real-life stimuli impact or generate responses in more than one sensory modality. Last but not least, various results from crossmodal experimentation, and specifically those related to the timing and location of crossmodal integration, have forced a reinterpretation or a revision of longstanding beliefs about sensorial, perceptual, and information processing in the brain.

Despite the latest methodological developments that allow us to accurately locate which brain areas are involved in crossmodal integration, the relation between this process and some other cognitive processes such as attention has been elusive. This has occurred perhaps owing to the fact that, at a neural level, both phenomena imply an increase in the neural activity associated with stimulus processing, and ultimately cause an improvement in the behavioral performance of the subject, for instance, a reduction in response times.

Event-related potentials (ERP), an EEG-derived technique, seems to be very well suited to address this problem and to study this kind of association. This is a noninvasive technique that can be used with no risks in humans where attention manipulations are easier to accomplish and more assessable. The most common way to explore if a particular process is affected by attention or not is to instruct the participants to pay attention to a specific type of stimuli or conditions and ignore others, or the participants are simply presented with task-relevant and task-irrelevant stimuli. Besides, ERPs have a great temporal resolution (on the order of milliseconds), allowing one to address the other great challenge of the topic: the timing of the relationship between attention and perceptual crossmodal integration. How these two processes interact during cognitive processing, what causes what and in what order are highly relevant and still partially unsolved questions in this field.

Multisensory Integration

Our current understanding of multisensory integration or multimodal coordination has been built upon empirical data and theoretical interpretations from several disciplines, particularly, on one hand, studies of anatomy and physiology of the nervous system in animal models and, on the other hand, psychophysical or "behavioral" studies in humans.

Evidences from Anatomical and Physiological Research in Animals

Sensory systems and related neural mechanisms have traditionally been studied as individual and not interrelated systems. That is, sensory processing in one modality has been assumed to occur in a modality-specific neural structure or area from which the information was later transferred to hierarchical higher-order areas where the integration with the information coming from other sensory systems would take place. These higher-order areas, named "association or multimodal areas," comprise several regions, including the parietal cortex, the superior temporal sulcus, and the prefrontal cortex. Several studies have confirmed the importance of these areas in the multisensory integration process (Barraclough, Xiao, Baker, Oram, & Perrett, 2005; Bernstein, Auer, Wagner, & Ponton, 2008; Calvert, Hansen, Iversen, & Brammer, 2001). However, several other studies have suggested that some areas classically considered as "modality-specific" also participate in this integration process (Driver & Noesselt, 2008; Macaluso, 2006; Shimojo & Shams, 2001).

One of the landmarks in the study of multisensory integration comes from the research conducted in the cat's superior colliculus (Stein & Meredith, 1993). This subcortical structure uses visual, auditory, and somatosensory information to initiate and control attention- and orientation-related behavior. In different neural layers some collicular neurons respond to one sensory modality, others receive input from two sensory modalities and others from three sensory modalities and have receptive fields in each modality. Such observations were the essential empirical evidence to reveal the existence of multisensory integration at an early stage of cognitive processing. Later studies, in both humans and nonhuman primates, demonstrated that stimulation of a sensory modality can elicit correlated neural activity in sensory cortical areas considered to be specific or exclusive to other sensory modality (Ghazanfar & Schroeder, 2006; Kayser & Logothetis, 2007). In one of the first studies in this context, Schroeder, Lindsley, Specht, Marcovici, Smiley et al. (2001), using current density measurements and multiunit recordings in macaque, showed that simultaneous presentation of auditory and somatosensory stimuli produces a convergent response in the auditory association cortex posterior to the primary auditory cortex (A1).

Neuroimaging studies in humans have shown results in the same line (Foxe, Wylie, Martinez, Schroeder, Javitt et al. 2002). More recently this convergent activity elicited by somatosensory stimulation has been shown to impact also A1, producing a phase resetting of slow-wave oscillations (around 10 Hz) (Lakatos, Chen, O'Connell, Mills, & Schroeder, 2007). Similarly, Kayser, Petkov, & Logothetis (2008) demonstrated that visual stimuli can modify the neural activity of the auditory cortex in monkeys. They utilized a set of audiovisual stimuli of two types: naturalistic (scenes of animals in their natural state) and artificial (filtered noise and flashes) and the same stimuli but in single modality. Their results show that audiovisual scenes (naturalistic or artificial) produce an increase in the activity of primary and secondary auditory cortex that can be clearly differentiated from the effects of single-modality auditory or visual stimuli. That effect is assessable in single neuron responses and in recordings of the field potential. They concluded that "the information communicated from early auditory cortices to higher processing stages not only reflects the acoustical environment but also depends on the multisensory context of a sound" (Kayser et al., 2008).

Standford & Stein (2007) and Kayser & Logothetis (2007) proposed three principles by which multisensory-related neural activity can be generated: spatial coincidence, temporal coincidence, and inverse effectiveness. The first one refers to the fact that receptive fields of multisensory neurons are generally overlapping and only those stimuli that are spatially located in these overlapped areas produce a real increase in the neural response. The second principle postulates that the increase in the neural response depends on the temporal proximity of the stimuli, and then only those stimuli that have a close temporal relation can generate an increased response. Otherwise they would produce a normal single-modality response. The third principle, "inverse effectiveness," suggests that multimodal integration is modulated by the strength of the neural response of single-modality stimuli. When the unimodal stimulus is strong (effectively offers relevant information about the environment) multimodal interactions rarely take place; in turn, when the response for the unimodal stimulus is weak, the probability of multimodal interactions increases if the second stimulus comply with at least one of the first two principles. These three explanatory principles may account for many of the descriptions of increased neural responses for multimodal stimuli compared with the response to the same stimuli when presented in a single modality, which is also known as the "superadditivity effect." This would also explain the absence of the effect in some experimental designs (Populin & Yin, 2002) or the contradictory results sometimes reported (Bizley, Nodal, Bajo, Nelken, & King, 2007). On the other hand, (Stanford & Stein (2007) advise about the potential risk of extending the "superadditivity effect" interpretation to results obtained with other methods, such as like neuroimages, ERPs, or behavioral measurements, owing to technical differences and also the possible existence of other, still unknown, neural mechanisms that account for multisensory integration.

Another important factor to necessary to understand multisensory integration is the description of the anatomical connections that may allow the occurrence of such integration. In the last few years several studies have shown direct cortical connections between areas belonging to different sensory modalities. Falchier, Clavagnier, Barone, & Kennedy (2002) unveiled unidirectional connections from primary and secondary auditory cortices to the periphery of the primary visual cortex in monkeys. A different study, using macaques, showed direct connections between the association auditory cortex and the secondary visual area (V2), with some projections reaching the primary visual area (V1) (Rockland & Ojima, 2003). Later, Cappe & Barone (2005) described the existence of three types of connections between unimodal sensory areas in marmoset: projections from visual areas (superior and medial temporal areas) to somatosensory cortex (areas 1/3b), projections from somatosensory sensory (S2) to auditory cortex (anterior bank of the lateral sulcus) and projections from visual areas (anterior to superior temporal sulcus) to primary auditory cortex. Recently a pattern of connections have been described between auditory and visual cortices in ferrets, including a limited number of them connecting the primary auditory cortex with the primary visual cortex but a considerable number connecting superior visual areas with the auditory cortex (Bizley et al., 2007).

In a study in rodents Budinger, Heil, Hess, & Scheich (2006) described direct connections between the primary auditory cortex and somatosensory and olfactory primary cortices, along with some others reaching secondary visual cortices. They also found that a great number of afferent connections to auditory cortices come from multisensory areas in the thalamus, suggesting possible multisensory integration even before the arrival at the cortex. These important multisensory afferent thalamic projections to auditory cortices were also found in macaques, where auditory association areas seem to receive more projections from thalamic multisensorial nuclei than the auditory primary area (Hackett, De La Mothe, Ulbert, Karmos, Smiley et al., 2007).

Evidences from Behavioral Studies

A great variety of behavioral studies have looked into the effects of multisensory integration. Together they have demonstrated that the perception of a stimulus in one modality influences the perception of another stimulus in other sensory modalities. Stimulation of a sensory modality can facilitate the perception of stimuli presented in a different modality when they coincide in spatial location or are presented in close temporal proximity. This is inferred from shorter response times, better performance in the task, or both (Doyle & Snowden, 2001; McDonald, Teder-Salejarvi, & Hillyard, 2000; Teder-Salejarvi, Di Russo, McDonald, & Hillyard, 2005; Vroomen & de Gelder, 2000; Zampini, Torresan, Spence, & Murray, 2007). Several models have been proposed trying to explain this phenomenon. The "race model" assumes that reduction in reaction times occurs because the response is triggered by the first cue detected (Raab, 1962). The "coactivation model" suggests that unimodal cues generate separate activations that are summated in the bimodal response (Miller, 1982). The "interactive coactivation model" states that the coactivation produced by a bimodal stimulus is greater than the sum of the activations produced by each of the two stimuli when presented in an independent manner (Miller, 1991); therefore, the processing of a stimulus in one modality necessarily impacts the processing of the other stimulus in a different modality.

On the other hand, there is a series of studies that concentrate on the processing of conflicting information between sensory modalities, which occurs when two sensory modalities receive incongruent information about a single event (De Gelder & Bertelson, 2003). In this context illusory phenomena can occur. The "McGurck effect" is a perceptual phenomenon in which vision alters the perception of spoken language. This effect takes place when, for example, a subject perceives the sound "ba" but is at the same time exposed to a visual image of a face articulating the incompatible syllable "ga," and the subject usually reports hearing the compatible syllable "da" (McGurk & MacDonald, 1976). This effect has recently been used to evaluate the recuperation of the capacity for multimodal integration in deaf people (who lost hearing after language acquisition) with cochlear implants. The results showed that they experienced the effect in the same way hearing people do but in their perceptions the information presented through visual images was clearly more predominant, perhaps owing to the known tendency of continuing reading the lips even after recovery of hearing (Rouger, Fraysse, Deguine, & Barone, 2008).

Another classic illusory effect is that known as the "ventriloquist effect" in which the simultaneous presentation of a visual and an auditory stimulus in different spatial locations tends to be perceived as coming from the same source. In the laboratory this effect is elicited by the simultaneous presentation of a series of light flashes and tones in separate locations. Subjects are instructed to ignore flashes when asked to report the spatial location of the sounds, but they tend to localize the sound in the places where the flashes occurred (Spence & Driver, 2000). A similar effect occurs when flashes and tones are presented in the same or different spatial locations but separated in time. Subjects are inclined to report a shorter interstimulus interval if the stimuli coincide in spatial location. This effect has been named "temporal ventriloquism" (Aschersleben & Bertelson, 2003; Bertelson & Aschersleben, 2003). A rather comparable effect occurs when tones and flashes are presented with an apparent movement (from left to right or right to left). Participants failed to determine the direction of the movement of auditory stimuli when they were presented simultaneously with visual stimuli moving in the opposite direction (Soto-Faraco, Lyons, Gazzaniga, Spence, & Kingstone, 2002)

Another classic effect due to bias in the multimodal integration process is that known as the "Colavita visual dominance effect" (Colavita & Weisberg, 1979). In this design subjects are confronted with auditory, visual, and audiovisual stimuli. When asked to respond to the auditory component of the audiovisual stimulus, subjects commit more errors, especially if the audiovisual stimulus was preceded by a visual stimulus. As a possible explanation, it has been suggested that visual characteristics of the stimuli attract more easily the attention of the participants than the auditory ones (Koppen & Spence, 2007).

Most of the known crossmodal effects are those in which vision influences the perceptual processing in other modalities. This is in line with the claim that visual cues are more important for humans than any other cue in a different modality (Eimer, 2004). Nevertheless, there are examples in which auditory stimuli modify some aspects of visual perception. It has been demonstrated, for example, that when a flash is presented accompanied by a series of auditory tones it tends to be perceived as multiple flashes (Shams, Kamitani, & Shimojo, 2000, 2002). Additional studies have unveiled the influence of auditory and tactile stimuli in the interpretation of the meaning of visual stimuli (Sekuler, Sekuler, & Lau, 1997; Watanabe & Shimojo, 1998, 2001); and other studies have concentrated on the influence of audition on tactile perception. The sound that emerges from touching a textured surface gives information about the texture of that surface. Guest, Catmur, Lloyd, & Spence (2002) manipulated the auditory feedback in a task where subjects were asked to discriminate the relative roughness of different surfaces. They discovered that perception of roughness was modulated by the frequency of the sound. High-frequency attenuation induced participants to perceive surfaces as softer.

Recent technological achievements allow the study of multisensory integration in chemical senses such as olfaction and taste. A clear example for the need for this integration in real life is the process of tasting a food or a beverage in which olfactory, gustatory, and somatosensory cues are necessary to construct a unique percept (Small & Prescott, 2005). White & Prescott (2007) conducted an experiment using a variant of the Stroop test (Stroop, 1935) in which participants had to discriminate between sweet or bitter taste while they were confronted with odorous substances that could be congruent or incongruent to the perceived taste. Congruency between taste and odor resulted in shorter reaction times.

Olfactory cues can also modulate the judgment about the attractiveness of a face. Dematte, Osterbauer, & Spence (2007) asked female participants to judge how attractive male faces were that were presented simultaneously with pleasant and unpleasant odors. Their results showed that the faces were evaluated as less attractive when paired with unpleasant odors. Other studies have pointed out the relevance of odors in face and word processing (for a review, see Walla, 2008).

Multisensory Integration and Event-Related Potentials

A rather large series of studies have focused on the neural activity related to multisensory integration. In a study of audiovisual integration in object recognition Giard & Peronnet (1999) applied a strategy, similar to that used in animal models (see earlier), to extract the neural activity purely related to multisensory integration. They subtracted the sum of the neural responses obtained by single-modality stimulation from the neural response elicited by bimodal stimulation. That is, AV − (A + V), where A is auditory, V is visual, and AV is audiovisual.

In this way they described various ERP components alleged to be elicited by multimodal interaction. Among them there is a very early crossmodal-interaction-related component that appears around 40 ms after stimulus presentation. These results were challenged by Teder-Salejarvi, McDonald, Di Russo, & Hillyard (2002), who interpreted this early effect as an artifact produced by the subtraction technique and the presence of anticipatory slow potentials beginning before the onset of the stimuli from which the time-locked response had been extracted. Nevertheless, Giard & Peronnet's results were corroborated in a follow-up study by the same group in which they also addressed the abovementioned criticisms (Fort, Delpuech, Pernier, & Giard, 2002), and in an independent study that used a different set of audiovisual stimuli (Molholm, Ritter, Murray, Javitt, Schroeder et al., 2002).

Recently other ERP studies have focused on the neural correlates of the illusory effects described in the previous section. Bonath, Noesselt, Martinez, Mishra, Schwiecker et al. (2007) studied the neural bases of the ventriloquist effect. They found two ERP components that could be related to this effect. The first one is a positivity that appears around 180 ms after stimulus onset (P180) which had a symmetric distribution in the scalp in all task conditions. The second component was a negativity that appears 260 ms after stimulus onset (N260) which was distributed symmetrically only if the illusion was not perceived. When the illusory effect occurred the N260 had greater amplitude over the hemisphere contralateral to the visual stimulation (and to the location where the apparent source of the sound was). With use of a version of this paradigm adapted for functional magnetic resonance imaging, (fMRI) the neural source for this activity was located to the temporal plane in the auditory cortex. In a different study Saint-Amour, De Sanctis, Molholm, Ritter, & Foxe (2007) investigated the changes in a well-known auditory ERP component, the mismatch negativity, during the McGurck effect. They observed a strong lateralization of this component over the left hemisphere starting around 175 ms, the component later exhibited a bilateral frontocentral distribution peaking at 290 ms, and finally between 350 and 400 ms its amplitude distribution was again lateralized over the left hemisphere. Source analysis localized this activity in the left temporal auditory cortex.

Finally, Mishra, Martinez, Sejnowski, & Hillyard (2007) studied the effects over the visual and the auditory cortices of the illusion that makes a single flash be perceived as two flashes when it is presented simultaneously with two briefs sounds. Their results showed an early modulation of the activity in the visual cortex 30–60 ms after the second sound. Trial-by-trial analysis showed a short latency ERP modification localized to the auditory cortex in the temporal lobe that occurred simultaneously with the appearance of gamma-band activation apparently coming out from the visual cortex. These results suggest that the second sound triggers a rapid and dynamic interaction between visual and auditory cortical areas that seems to be the real neural correlate for perception of the particular illusion.

Attention Mechanisms

Attention is a cognitive process difficult to define without incurring important omissions. It is convenient, however, to state at least an operational definition to delimit the meaning of the term in the present text. Attention refers here to "selective attention," which is defined as the "cognitive mechanism of the brain that allows processing relevant sensory inputs, thoughts or actions while ignoring the irrelevant ones" (Gazzaniga, Ivry, & Mangun, 2002).

Studies about attention have been marked by some dichotomous definitions that still today are, to some extent, clearly found in the attention literature. First the process can be divided into voluntary or reflexive attention. Voluntary attention refers to the capacity of intentionally paying attention to an external stimulus, or an internal thought. Involuntary or reflexive attention refers to the capacity of the nervous system to automatically orient or redirect attention to unexpected, novel, or dangerous stimuli (Gazzaniga et al., 2002). Several factors of diverse nature modulate or control attention. There are cognitive factors such as previous knowledge, expectancy, or intention that could incline us to voluntarily select and pay attention to certain parts or characteristic of the stimulus (top-down processing). On the other hand, there are factors pertaining to the stimulus by itself, such as intensity, for example, that determine a reflexive or involuntary capture of attention (bottom-up processing) (Corbetta & Shulman, 2002). Despite the clear correspondence between voluntary attention and top-down processing, and between reflexive attention and bottom-up processing, there is evidence suggesting that they are different phenomena. Sussman, Winkler, & Schroger (2003) studied reflexive changes in the attention to auditory stimuli and found that this "passive" capture of attention could be modulated by top-down factors. In their design only those changes in the auditory stimulus that were really unexpected captured attention. If the modification of the auditory stimuli was somehow predictable, the reflexive capture of attention did not play an important role, although the so-called automatic system for change detection continued operating. Several other findings support the distinction between these phenomena (Escera, Yago, Corral, Corbera, & Nunez, 2003; Sussman, Winkler, Huotilainen, Ritter, & Naatanen, 2002).

Another pseudo-dichotomous distinction relevant in the attention literature is that between space-oriented attention and object-oriented attention (Scholl, 2001). Spatially defined attention refers to the capacity of paying attention (voluntarily or involuntarily) to a specific area of the space, among other multiple possible locations. Stimuli that appear in that area will be preferentially processed. Object-oriented attention refers to the capacity of paying attention primarily to an object and disregarding its spatial location (Herrmann & Knight, 2001). Apparently human attention mechanisms use both strategies depending on the specific situation. Spatial location seems to be very important in early stages of attention processing, while surface- and object-oriented attention are more important in later stages (Luck, Woodman, & Vogel, 2000).

The temporal course of selective attention is another important point of debate in the attention literature. Two opposing views have been exposed: early- versus late-selection hypotheses. Early selection entails that stimuli are selected long before the end of a complete perceptual processing (Posner & Dehaene, 2000). The late-selection hypothesis suggests that both attended and unattended stimuli receive the same early processing and then selection occurs at a later stage, influencing mainly memory encoding, conscious perception, and response initiation (Gazzaniga et al., 2002). It is important to consider that most of these distinctions appear owing to the conceptualization of attention in the context of the information processing paradigm that assumes a hierarchical and serial order in cognition that cannot always be easily applied in real life.

Crossmodal Attention

Attention, as most cognitive processes, has been traditionally studied in single-sensory-modality designs, in an independent approach for each sense, and disregarding any interaction. Nevertheless, a growing corpus of evidence suggests that attention is really a crossmodal process. Some of this evidence will be summarized here.

Spatial Attention

Spatial location is perhaps the most studied factor in multimodal interactions. Several studies have shown that spatial coincidence between stimuli presented to different sensory modalities can modulate attentional allocation (Eimer & Driver, 2001; McDonald, Teder-Salejarvi, Di Russo, & Hillyard, 2003; Teder-Salejarvi et al., 2002). Schroger & Widmann (1998) used visual and auditory stimuli in a bimodal design that replicated the classic experiment of spatial cued attention described in the visual modality (Posner, 1980). They found that when auditory and visual stimuli had a similar spatial location reaction times were shorter and a negative component became evident in the ERP waveform at around 100–300 ms (negative difference, Nd). This effect was not attenuated when subjects were instructed to direct attention to the side opposite that indicated by the cue. Their results showed that stimuli from different sensory modalities but coincident in spatial location are preferentially attended and generate faster responses and larger negativities in the ERP. In the same line Teder-Salejarvi, Munte, Sperlich, & Hillyard (1999) used flashes and noises in a design where subjects were separated into two groups: those that attended only visual stimuli and those that attended the auditory ones. All subjects were instructed to pay attention to the right or to the left side in different blocks of the experiment. In the auditory attending group, sounds in the attended location elicited a negative biphasic component (Nd) starting around 70 ms after the stimulus onset that was not present in the responses elicited by sounds presented in unattended locations. In the visual attending group, flashes presented in attended locations elicited larger early visual components between 100 and 200 ms, and also larger late components (between 200 and 350 ms). Regarding the unattended modalities, in the auditory attending group, visual stimuli presented at the cued locations exhibited a slight but significant increment in the amplitude of the early components. In the same way, in the visual attending group, auditory stimuli at the cued location elicited the Nd component, although with a reduction in amplitude compared with that seen in the auditory attending group. These results led to the conclusion that spatial attention selection in a sensory modality affects the processing of stimuli in other sensory modalities even when these stimuli are irrelevant to the task. Similar results have been reported for interactions between visual and tactile stimuli (Eimer & Driver, 2000).

In a different study (Eimer, van Velzen, & Driver 2002) visual cues were used to direct attention trial by trial to the left or to the right side of the space. Visual, auditory, or tactile stimuli were presented equally distributed between the cued and the unattended location. Subjects were also instructed to pay attention to a "relevant" modality that could be auditory or tactile. Cue-related activity exhibited the same ERP components described in unimodal designs (Harter & Anllo-Vento, 1991). A component that resembles the late directing attention positivity contralateral to the cued side was present independently of the instructed relevant modality. The analysis of the activity elicited by target stimuli (relevant or irrelevant to the task) showed very similar results to those previously described. In line with these results very similar effects have been found in several other experiments: crossmodal modulation by endogenous or exogenous attention (Eimer & Driver, 2001), under conditions of light or darkness (Eimer, van Velzen, Forster, & Driver, 2003), studying separately the effects produced by stimulation of different hemifields (Eimer, van Velzen, & Driver, 2004), etc. This has led to the widely accepted notion that spatial attention uses a supramodal system that controls and coordinates attention selection in different sensory modalities (Eimer & Van Velzen, 2002).

In a fMRI study using simultaneous visual and tactile stimuli Macaluso, Frith, & Driver (2000) found that visuospatial attention was facilitated in the proximity of the hand where the tactile stimulus was presented. They demonstrated that cortical areas participating in this process included not only intermodal association areas of the parietal cortex. The activity in the primary visual area was clearly incremented when visual and tactile stimuli coincided on the same side. McDonald et al. (2003) reported similar attention modulation effects in the audiovisual integration. They proposed a feedback mechanism from crossmodal to unimodal areas as the neural substrate of these processes that involved the intermodal cortex of the superior temporal sulcus, the fusiform occipitotemporal area, and finally the extrastriate visual cortex. Other fMRI studies have reported that the visualization of facial movements related to language articulation generates an increment in the activity of primary auditory cortices that is not present when the facial movement is not language-related (Calvert, Bullmore, Brammer, Campbell, Williams et al., 1997; Puce, Allison, Bentin, Gore, & McCarthy, 1998).

A posterior study demonstrated that these interactions between cortical areas were not exclusive of the audiovisual integration necessary for language comprehension. The emerging network of cortical interactions depends, to a great extent, on the nature of the stimuli and the congruency between them (for example, in terms of space, time, and/or shape). These networks include subcortical structures such as the superior colliculus and cortical ones such as the intraparietal sulcus and the superior and ventromedial frontal cortices (Calvert, 2001).

The insular cortex, particularly in the right hemisphere, may play an important role in the detection of temporal coincidence between visual and auditory stimuli, information that is crucial for the ulterior processing and perceptual integration (Bushara, Grafman, & Hallett, 2001). Macaluso, George, Dolan, Spence, & Driver (2004) investigated the temporal and spatial factors involved in the audiovisual processing of speech. Synchronous stimuli produced an increased activation of association multimodal areas such as the superior temporal sulcus. Stimulus location was also important. Lateral and dorsal occipital areas were selectively activated when the synchronous bimodal stimulation was also congruent regarding spatial location. Right inferior parietal lobe was consistently activated when stimuli were synchronous but incongruent in spatial location, a condition clearly associated with the ventriloquist effect. As a whole, ventral areas seem to be related to the analysis of stimuli synchrony, while dorsal areas are more related to the detection of spatial coincidences in multisensory interactions.

Object-Oriented Attention

Although there are fewer studies of object-oriented attention in comparison with spatial attention, there is evidence pointing to the fact that attention allocated to a specific characteristic of the object in one modality can be extended to characteristics of the same object in other modalities, even when these characteristics are task-irrelevant or spatially distant from the attended area. Busse, Roberts, Crist, Weissman, & Woldorff (2005) used an experimental design based on the ventriloquist effect, assuming that a light and a sound simultaneously presented are processed as a single object even when they have different spatial locations. They presented a series of visual stimuli that could be accompanied or not by sounds, and found a significant improvement in visual target detection when the visual stimulus was presented simultaneously with an auditory stimulus. They used a substraction technique to isolate the activity purely elicited by bimodal stimulation. In the resulting waveform they found a negative component starting around 220 ms and extending beyond 700 ms. In a fMRI version of this paradigm this activity was located to the auditory cortex in the temporal cortex. Molholm, Martinez, Shpaner, & Foxe (2007) used two audiovisual stimuli: a drawing of a guitar with the sound of a strum and a drawing of a dog with the sound of a bark. Stimuli could appear as single-modality stimuli or as bimodal stimuli. Participants were asked to pay attention to one modality or to both of them and to one of the two bimodal stimuli. In this way, Molholm et al. were able to assess two types of crossmodal facilitation: the one that occurred when congruent stimuli appeared sequentially (in different modalities) and the one produced by the simultaneous presentation of congruent stimuli as a single bimodal construct.

Molholm et al. compared the effect of attention within a sensory modality (attended object versus unattended object in the attended sensory modality) and the crossmodal effects described above (attended object versus unattended object in the unattended sensory modality). Visual attention selection produced a negative component (selection negativity, SN) peaking around 270 ms and with a scalp amplitude distribution characterized by a maximum over the lateral occipital region. The amplitude of this component differentiated attended from unattended visual stimuli when the subject was paying attention to the visual modality. When attention was allocated to the auditory modality, crossmodal effects were seen in the visual one: the same component with the same peak latency (270 ms) for the simultaneous presentation but with a peak latency of 310 ms in the sequential presentation.

Auditory attention selection was reflected by the amplitude modulation of the Nd component. This negativity had a peak at 220 ms and maximum amplitude over the frontocentral region and became evident when the response to the attended stimulus was compared with that elicited by the unattended one (when subjects were paying attention to the auditory modality). Crossmodal interactions also elicited an Nd component with a more frontal scalp distribution and lateralized to the left hemisphere. The latency was preserved in the simultaneous presentation (220 ms) but was delayed in the sequential presentation (300 ms). These results suggest that attention to one attribute of the stimulus can expand to other attributes of the same object but in a different sensory modality if stimuli are presented simultaneously or in close temporal relation.

The study of Talsma, Doty, & Woldorff (2007) differs from those described above. They studied crossmodal attention to objects when both visual and auditory attributes have to be attended at the same time. They used a rapid serial visual presentation (RSVP) paradigm in conjunction with visual, auditory, and audiovisual stimuli. Subjects were instructed to pay attention to the visual part, the auditory part, or both parts of the audiovisual objects. At a different moment they had to attend to the isolated visual or auditory stimuli or the sequence of letters in the RSVP. Talsma et al. found a very early crossmodal attention-modulated effect expressed in a positive component with a peak around 50 ms (P50) and a scalp distribution showing maximum amplitude over the medial central region. This component had a larger amplitude when attributes in both sensory modalities were attended, even larger than the sum of the activity elicited by isolated visual and auditory stimuli when attention was allocated to them. The effect reverted when stimuli in both modalities were ignored (attention allocated to the RSVP). These results are a relevant contribution to the debate about how early attention can modulate crossmodal interactions.

Discussion

The present text is not intended to exhaustively revise the growing field of multisensory integration and crossmodal attention, but rather to selectively summarize some of the aspects that we consider very novel or relevant to orient future research. In this context we will present here a very brief commentary about what we consider milestones in the current state of the research.

Many authors consider multisensory integration and crossmodal attention as completely separate processes (McDonald, Teder-Salejarvi, & Ward, 2001). Basically, they support their view with the fact that some crossmodal illusory effects (such as the ventriloquist effect) are not modulated by attention according to what behavioral data have suggested (Bertelson, Vroomen, de Gelder, & Driver,

2000; Driver, 1996; Vroomen, Bertelson, & de Gelder, 2001). It has also been claimed that the early stages of multisensory integration seem to be automatic and attention-independent. They suggest that attention would have an impact only in later stages of the process. In line with this, ERP studies have shown very early crossmodal integration effects, before 100ms (Fort et al., 2002; Giard & Peronnet, 1999; Mishra et al., 2007; Molholm et al., 2002), but the effects of crossmodal attention modulation has been found at later stages (Eimer & Driver, 2001; Talsma & Woldorff, 2005). This position has strong empirical support. Nevertheless, recent studies have challenged this view and its generalization to other situations. This evidence emerged in behavioral studies (Alsius, Navarra, Campbell, & Soto-Faraco, 2005; Alsius, Navarra, & Soto-Faraco, 2007; Koppen & Spence, 2007), and also in studies using fMRI (Pekkola et al., 2006; van Atteveldt, Formisano, Goebel, & Blomert, 2007) and ERP (Talsma et al., 2007). This makes even more complicated the already-entangled field of research. Small variations in the experimental paradigm can lead to big changes in the results and their interpretation. Further research is needed to elucidate the relationship between crossmodal attention and multisensory integration, especially concerning the boundaries of each phenomenon and how to distinguish the effects from one or the other.

In many of the studies presented here it is evident that modality specific cortical areas participate in multisensory integration and crossmodal attention. Although most of these activations could be explained by feedback or reentry mechanisms coming from superior integration areas, the possibility of a feed-forward phenomenon cannot be ruled out by the current corpus of evidence (Foxe & Schroeder, 2005), especially in those cases where crossmodal interactions occur at very early stages (Giard & Peronnet, 1999; Mishra et al., 2007; Molholm et al., 2002; Talsma et al., 2007). This makes it difficult to consider the involvement of superior association cortices given the known pattern of corticocortical connections of these regions. The description of anatomical connections between cortical areas up to now considered modality-specific (Bizley et al., 2007; Budinger et al., 2006; Cappe & Barone, 2005; Falchier et al., 2002; Rockland & Ojima, 2003) emphasizes the necessity for a revision of the predominant model of hierarchical organization and cortical integration processes. This is especially important if we consider that the long history of unimodal studies is more related to experimental limitations than to real experience and that multisensory integration is a basic fact for brain processes in real life.

The evidence presented here about attention mechanisms supports the existence of a supramodal attention system that controls and distributes attention resources across all sensory modalities (Eimer & Driver, 2001; Eimer & Van Velzen, 2002; Macaluso, Eimer, Frith, & Driver, 2003). The observations described here suggest that the function of this system goes well beyond the interactions related to the location of the stimuli spatial in the different sensory modalities involved. Attention can affect sensory-perceptual processing at different stages, and this varies depending on the task complexity, the relevant characteristics of stimuli, and the attentional load (Luck et al., 2000). In any event, attention seems to be crucial for the meaningful integration of object attributes captured by different sensory channels, and therefore for the perception of objects as unique multimodal percepts (Busse et al., 2005; Molholm et al., 2007). In this sense it is absolutely essential that the attention a stimulus is receiving can be continuously modified by the ongoing activity in other cortical areas pertaining to different sensory systems. Top-down modulations of attention seem to be present along the whole cognitive processing of a stimulus and operate independently of the supposed location in the chain of the "attentional filter". In this case, the alleged automatism of the early stages of multisensory integration would be only "relative" and certainly these early stages of processing can be influenced by the activity in other sensory modalities or in superior association cortices.

In the same way, the so-called late selection could make use of relevant information extracted from the stimuli at earlier stages, and then, more than the effect of a late attentional filter, the late facilitation or suppression of stimulus processing should be interpreted as the effect of an extended modulatory system that operates throughout the entire cognitive processing. These mechanisms could operate independently of the subjective awareness and intentionality; therefore, they could be involved both in voluntary attention and in reflexive attention phenomena.

Even though the last decade has been the most important period for the comprehension of the neural bases of crossmodal attention, this phenomenon is still far from being completely understood and further important research and paradigm-changing evidence are to be expected in the relatively near future.

References

Alsius, A., Navarra, J., Campbell, R., & Soto-Faraco, S. (2005). Audiovisual integration of speech falters under high attention demands. *Current Biology, 15*(9), 839–843.

Alsius, A., Navarra, J., & Soto-Faraco, S. (2007). Attention to touch weakens audiovisual speech integration. *Experimental Brain Research, 183*(3), 399–404.

Aschersleben, G., & Bertelson, P. (2003). Temporal ventriloquism: Crossmodal interaction on the time dimension. 2. Evidence from sensorimotor synchronization. *International Journal of Psychophysiology, 50*(1–2), 157–163.

Barraclough, N. E., Xiao, D., Baker, C. I., Oram, M. W., & Perrett, D. I. (2005). Integration of visual and auditory information by superior temporal sulcus neurons responsive to the sight of actions. *Journal of Cognitive Neuroscience, 17*(3), 377–391.

Bernstein, L. E., Auer, E. T., Jr., Wagner, M., & Ponton, C. W. (2008). Spatiotemporal dynamics of audiovisual speech processing. *Neuroimage, 39*(1), 423–435.

Bertelson, P., & Aschersleben, G. (2003). Temporal ventriloquism: Crossmodal interaction on the time dimension. 1. Evidence from auditory-visual temporal order judgment. *International Journal of Psychophysiology, 50*(1–2), 147–155.

Bertelson, P., Vroomen, J., de Gelder, B., & Driver, J. (2000). The ventriloquist effect does not depend on the direction of deliberate visual attention. *Perception and Psychophysics, 62*(2), 321–332.

Bizley, J. K., Nodal, F. R., Bajo, V. M., Nelken, I., & King, A. J. (2007). Physiological and anatomical evidence for multisensory interactions in auditory cortex. *Cerebral Cortex, 17*(9), 2172–2189.

Bonath, B., Noesselt, T., Martinez, A., Mishra, J., Schwiecker, K., Heinze, H. J., et al. (2007). Neural basis of the ventriloquist illusion. *Current Biology, 17*(19), 1697–1703.

Budinger, E., Heil, P., Hess, A., & Scheich, H. (2006). Multisensory processing via early cortical stages: Connections of the primary auditory cortical field with other sensory systems. *Neuroscience, 143*(4), 1065–1083.

Bushara, K. O., Grafman, J., & Hallett, M. (2001). Neural correlates of auditory-visual stimulus onset asynchrony detection. *Journal of Neuroscience, 21*(1), 300–304.

Busse, L., Roberts, K. C., Crist, R. E., Weissman, D. H., & Woldorff, M. G. (2005). The spread of attention across modalities and space in a multisensory object. *Proceedings of the National Academy of Sciences of the United States of America, 102*(51), 18751–18756.

Calvert, G. A. (2001). Crossmodal processing in the human brain: Insights from functional neuroimaging studies. *Cerebral Cortex, 11*(12), 1110–1123.

Calvert, G. A., Bullmore, E. T., Brammer, M. J., Campbell, R., Williams, S. C., McGuire, P. K., et al. (1997). Activation of auditory cortex during silent lipreading. *Science, 276*(5312), 593–596.

Calvert, G. A., Hansen, P. C., Iversen, S. D., & Brammer, M. J. (2001). Detection of audio-visual integration sites in humans by application of electrophysiological criteria to the BOLD effect. *Neuroimage, 14*(2), 427–438.

Cappe, C., & Barone, P. (2005). Heteromodal connections supporting multisensory integration at low levels of cortical processing in the monkey. *European Journal of Neuroscience, 22*(11), 2886–2902.

Colavita, F. B., & Weisberg, D. (1979). A further investigation of visual dominance. *Perception and Psychophysics, 25*(4), 345–347.

Corbetta, M., & Shulman, G. L. (2002). Control of goal-directed and stimulus-driven attention in the brain. *Nature Reviews Neuroscience, 3*(3), 201–215.

De Gelder, B., & Bertelson, P. (2003). Multisensory integration, perception and ecological validity. *Trends in Cognitive Science, 7*(10), 460–467.

Dematte, M. L., Osterbauer, R., & Spence, C. (2007). Olfactory cues modulate facial attractiveness. *Chemical Senses, 32*(6), 603–610.

Doyle, M. C., & Snowden, R. J. (2001). Identification of visual stimuli is improved by accompanying auditory stimuli: The role of eye movements and sound location. *Perception, 30*(7), 795–810.

Driver, J. (1996). Enhancement of selective listening by illusory mislocation of speech sounds due to lip-reading. *Nature, 381*(6577), 66–68.

Driver, J., & Noesselt, T. (2008). Multisensory interplay reveals crossmodal influences on 'sensory-specific' brain regions, neural responses, and judgments. *Neuron, 57*(1), 11–23.

Driver, J., & Spence, C. (1998). Crossmodal attention. *Current Opinion in Neurobiology, 8*(2), 245–253.

Eimer, M. (2004). Multisensory integration: How visual experience shapes spatial perception. *Current Biology, 14*(3), R115–117.

Eimer, M., & Driver, J. (2000). An event-related brain potential study of cross-modal links in spatial attention between vision and touch. *Psychophysiology, 37*(5), 697–705.

Eimer, M., & Driver, J. (2001). Crossmodal links in endogenous and exogenous spatial attention: Evidence from event-related brain potential studies. *Neuroscience Biobehavioral Review, 25*(6), 497–511.

Eimer, M., & Van Velzen, J. (2002). Crossmodal links in spatial attention are mediated by supramodal control processes: Evidence from event-related potentials. *Psychophysiology, 39*(4), 437–449.

Eimer, M., van Velzen, J., & Driver, J. (2002). Cross-modal interactions between audition, touch, and vision in endogenous spatial attention: ERP evidence on preparatory states and sensory modulations. *Journal of Cognitive Neuroscience, 14*(2), 254–271.

Eimer, M., van Velzen, J., & Driver, J. (2004). ERP evidence for cross-modal audiovisual effects of endogenous spatial attention within hemifields. *Journal of Cognitive Neuroscience, 16*(2), 272–288.

Eimer, M., van Velzen, J., Forster, B., & Driver, J. (2003). Shifts of attention in light and in darkness: An ERP study of supramodal attentional control and crossmodal links in spatial attention. *Brain Research. Cognitive Brain Research, 15*(3), 308–323.

Escera, C., Yago, E., Corral, M. J., Corbera, S., & Nunez, M. I. (2003). Attention capture by auditory significant stimuli: Semantic analysis follows attention switching. *European Journal of Neuroscience, 18*(8), 2408–2412.

Falchier, A., Clavagnier, S., Barone, P., & Kennedy, H. (2002). Anatomical evidence of multimodal integration in primate striate cortex. *Journal of Neuroscience, 22*(13), 5749–5759.

Fort, A., Delpuech, C., Pernier, J., & Giard, M. H. (2002). Dynamics of cortico-subcortical crossmodal operations involved in audio-visual object detection in humans. *Cerebral Cortex, 12*(10), 1031–1039.

Foxe, J. J., & Schroeder, C. E. (2005). The case for feedforward multisensory convergence during early cortical processing. *Neuroreport, 16*(5), 419–423.

Foxe, J. J., Wylie, G. R., Martinez, A., Schroeder, C. E., Javitt, D. C., Guilfoyle, D., et al. (2002). Auditory-somatosensory multisensory processing in auditory association cortex: An fMRI study. *Journal of Neurophysiology, 88*(1), 540–543.

Gazzaniga, M. S., Ivry, R. B., & Mangun, G. R. (2002). *Cognitive neuroscience: The biology of the mind* (2nd ed.). New York: Norton.

Ghazanfar, A. A., & Schroeder, C. E. (2006). Is neocortex essentially multisensory? *Trends in Cognitive Science, 10*(6), 278–285.

Giard, M. H., & Peronnet, F. (1999). Auditory-visual integration during multimodal object recognition in humans: A behavioral and electrophysiological study. *Journal of Cognitive Neuroscience, 11*(5), 473–490.

Guest, S., Catmur, C., Lloyd, D., & Spence, C. (2002). Audiotactile interactions in roughness perception. *Experimental Brain Research, 146*(2), 161–171.

Hackett, T. A., De La Mothe, L. A., Ulbert, I., Karmos, G., Smiley, J., & Schroeder, C. E. (2007). Multisensory convergence in auditory cortex, II. Thalamocortical connections of the caudal superior temporal plane. *The Journal of Comparative Neurology, 502*(6), 924–952.

Harter, M. R., & Anllo-Vento, L. (1991). Visual-spatial attention: Preparation and selection in children and adults. *Electroencephalography and Clinical Neurophysiology Supplement, 42*, 183–194.

Herrmann, C. S., & Knight, R. T. (2001). Mechanisms of human attention: Event-related potentials and oscillations. *Neuroscience Biobehavior Review, 25*(6), 465–476.

Kayser, C., & Logothetis, N. K. (2007). Do early sensory cortices integrate cross-modal information? *Brain Structure and Function, 212*(2), 121–132.

Kayser, C., Petkov, C. I., & Logothetis, N. K. (2008). Visual Modulation of Neurons in Auditory Cortex. *Cerebral Cortex, 18*(7), 1560–1574.

Koppen, C., & Spence, C. (2007). Audiovisual asynchrony modulates the Colavita visual dominance effect. *Brain Research, 1186*, 224–232.

Lakatos, P., Chen, C. M., O'Connell, M. N., Mills, A., & Schroeder, C. E. (2007). Neuronal oscillations and multisensory interaction in primary auditory cortex. *Neuron, 53*(2), 279–292.

Luck, S. J., Woodman, G. F., & Vogel, E. K. (2000). Event-related potential studies of attention. *Trends in Cognitive Science, 4*(11), 432–440.

Macaluso, E. (2006). Multisensory processing in sensory-specific cortical areas. *Neuroscientist, 12*(4), 327–338.

Macaluso, E., Eimer, M., Frith, C. D., & Driver, J. (2003). Preparatory states in crossmodal spatial attention: Spatial specificity and possible control mechanisms. *Experimental Brain Research, 149*(1), 62–74.

Macaluso, E., Frith, C. D., & Driver, J. (2000). Modulation of human visual cortex by crossmodal spatial attention. *Science, 289*(5482), 1206–1208.

Macaluso, E., George, N., Dolan, R., Spence, C., & Driver, J. (2004). Spatial and temporal factors during processing of audiovisual speech: A PET study. *Neuroimage, 21*(2), 725–732.

McDonald, J. J., Teder-Salejarvi, W. A., Di Russo, F., & Hillyard, S. A. (2003). Neural substrates of perceptual enhancement by cross-modal spatial attention. *Journal of Cognitive Neuroscience, 15*(1), 10–19.

McDonald, J. J., Teder-Salejarvi, W. A., & Hillyard, S. A. (2000). Involuntary orienting to sound improves visual perception. *Nature, 407*(6806), 906–908.

McDonald, J. J., Teder-Salejarvi, W. A., & Ward, L. M. (2001). Multisensory integration and crossmodal attention effects in the human brain. *Science, 292*(5523), 1791.

McGurk, H., & MacDonald, J. (1976). Hearing lips and seeing voices. *Nature, 264*(5588), 746–748.

Miller, J. (1982). Divided attention: Evidence for coactivation with redundant signals. *Cognitive Psychology, 14*(2), 247–279.

Miller, J. (1991). Channel interaction and the redundant-targets effect in bimodal divided attention. *Journal of Experimental Psychology: Human Perception and Performance, 17*(1), 160–169.

Mishra, J., Martinez, A., Sejnowski, T. J., & Hillyard, S. A. (2007). Early cross-modal interactions in auditory and visual cortex underlie a sound-induced visual illusion. *Journal of Neuroscience, 27*(15), 4120–4131.

Molholm, S., Martinez, A., Shpaner, M., & Foxe, J. J. (2007). Object-based attention is multisensory: co-activation of an object's representations in ignored sensory modalities. *European Journal of Neuroscience, 26*(2), 499–509.

Molholm, S., Ritter, W., Murray, M. M., Javitt, D. C., Schroeder, C. E., & Foxe, J. J. (2002). Multisensory auditory-visual interactions during early sensory processing in humans: a high-density electrical mapping study. *Brain Research. Cognitive Brain Research, 14*(1), 115–128.

Pekkola, J., Ojanen, V., Autti, T., Jaaskelainen, I. P., Mottonen, R., & Sams, M. (2006). Attention to visual speech gestures enhances hemodynamic activity in the left planum temporale. *Human Brain Mapping, 27*(6), 471–477.

Populin, L. C., & Yin, T. C. (2002). Bimodal interactions in the superior colliculus of the behaving cat. *J Neurosci, 22*(7), 2826–2834.

Posner, M. I. (1980). Orienting of attention. *Quartely Journal of Experimental Psychology, 32*(1), 3–25.

Posner, M. I., & Dehaene, S. (2000). Attentional networks. In M. S. Gazzaniga (Ed.), *Cognitive neuroscience a reader* (pp. 156–164). Oxford: Blackwell Publishers Ltd.

Puce, A., Allison, T., Bentin, S., Gore, J. C., & McCarthy, G. (1998). Temporal cortex activation in humans viewing eye and mouth movements. *Journal of Neuroscience, 18*(6), 2188–2199.

Raab, D. H. (1962). Statistical facilitation of simple reaction times. *Transaction of New York Academy of Science, 24,* 574–590.

Rockland, K. S., & Ojima, H. (2003). Multisensory convergence in calcarine visual areas in macaque monkey. *International Journal of Psychophysiology, 50*(1–2), 19–26.

Rouger, J., Fraysse, B., Deguine, O., & Barone, P. (2008). McGurk effects in cochlear-implanted deaf subjects. *Brain Research, 1188,* 87–99.

Saint-Amour, D., De Sanctis, P., Molholm, S., Ritter, W., & Foxe, J. J. (2007). Seeing voices: High-density electrical mapping and source-analysis of the multisensory mismatch negativity evoked during the McGurk illusion. *Neuropsychologia, 45*(3), 587–597.

Scholl, B. J. (2001). Objects and attention: The state of the art. *Cognition, 80*(1–2), 1–46.

Schroeder, C. E., Lindsley, R. W., Specht, C., Marcovici, A., Smiley, J. F., & Javitt, D. C. (2001). Somatosensory input to auditory association cortex in the macaque monkey. *Journal of Neurophysiology, 85*(3), 1322–1327.

Schroger, E., & Widmann, A. (1998). Speeded responses to audiovisual signal changes result from bimodal integration. *Psychophysiology, 35*(6), 755–759.

Sekuler, R., Sekuler, A. B., & Lau, R. (1997). Sound alters visual motion perception. *Nature, 385*(6614), 308.

Shams, L., Kamitani, Y., & Shimojo, S. (2000). Illusions. What you see is what you hear. *Nature, 408*(6814), 788.

Shams, L., Kamitani, Y., & Shimojo, S. (2002). Visual illusion induced by sound. *Brain Research. Cognitive Brain Research, 14*(1), 147–152.

Shimojo, S., & Shams, L. (2001). Sensory modalities are not separate modalities: Plasticity and interactions. *Current Opinion in Neurobiology, 11*(4), 505–509.

Small, D. M., & Prescott, J. (2005). Odor/taste integration and the perception of flavor. *Experimental Brain Research, 166*(3–4), 345–357.

Soto-Faraco, S., Lyons, J., Gazzaniga, M., Spence, C., & Kingstone, A. (2002). The ventriloquist in motion: Illusory capture of dynamic information across sensory modalities. *Brain Research. Cognitive Brain Research, 14*(1), 139–146.

Spence, C., & Driver, J. (2000). Attracting attention to the illusory location of a sound: Reflexive crossmodal orienting and ventriloquism. *Neuroreport, 11*(9), 2057–2061.

Stanford, T. R., & Stein, B. E. (2007). Superadditivity in multisensory integration: Putting the computation in context. *Neuroreport, 18*(8), 787–792.

Stein, B. E., & Meredith, M. A. (1993). *The merging of the senses*. Cambridge, MA: MIT.

Stroop, J. R. (1935). Studies of interference in serial verbal reactions. *Journal of Experimental Psychology, 18,* 643–661.

Sussman, E., Winkler, I., Huotilainen, M., Ritter, W., & Naatanen, R. (2002). Top-down effects can modify the initially stimulus-driven auditory organization. *Brain Research. Cognitive Brain Research, 13*(3), 393–405.

Sussman, E., Winkler, I., & Schroger, E. (2003). Top-down control over involuntary attention switching in the auditory modality. *Psychonomic Bulletin and Review, 10*(3), 630–637.

Talsma, D., Doty, T. J., & Woldorff, M. G. (2007). Selective attention and audiovisual integration: Is attending to both modalities a prerequisite for early integration? *Cerebral Cortex, 17*(3), 679–690.

Talsma, D., & Woldorff, M. G. (2005). Selective attention and multisensory integration: Multiple phases of effects on the evoked brain activity. *The Journal of Cognitive Neuroscience, 17*(7), 1098–1114.

Teder-Salejarvi, W. A., Di Russo, F., McDonald, J. J., & Hillyard, S. A. (2005). Effects of spatial congruity on audio-visual multimodal integration. *The Journal of Cognitive Neuroscience, 17*(9), 1396–1409.

Teder-Salejarvi, W. A., McDonald, J. J., Di Russo, F., & Hillyard, S. A. (2002). An analysis of audio-visual crossmodal integration by means of event-related potential (ERP) recordings. *Brain Research. Cognitive Brain Research, 14*(1), 106–114.

Teder-Salejarvi, W. A., Munte, T. F., Sperlich, F., & Hillyard, S. A. (1999). Intra-modal and cross-modal spatial attention to auditory and visual stimuli. An event-related brain potential study. *Brain Research. Cognitive Brain Research, 8*(3), 327–343.

van Atteveldt, N. M., Formisano, E., Goebel, R., & Blomert, L. (2007). Top-down task effects overrule automatic multisensory responses to letter-sound pairs in auditory association cortex. *Neuroimage, 36*(4), 1345–1360.

Vroomen, J., Bertelson, P., & de Gelder, B. (2001). The ventriloquist effect does not depend on the direction of automatic visual attention. *Perception and Psychophysics, 63*(4), 651–659.

Vroomen, J., & de Gelder, B. (2000). Sound enhances visual perception: Cross-modal effects of auditory organization on vision. *Journal of Experimental Psychology: Human Perception and Performance, 26*(5), 1583–1590.

Walla, P. (2008). Olfaction and its dynamic influence on word and face processing: Cross-modal integration. *Progress in Neurobiology, 84*(2), 192–209.

Watanabe, K., & Shimojo, S. (1998). Attentional modulation in perception of visual motion events. *Perception, 27*(9), 1041–1054.

Watanabe, K., & Shimojo, S. (2001). When sound affects vision: Effects of auditory grouping on visual motion perception. *Psychology Science, 12*(2), 109–116.

White, T. L., & Prescott, J. (2007). Chemosensory cross-modal stroop effects: Congruent odors facilitate taste identification. *Chemical Senses, 32*(4), 337–341.

Zampini, M., Torresan, D., Spence, C., & Murray, M. M. (2007). Auditory-somatosensory multi-sensory interactions in front and rear space. *Neuropsychologia, 45*(8), 1869–1877.

Chapter 7
Measuring and Modulating Hemispheric Attention

A. Hill, A. Barnea, K. Herzberg, A. Rassis-Ariel, S. Rotem, Y. Meltzer, Y. H. Li, and E. Zaidel

Abstract Studies combining electrophysiological and behavioral laterality measures hold great potential to illuminate hemispheric relations in attention. However, data from event-related potentials as well as spectral analyses (quantitative EEG and band power) are conflicting, and do not support a coherent theory of the electrophysiology of hemispheric attention. At the same time, a definitive behavioral measure of attention does not currently exist. To remedy these lacunae, we carried out the following experiment. Four groups of learning-disabled young adults received the same EEG biofeedback (EEGBF) protocol, consisting of training theta (4–8 Hz) down and training sensorimotor rhythm (12–15 Hz) up at four different electrode sites: C3 (seven subjects), C4 (ten subjects), Cz (nine subjects), and Fz (eight subjects). The C3 site is over the left sensory motor cortex, C4 is over the right sensory motor cortex, and Fz and Cz are over the front and middle regions of the central strip, respectively. Attention in each hemisphere was measured before and after EEGBF training using the computerized Lateralized Attention Network Test (LANT). The LANT estimates four separate networks of attention: executive-frontal Conflict resolution, the benefit and cost of spatial Orienting, and Alerting, or sustained attention. EEGBF affected different networks maximally at different sites: Training at C3 reduced Conflict in the right hemisphere, training at C4 improved Alerting bilaterally and training at Cz increased the benefit of spatial Orienting bilaterally. Generally, C3 training improved attention in the right hemisphere and C4 training improved attention in the left hemisphere. This suggests that training at C3 and C4 activates a metacognitive control system which is contralaterally organized. We concluded that EEGBF has site-specific and function-specific effects on attention. Further, unilateral training can have bilateral effects or even predominantly contralateral ones. This procedure suggests a new way to probe discrete physiological correlates of discrete behavioral changes following EEGBF. Such data inform functional theories of attention and clinical interventions in disorders of attention.

E. Zaidel
Department of Psychology, University of California at Los Angeles, Los Angeles, CA 90095-1563, USA, e-mail: ezaidel@psych.ucla.edu

F. Aboitiz and D. Cosmelli (eds.) *From Attention to Goal-Directed Behavior.*
© Springer-Verlag Berlin Heidelberg 2009

125

Abbreviations

ADHD   Attention deficit–hyperactivity disorder
ANT   Attention Network Task
EEGBF   EEG biofeedback
ERP   Event-related potential
LANT   Lateralized Attention Network Task
LH   Left hemisphere
LVF   Left visual field
RH   Right hemisphere
RT   Reaction time
RVF   Right visual field
SD   Standard deviation
SMR   Sensorimotor rhythm
TVF   Target visual field

Introduction

Electrophysiological Wars

There is overwhelming EEG evidence suggesting functional hemispheric asymmetry in attention, although contradictory data are reported for specific EEG frequency bands. Changes in hemispheric specialization are also seen in pathological conditions, with atypical frontal activation in attention deficit–hyperactivity disorder (ADHD) (Baving, Laucht & Schmidt, 1999). These authors found that boys with ADHD exhibited a *reduced* right-lateralized frontal activation when compared with normal controls (as indicated by relatively higher right-frontal alpha power), while girls with ADHD exhibited *increased* right-frontal lateralization (as indicated by relatively higher left-frontal alpha power). At the same time, Ciçek & Nalçaci (2001) suggest that in normal participants greater resting alpha power as well as decreased alpha power during a task can be positively correlated with better performance on the Wisconsin Card Sorting Test. In addition, Clarke, Barry, McCarthy, Selikowitz, Johnstone, Hsu, Magee, Lawrence, & Croft (2007) found that ADHD children showed increased interhemispheric EEG coherence (in theta and beta bands) in frontal and parietal regions, suggesting reduced functional laterality. Furthermore, Valentino & Dufresne (1991) demonstrated that directed attention produced a shift in alpha and beta asymmetry towards right-hemisphere (RH) lateralization. On the basis of such evidence, cogent theories of right-hemisphere systems of attention have been advanced (Fan, McCandliss, Sommer, Raz, & Posner, 2002), although it is not clear whether right lateralization is due to cognitive specialization or is simply a result of interhemispheric interaction in early visual perceptual processes (Jutai, 1984). The relation of hemifield attention to hemispheric activation is also not clear, as research has found that both hemispheres respond to selective attention demands related to stimuli in either visual field (Heilman & Van Den Abell, 1980).

The findings of event-related potential (ERP) studies are also often contradictory. Among the most interesting ERP studies, Miniussi, Rao & Nobre (2002) demonstrated that foveal attention increased the evoked N1 component bilaterally, when attention was directed to the right visual field (RVF), but increased the N1 component only in the RH when attention was directed to the left visual field (LVF). This suggests that the RH systems of directed attention affect the hemispheres asymmetrically. In contrast, Yamaguchi, Yamagata & Kobayashi (2000) found that the N1 component was increased in the posterior temporal lobe contralateral to the stimulus field and was reduced over the ipsilateral hemisphere.

Combined evidence from spectral EEG and ERP demonstrates the effect of attention on gamma-band EEG frequencies. Müller, Gruber & Keil (2000) have shown that gamma is increased in the hemisphere contralateral to a presented stimulus, but only when attention is paid to the stimulus. Other researchers (Gobbelé, Waberski, Schmitz, Sturm & Buchner, 2002) have shown that the gamma band synchronizes in the RH in response to selective attention (suggesting a temporoparietal network driving selective spatial attention). However, Doesburg et al. (2007) have suggested that gamma synchronization occurs not only contralateral to the attended stimuli but also throughout the cortex. Indeed, Senkowski, Talsma, Herrmann & Woldorff (2005) demonstrated gamma changes in medial-frontal areas in response to attended (but not unattended) stimuli.

Many of these discrepancies may be related to methodological choices in data collection, artifacting, or analysis, although most papers specify similar procedures. It is likely that additional, yet unknown causes of discrepancy exist. Regardless of the cause, the current literature the functional hemispheric asymmetry often does not fit common theories that tie cortical activity to behavior.

The Lateralized Attention Network Test

Recently, Posner and associates developed a brief computerized battery designed to measure executive Conflict, spatial Orienting, and Alerting (Fan et al., 2002). These networks are claimed to be independent of each other, localized in different brain regions, and mediated predominantly via different neurotransmitters. Behavioral and physiological data suggest that Conflict is controlled by both dorsolateral prefrontal and anterior cingulate cortices and is predominantly dopaminergic. Orienting is localized in the right parietal cortex and is mediated predominantly by acetylcholine. Alerting engages the parietal/prefrontal cortex and it is predominantly adregenergic.

We adapted the Attention Network Test (ANT) to lateralized presentation by tachistoscopically presenting to the LVF and the RVF "up arrows" and "down arrows," flanked vertically by additional arrows. Target arrows were preceded by cues to target location; cues presented were valid, invalid, central, double, or absent (not presented). The inclusion of invalid cues (not included in the ANT) permitted separate measurement of the cost and benefit of spatial Orienting. These two components are thought to be associated with the N1 and N2 deflections of the ERP, respectively (Luck, Hillyard, Mouloua, Woldorff, Clark, & Hawkins 1994). Further, split brain and normal data suggest that the cost of spatial Orienting is larger in the LVF or RH, whereas the benefit of spatial Orienting is larger in the RVF or left hemisphere (LH) (Zaidel, 1995). Studies of the Lateralized Attention Network Test (LANT) in normal subjects suggest that the two cerebral hemispheres have largely separate and comparable networks of attention (Barnea et al., in preparation; Greene, 2008). Occasionally, Orienting is found to be larger in the RH, whereas Conflict and Alerting may be larger in the LH. In turn, clinical and imaging studies confirm that Alerting is specialized in the RH, but there is some evidence that Conflict selectively engages the right anterior cingulate cortex (Kaplan & Zaidel, 2001).

The three networks are independent of each other in each hemisphere, but they may cross-correlate significantly between the hemispheres. These results are consistent with the view that each normal hemisphere has a complete cognitive repertoire for interacting with the environment, including its own attentional control system. Thus, attention orchestrates the dynamics of hemispheric information processing in both the normal and the split brain by assigning resources to, and enabling, specific cognitive operations.

Attention and Laterality in Learning Disability

Attention

Probably the most studied subgroup of learning disabilities consists of individuals with ADHD. Although several reports have identified attention disorders in ADHD, some have not. Thus, the covert Orienting of spatial attention paradigm adapted from Posner shows conflicting results. Sustained attention, assessed using the classic Continuous Performance Test paradigm and other paradigms, also produced inconsistent results. Perhaps the most consistent finding is evidence of impaired performance in conflict-producing tasks. The evidence is both behavioral and physiological. These results are consistent with the notion that ADHD involves executive deficits.

Laterality

Several cognitive studies have supported abnormal cerebral laterality in ADHD, but the evidence is largely circumstantial and few studies have been explicitly designed to address laterality directly. Some have found spatial deficits in children with ADHD (Sheppard, Bradshaw & Mattingley, 1999; Garcia-Sanchez, Estevez-Gonzolas, Suarez-Romero & Junque, 1997), whereas others have found evidence for LH dysfunction (Cohen et al., 2000; McInnes et al., 2003; Rucklidge and Tannock, 2002; Seidman et al., 1998; Semrud-Clikeman, 2000).

Both structural and functional imaging studies have also reported laterality differences in ADHD individuals. In particular, a number of baseline functional activation studies have indicated abnormal asymmetries (RH > LH) in subjects with ADHD, while others looking at activation during cognitive challenges have indicated decreased right-sided activation.

McCracken (1991) has suggested that ADHD may be associated with a dysregu-lation of the locus ceruleus nucleus in the brainstem, which is thought to produce 80% of the largely right-lateralized norepinephrine projections to the cerebral cor-tex. This system has been shown to play a key role in regulating arousal, vigilance, sleep, autonomic function, and emotion. This account implies a dysregulation of RH function.

Recent data from our laboratory are directly relevant and more definitive. Adults with ADHD were compared with age-matched controls in tests of lateralized lexical decision as well as of dichotic listening to words (specialized in the LH) and to emotional prosody (specialized in the RH). There seemed to be a selective deficit in LH control of processing strategies. Adults with ADHD appeared to emphasize a RH processing strategy during challenges that did not overtly tax executive proc-esses. This RH emphasis in adult ADHD is associated with deficits for linguistic functions and advantages for processes specialized to the RH. However, processing could be normalized in adults with ADHD under certain attentional states.

EEG Biofeedback

EEG biofeedback (EEGBF) is a procedure whereby an individual modifies the amplitude, frequency, synchrony, or other derived measures of the electrical activity of his/her own brain. Many authors have demonstrated control of various electro-encephalographic parameters in animals and humans (Birbaumer, 1977; Birbaumer, 1984; Birbaumer et al., 1981; Kamiya, 1969; Plotkin, 1976; Sterman, 1977). Work by Monastra et al. (Monastra, 2003; Monastra & Lubar, 2000; Monastra et al., 1999, 2002) concluded that EEGBF is effective in treating attentional disorders, but some methodological issues persist (Loo, 2003). Foremost is the ubiquitous prob-lem of a proper control group. It is critical to exclude experimenter bias, simple repetition, or incidental context/attention effects of the EEGBF training protocol. Although it is possible to demonstrate that EEGBF works by showing the differen-tial effect of different protocols, determination of the degree of change, relative to a neutral or a baseline condition, is problematic. A double-blind sham control, where the subject engages in the same kind of feedback condition as with EEGBF but without veridical feedback, is expensive and often detectable by the subject. Instead, we will contrast several alternative EEGBF protocols, which will enable comparison of the results of lateralized training at C3 and C4 in the LANT. This permits an assessment of the degree to which the laterality of the training electrode determines the laterality of the affected network. A significant interaction of the site of the training electrode by the network by EEGBF would demonstrate the efficacy of training.

There have been few controlled experiments on the effects of EEGBF training on hemispheric specialization and interhemispheric interaction in the normal brain, mostly using the slow cortical potential shift approach (Hardman et al., 1997; Kotchoubey et al., 1996; Rockstroh et al., 1993; Pulvermuller et al., 2000). However, the evidence for hemispheric engagement is at best indirect. Our approach addresses these shortcomings.

Effects of EEGBF on Attention

Primary improvement on attentional symptoms has shown the effectiveness of EEGBF in addressing ADHD symptoms (Monastra, 2005; Gruzelier & Egner, 2005), although much research still needs to be done to validate the methods of action of EEGBF. Clinically, much EEGBF for attentional symptoms is performed with an "active" electrode placed at C4, Cz, or C3. (We use the term "active" here loosely, as any EEGBF training circuit consists of the difference between two electrodes and thus the "reference" electrode will most likely measure an electrically active area.) As early as 1970, Sterman showed that C4 sensorimotor rhythm (SMR) training produces motorically "calm" cats, with a trained increase of SMR (12–15 Hz) on cortical motor strip. Similar motor stilling has been assumed to have a beneficial effect on attention and impulsivity and most clinicians still choose to train attentional symptoms on the motor strip, at C3, Cz, or C4, even though there is strong functional neuroimaging support for a frontal hypoarousal model of ADHD (Liotti, et al., 2005; Max, et al., 2005) along with central and midline cortical slowing (Mann, et al., 1992; Chabot, et al., 1996; Monastra, et al., 1999; Clarke, et al., 2001).

For training-frequency selection, inhibiting slower frequency ranges including theta (4–8 Hz) is typical, as we replicated here. Reward frequency ranges are usually SMR (12–15 Hz) on the motor strip, as SMR is considered a "resting" rhythm, perhaps analogous in function to alpha elsewhere in the brain. Other frequency ranges are also sometimes chosen for reward by clinicians. While SMR is typically trained at C4, either SMR or low beta (15–18 Hz) may be typically chosen as a reward frequency at C3. Standard Cz reward frequencies could be either SMR or low beta. In addition, clinicians often reward slower frequencies posterior to the motor strip, and faster frequencies anterior to the central motor strip. Interhemispheric training (C3 active, C4 reference, for example) is also used, rewarding frequencies in the 12–18-Hz range. The subjective nature and variety of these choices is one of the large confounding factors in comparing clinical efficacy of training techniques. For the purpose of comparing hemisphere and site specificity, we rewarded all training sites at 12–15 Hz.

The main goal of this chapter is to contrast the effects of four EEGBF training protocols on the attention networks in each cerebral hemisphere of learning-disabled young adults. This was accomplished by probing the three networks of attention, namely, Conflict resolution, spatial Orienting, and Alerting/vigilance. These networks were assessed using a lateralized version of Posner and associates' ANT. The training protocols rewarded the subject for decreasing the power in the theta band (4–8 Hz) and for increasing power in the "SMR" 12–15-Hz band. The training electrode varied systematically between C3, C4, Cz, and Fz, using the standard 10:20 placement system. We addressed the following question: What is the relationship between training site and affected attentional process as a function of inter- and intrahemispheric training and assessment location?

Methods

Lateralized Attention Network Test

Stimuli were presented using E-Prime software, on an IBM-compatible PC. Participants viewed the screen from a distance of 57 cm. Targets consisted of an upward or a downward arrow centered 1.06 cm to the left or to the right of the fixation. This target was flanked above and below (1) by two arrows in the same direction (congruent condition), (2) by two arrows in the opposite direction (incongruent condition), or (3) by two lines without arrowheads (neutral condition). The participants' task was to identify the direction of the middle arrow by pressing the up or down keys on the mouse. We used two slightly different versions of the test. Version 1 was presented in four blocks of 96 trials each, and version 2 was presented in four blocks of 108 trials each. The response hand alternated between blocks in the following order: right hand, left hand, right hand, left hand.

A single arrow or line subtended 0.55 cm (version 1) or 0.68 cm (version 2), and the contours of adjacent arrows or lines were separated by 0.08 cm (version 1) or 0.2 cm (version 2). The stimuli (one central arrow plus four flankers) subtended a total of 3.07 cm (version 1) or 4.20 cm (version 2). Performance in the congruent condition minus performance in the incongruent condition defines the Conflict component of attention. Targets were preceded by one of six types of cues, used to define the cost of spatial Orienting, the benefit of spatial Orienting, and Alerting: no cue, center cue, double cue, a valid spatial cue, and an invalid spatial cue. For the double-cue trials, there were two warning cues corresponding to the two possible target positions – left and right. For the valid-cue trials, the cue was at the target position, and for the invalid-cue trials, the cue appeared in the visual field opposite the target. Alerting was defined as performance in the double-cue condition minus performance in the no-cue condition. Benefit of spatial Orienting was defined as performance in the valid-cue condition minus performance in the center-cue condition. Cost of spatial Orienting was defined as performance in the center-cue condition minus performance in the invalid-cue condition. A variable duration of the first fixation was used to produce additional uncertainty about cue onset.

Each trial consisted of five events. First, there was a fixation period for a random, variable duration between 400 and 1,600 ms. Next, a warning cue was presented for 400 ms. There was a short fixation period for 400 ms after the warning cue and then the targets and flankers were presented simultaneously for 170 ms. The participants had a window of 1,500 ms (version 1) or 1,000 ms (version 2) to make the response. After the response, in version 1, there was a delay between the response and initiation of the next trial equal to 3,500 ms minus duration of the first fixation minus reaction time (RT). In version 2, there was no delay between the response and the beginning of the next trial. After this interval, the next trial began. Each trial in which a response was made lasted from 3,170 to 4,170 ms for version 1 and a total of 670–2,770 ms for version 2. Trials to which the subject did not respond lasted on average an extra 1,000 ms for version 1 and an extra 500 ms for version 2. The fixation cross appeared at the center of the screen throughout the whole trial. Target location was uncertain except when a valid spatial cue preceded it. Thus, the first two experiments (C3 and C4 training) used smaller stimuli, included no neutral flankers, and used a longer interstimulus interval (3,170–4,170 ms), whereas the last two experiments (Cz and Fz training), used the larger stimuli, included neutral flankers, and used a shorter interstimulus interval (670–2,770 ms). More critically, the first two experiments included a smaller proportion of invalid cues (one quarter of the valid cues) than the second two experiments (half of the valid cues). The two sets of experiments are part of a series designed to produce a progressively more sensitive clinical tool for measuring attention. They were carried out 1 year apart and were parts of larger studies. Thus, the variation in the LANT was a result of improving the clinical sensitivity of the test over time.

Responses were made unimanually on a mouse placed at the midline on its side, facing the responding hand. Thus, "up" responses were made with both index fingers, and "down" responses were made with both middle fingers.

Participants

Thirty young Israeli adults with heterogeneous learning disabilities (mixed ADHD, dyslexia, alcalculia, etc.) were recruited from remedial college preparatory classes at the Academic College of Tel Chai in Israel during July and August of 2004 and 2005. All participants were between 21 and 26 years old, with a mean age of 23 years 8 months. The participants were partitioned into four groups. Each group was assigned to a different training site from the set, C3, C4, Cz, and Fz.

EEGBF Training

EEGBF training was conducted over a period of 8 weeks; each participant received 20 training sessions. Each session consisted of ten segments of 3 min each, for 30-min total session training time. EEG signal was recorded and relevant frequency components were extracted. Feedback was provided in the form of visual and auditory videogame responses. The amplitude of the target frequency was represented by the size or speed of the object in the game. The participants' task was to increase the size and accelerate the speed of those objects. When reward conditions were satisfied for a minimum of 0.5 s, an auditory beep and a visual incentive stimulus (e.g., highway stripe, star in sky) were provided as reinforcement. The participants were instructed to maximize their point scores as well.

EEGBF was administered using the Deymed TruScan 32 system. Signal was acquired at 256-Hz sampling rate, converted from analog to digital and band-filtered to extract delta (0–4 Hz), theta (4–8 Hz), alpha (8–12 Hz), SMR (12–15 Hz), and beta (18–22 Hz) components. An active scalp electrode was placed at C3, C4, Fz, or Cz for the four groups, respectively, according to the standard 10:20 system. The reference electrode was placed on the ipsilateral ear for the C3 and C4 groups and on the left ear for the Cz and Fz groups, and the ground electrode was placed on the contralateral earlobe. Impedance was kept below 5 kΩ, and artifact-rejection thresholds were set individually for each participant so as to interrupt feedback during eye and body movements that produced gross EEG fluctuations.

Results

We carried out a within-subject repeated-measures analysis of variance (ANOVA), site of training (C3, C4, Cz, Fz) × network (Conflict, cost of spatial Orienting, benefit of spatial Orienting, Alerting) × target visual field (TVF) (left, right). The dependent variable was the change in z score following EEGBF for each network for each experimental group relative to the respective control group. For Conflict and for cost of spatial Orienting the dependent variable was based on the RT such that it was (RT pre training-RT post training)/SD of the control group for that condition and for benefit of spatial Orienting and Alerting the dependent variable was (RT post training-RT pre training)/SD of the control group for that condition. In this way, positive change indicates improved attention following EEGBF.

The results showed a significant site × network interaction, $F(9,90) = 5.064$, $p = 0.0157$ (Fig. 7.1a). This shows that Conflict improved the most following C3 training; benefit of spatial Orienting improved the most after Cz training, Alerting improved the most after C4 training, and cost of spatial Orienting did not improve at all.

Given our a priori focus on the separate attention networks in each hemisphere, we carried out two separate ANOVAs of network × site for each visual field. The analysis for LVF trials showed a significant network × site interaction, $F(9,90) = 2.25$, $p = 0.0256$, reflecting a selective and massive improvement in Conflict in the RH following C3 training (Fig. 7.1b). The network × site interaction for RVF trials was not significant ($p = 0.374$) (Fig. 7.1c).

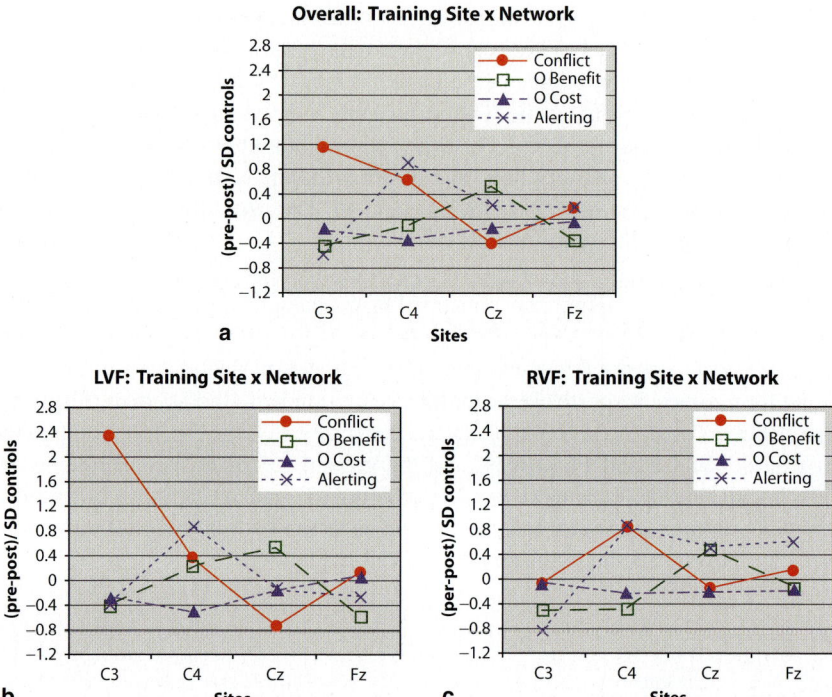

Fig. 7.1 The interaction between the site of the training electrode and the attention networks showing the effect of EEG biofeedback in (**a**) the two visual fields combined, (**b**) left visual field (*LVF*) targets, and (**c**) right visual field (*RVF*) targets. The dependent variable is the effect of EEG biofeedback in terms of standard deviation units of normal controls for the same networks. *C3, C4, Cz, Fz*: location of training electrode for the four experimental groups. *O*: Orienting

We next followed up with separate ANOVAs of site × TVF for each network. The result for Conflict disclosed a significant site × TVF interaction, $F(3, 30) = 3.261$, $p = 0.0351$. In particular, the site (C3, C4) × TVF interaction was significant, $F(1, 15) = 4.985$, $p = 0.0412$, showing that the LVF improvement in Conflict occurred for C3 but not for C4 (Fig. 7.2). Similarly, site (C3, Cz) × TVF was significant, $F(1,14) = 4.905$, $p = 0.0439$, showing that the selective LVF improvement in Conflict following C3 training, did not apply to Cz training. Finally, the site (C4, Cz) × TVF analysis disclosed a significant main effect of site, $F(1,17) = 4.665$, $p = 0.0454$, showing that C4 training resulted in an overall improvement (reduction) in Conflict ($\Delta z = 0.63$), whereas training at Cz resulted in an increase in Conflict ($\Delta z = -0.406$).

For benefit of spatial Orienting, the ANOVA site (Cz, Fz) × TVF exhibited a significant main effect of site, $F(1,15) = 5.817$, $p = 0.0291$, showing an improvement in benefit of spatial Orienting following Cz training (0.524), but a decrease follow-

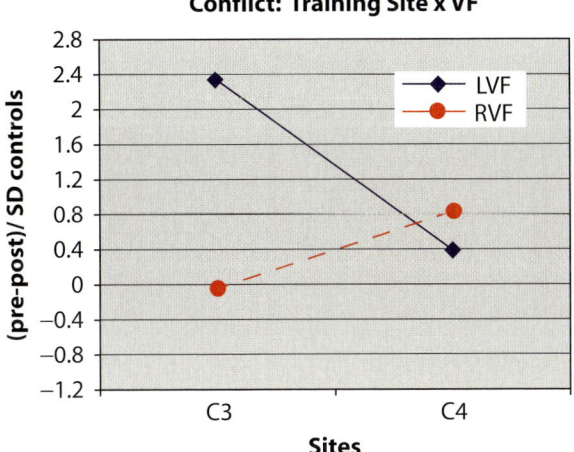

Fig. 7.2 The interaction between the site of the training electrode and the visual field (*VF*) of the target for Conflict in the C3 and C4 groups

ing Fz training (−0.341). Finally, the analysis site (Cz, C3) × TVF showed a trend to significance, $F(1, 14) = 3.644$, $p = 0.077$, contrasting with the improvement in benefit of spatial Orienting following Cz training and a decrease following C3 training (−0.455).

For Alerting, the analysis site (C3, C4, Cz, Fz) × TVF yielded a significant main effect of site, $F(3, 30) = 3.786$, $p = 0.02$, showing a large improvement following C4 training (0.89), a moderate improvement following Cz (0.219) and Fz (0.201) training, and a decrease following C3 training (−0.586). The analysis site (C3, C4) × TVF confirmed the results that C3 training differed significantly from C4 training, $F(1,15) = 7.947$, $p = 0.013$. Interestingly, the analysis site (Cz, Fz) × TVF disclosed a significant main effect of TVF, $F(1, 15) = 4.6$, $p = 0.05$, showing that there was an improvement in Alerting in the LH (0.576) but a decrease in Alerting in the RH (−0.155).

Discussion

The significant interaction of site × network shows, first, that EEGBF is effective, second, that training is site-specific, and, third, that training is function (network)-specific. In particular, the theta down–SMR up protocol applied at C3 selectively improved Conflict, whereas training at C4 selectively improved Alerting, and training at Cz selectively improved the benefit of spatial Orienting. In fact, C3 selectively improved Conflict in the LVF (Fig. 7.3a). Thus, an electrode placed over the left

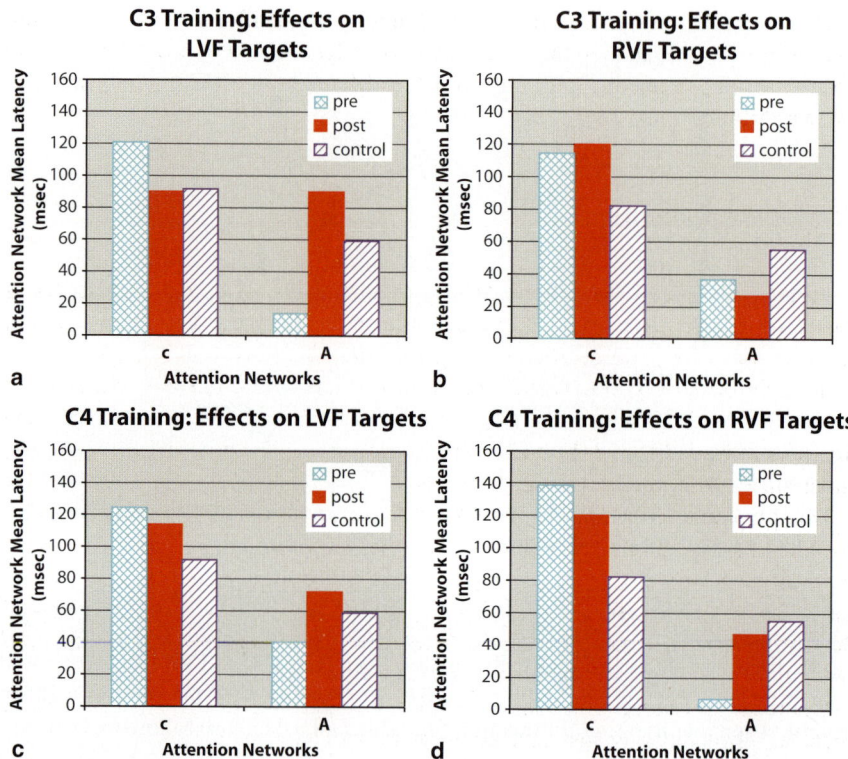

Fig. 7.3 The effect of training on Conflict (*C*) and Alerting (*A*) compared with normal controls
(**a**) on LVF targets in the C3 group, (**b**) on RVF targets in the C3 group, (**c**) on LVF targets in the
C4 group, and (**d**) on RVF targets in the C4 group

motor cortex selectively affected the RH. By the same token, C4, placed over right
motor cortex, improved Conflict in the LH (Fig. 7.3d) and Alerting in both hemispheres (Fig. 7.3c, d). Similarly, training with the centrally placed electrodes, Cz
and Fz, improved Alerting in the LH. Thus, there is no selective effect of the training electrode on the hemisphere underneath it. We found that C3 training and C4
training had greater effects on the attention networks than Cz training and Fz training. Furthermore, C3 training improved attention (Conflict and spatial Orienting)
selectively in the RH (Fig. 7.3a), whereas C4 training improved attention selectively in the LH (Conflict, benefit of spatial Orienting, Alerting) (Fig. 7.3d).

The results are consistent with a model in which C3 and C4 training engages a
metacognitive network that is contralaterally organized (Sterman, personal communication, April 29, 2005). Activation of the metacognitive network enables a change
in the opposite hemisphere, but training does not affect the organization of the metacognitive network itself. Consequently, the C3 and C4 electrodes need not change
their own physiological profile and can reflect the same theta to SMR ratio before
and after training. More stable changes in the theta to SMR ratio are likely to be
observed in electrodes that reflect the operation of each hemisphere separately.

It is also noteworthy that although Conflict and cost of spatial Orienting both involve interference and are known to engage the dorsolateral prefrontal cortex and the anterior cingulate cortex, they are affected differently by EEG training, suggesting that they are based on different structures.

Clinical Implications

In clinical EEGBF, C3 is often chosen for inattention symptoms, while C4 is chosen for symptoms of impulsivity and restlessness (cf. Egner & Gruzelier, 2004), but there has been little justification beyond clinical observation to support these choices. Gruzelier suggests that C3 training decreases errors of omission, whereas we found that C3 is *least* effective at improving Alerting. However, Egner and Gruzelier rewarded 15–18 Hz at C3, while we rewarded 12–15 Hz, underscoring the importance of different reward frequencies.

For symptoms of inattention as well as for perseveration or problems of switching attention, Cz and Fz are also sometimes selected as training locations, instead of motor strip sites. For Fz and Cz our results match clinical practice, improving the Orienting of attention, shown by the improvement of the benefit of spatial Orienting measure. However, our results suggest that typical ADHD symptoms of inattention and impulsivity would be best served by training at C4.

Our results converge with clinical practice, in suggesting that C4 training is most effective for modulating sustained attention in both hemispheres.

Methodological Considerations

Our study was limited in several ways. First, let us consider the choice of the reference electrode. This is often the ear ipsilateral to the active electrode, i.e., C3-A1 or C4-A2. Some clinicians also use a contralateral ear reference, and some use a linked ear reference. These choices may significantly alter the training protocol. Specifically, our C3 and C4 groups used ipsilateral references and thus localized the feedback to laterally measured changes, while our Fz and Cz groups used the left ear reference for both protocols. While our results at Fz and Cz *do* support the clinical belief that these areas affect switching of attention, it may not be the case that Fz-A2 and Cz-A2 would have effects similar to those seen here with Fz-A1 and Cz-A1.

One must also use caution when generalizing from our sample to all ADHD or learning-disabled individuals. Not only was there considerable variation among individual electroencephalograms, but also our subject pool may have exhibited EEG profiles not typical of adults with attention and learning disabilities. For example, Monastra et al. (1999) used a single EEG measurement at the vertex (Cz) to assess 482 individuals between the ages of 6 and 30 years, and found that a ratio of band power (theta/beta) was able to confirm ADHD diagnoses based on more traditional measures, including a clinical interview, continuous performance test, and behavior survey. This discriminant criterion predicted the diagnoses with a specificity of 98%

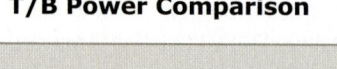

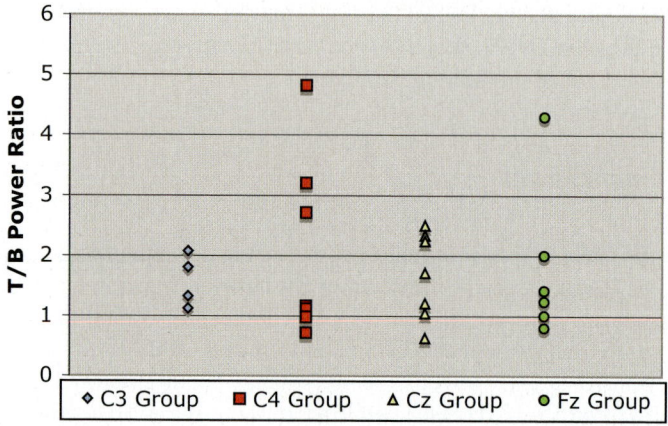

Fig. 7.4 The pretraining theta/beta (*T/B*) ratio in each participant group. All groups of mixed attention deficit hyperactivity disorder/learning-disabled individuals showed a wide range of T/B scores, without the expected high T/B ratios for all individuals

and a sensitivity of 86%. We applied a similar metric to our subject group, and found their theta/beta ratio scores to be widely distributed (Fig. 7.4.) This discrepancy may be more apparent than real. Monastra et al. averaged the EEG signal during eyes open, eyes closed, and task conditions, while we averaged EEG during eyes open and closed conditions only. We do suggest that electrophysiological criteria for inclusion may be useful in future studies.

Future Directions

We have shown here that attention is duplicated in the two hemispheres and that attention in one hemisphere can be affected differentially by lateralized training. C3 and C4 both seem to reduce Conflict in the contralateral hemisphere, raising the possibility that Conflict is mediated by a competition or interference mechanism. This model contrasts with the metacognitive model proposed in "Discussion" and further studies are necessary to decide between them. Our findings also raise the question of what happens following interhemispheric training aiming at increasing or decreasing interhemispheric coherence. In future studies it will be important to examine the effect of C3-C4 interhemispheric training in the LANT.

Furthermore, given that our results with C4 training have a strong LH effect, and given that C4-A2 training is common in clinical use, the question arises of whether there is a bilateral training protocol that would have similar or superior effects. In particular, what is the effectiveness and mechanism of action of two-channel (bilat-eral) training protocols involving hemispheric ratios, such as the asymmetry indices suggested by Davidson (2004)?

Finally, the version of the LANT that we used may be too complex, since it mixed components of both automatic and controlled orienting. In particular, in this study we used *peripheral* cues (automatic orienting) that were *informative*, i.e., mostly valid (controlled orienting). Instead, one can contrast the effect of automatic orienting, using *uninformative peripheral* cues, with controlled orienting, using *informative, central*, symbolic, cues. In this case, one would expect a reversal of the benefit of spatial orienting (inhibition of return) in automatic orienting at cue-to-target intervals longer than 300 ms. No such reversal, however, should occur in controlled orienting (Posner & Cohen, 1984). In a recent study, we administered four such conditions of the LANT to 55 young adult participants (Fig. 7.5). Conditions 1 and 2 included uninformative (50% valid) peripheral cues consisting of asterisks, presented 150 and 500 ms, respectively, before the target. Conditions 3 and 4 included informative (75%

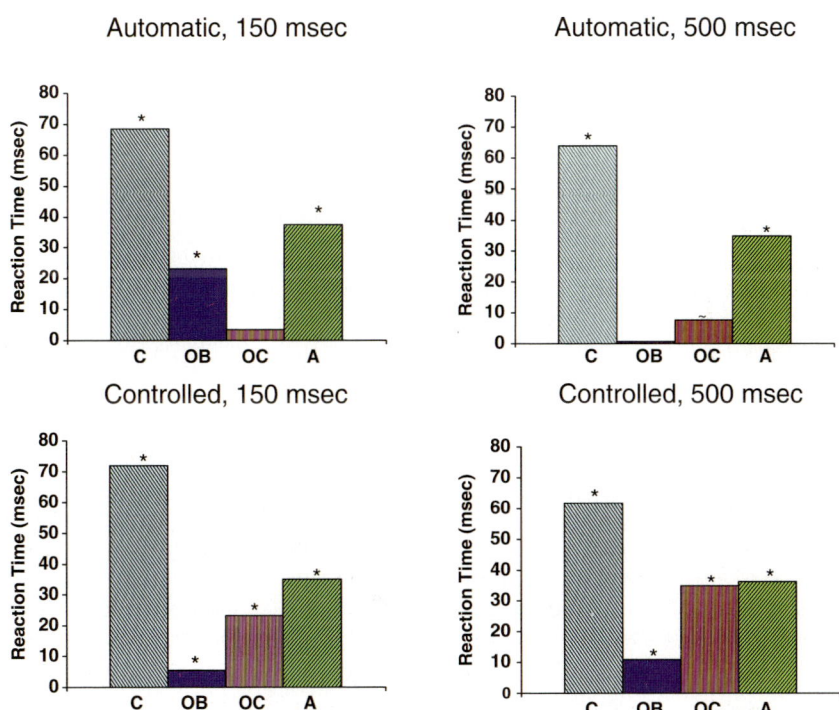

Fig. 7.5 Results of testing 60 young adults with two versions of the LANT at two different cue-to-target intervals (150 and 500 ms). The "automatic version" employs uninformative peripheral cues, while the "controlled version" employs informative bilateral symbolic cues. Only the auto-matic version shows inhibition of return, at 500-ms cue-to-target interval. *OB* benefit of spatial Orienting, *OC* cost of spatial Orienting

valid) central, symbolic cues consisting of hands pointing left or right in both visual fields, presented at 150 ms and 500 s, respectively, before the target.

The results showed significant networks in all conditions with the exception of automatic benefit of spatial Orienting at 500 ms. This reflects the expected Inhibition Of Return at the longer cue-to-target interval. Moreover, cue condition (automatic, controlled) interacted with TVF (left, right): there was an LVF advantage for the automatic conditions and an RVF advantage for the controlled conditions.

In conclusion, we have shown evidence for the existence of lateralized networks of attention and for the ability to affect these networks differentially through different EEGBF protocols. However, many questions remain about the roles of training site and training frequency in affecting attention.

Acknowledgements Thanks to Rachel Tamir for helping to coordinate the experiment and to S. Aakash Kishore and Nina Azer for assistance in preparing the manuscript. Thanks to Sigi Hale for help with reviewing the literature. Most of all, we would like to thank the participants for their enthusiastic and dedicated involvement in these experiments. This work was supported by the Center for Learning Disabilities at the Academic College of Tel Chai, Israel, and by USPHS NIH grant NS20187.

References

Barnea, A., Rassis, A., Raz, A., Neta, M., Herzberg, K. & Zaidel, E. (in preparation). The Lateralized Attention Network Test (LANT).

Baving, L., Laucht, M., & Schmidt, M. H. (1999). Atypical frontal brain activation in ADHD: Preschool and elementary school boys and girls. *Journal of the American Academy of Child and Adolescent Psychiatry, 38*, 1363–71.

Birbaumer, N. (1977). Operant enhancement of EEG-Theta activity. In J. Beatty & H. Legewie (Eds.), *Biofeedback and behavior* (pp. 135–145). New York: Plenum Press.

Birbaumer, N., Elbert, T., Rockstroh, B., & Lutzenberger, W. (1981). Biofeedback of event-related potentials of the brain. *International Journal of Psychology, 16*, 389–415.

Birbaumer, N. (1984). Operant control of slow brain potentials: A tool in the investigation of the potential's meaning and its relation to attentional dysfunction. In T. Elbert, B. Rockstroh, W. Lutzenberger & N. Birbaumer (Eds.), *Self-regulation of the brain and behaviour* (pp. 227–239). Berlin: Springer.

Bush, G., Frazier, J. A., Rauch, S. L., Seidman, L. J., Whalen, P. J., Jenike, M. A., Rosen, B. R., Biederman, J. (1999). Anterior cingulate cortex dysfunction in Attention- Deficit Hyperactivity/ Disorder revealed by fMRI and the Counting Stroop. *Biological Psychiatry, 45*, 1542–1552.

Clarke, A. R., Barry, R. J., McCarthy, R., Selikowitz, M., Johnstone, S. J., Hsu, C., Magee, C., Lawrence, C., & Croft, R. (2007). Coherence in children with Attention-Deficit/Hyperactivity Disorder and excess beta activity in their EEG. *Clinical Neurophysiology: Official Journal of the International Federation of Clinical Neurophysiology, 118*, 1472–9.

Clarke, A. R., Barry, R. J., McCarthy, R. & Selikowitz, M. (2001). Electroencephalogram differences in two subtypes of attention-deficit hyperactivity/disorder. *Psychophysiology, 38*(2), 212–221.

Cohen, N. J., Vallance, D. D., Barwick, M., Im, N., Menna, R., Horodezky, N., & Isaacson, L. (2000). The interface between ADHD and language impairment: An examination of language, achievement, and cognitive processing. *Journal of Child Psychology and Psychiatry, 43,* 353–362.

Davidson, R. J. (2004). What does the prefrontal cortex "do" in affect: Perspectives on frontal EEG asymmetry research. *Biological Psychology, 67*(1–2), 219–33.

Doesburg, S. M., Roggeveen, A. B., Kitajo, K., & Ward, L. M. (2007). Large-Scale Gamma-Band Phase Synchronization and Selective Attention. *Cerebral Cortex.* Egner, T. and J. H. Gruzelier (2004). EEG biofeedback of low beta band components: Frequency-specific effects on variables of attention and event-related brain potentials. *Clinical Neurophysiology, 115*(1), 131–9.

Fan, J., McCandliss, B. D., Sommer, T., Raz, A., & Posner, M. I. (2002). Testing the efficiency in independence of attentional networks. *Journal of Cognitive Neuroscience, 14,* 340–347.

Garcia-Sanchez, C., Estevez-Gonzolas, A., Suarez-Romero, E., & Junque, C. (1997). Right hemisphere dysfunction in subjects with Attention-Deficit disorder with and without hyperactivity. *Journal of Child Neurology, 12*(2), 107–115.

Greene, D.J., Barnea, A., Herzberg, K., Rassis, A., Neta, M., Raz, A., & Zaidel, E. (2008). Measuring attention in the hemispheres: the lateralized attention network test (LANT). *Brain and Cognition, 66*(1), 21–31.

Gobbelé, R., Waberski, T. D., Schmitz, S., Sturm, W., & Buchner, H. (2002). Spatial direction of attention enhances right hemispheric event-related gamma-band synchronization in humans. *Neuroscience Letters, 327,* 57–60.

Gruzelier, J. and T. Egner (2005). Critical validation studies of neurofeedback. *Child and Adolescent Psychiatry Clinics of North America, 14*(1), 83–104.

Hardman, E., Gruzelier, J., Cheesman, K., Jones, C., Liddiard, D., Schleichert, H., Birbaumer, N. (1997). Frontal interhemispheric asymmetry: Self regulation and individual differences in humans. *Neuroscience Letters, 221,* 117–120.

Heilman, K. M., & Van Den Abell, T. (1980). Right hemisphere dominance for attention: The mechanism underlying hemispheric asymmetries of inattention (neglect). *Neurology, 30,* 327–30.

Jutai, J. W. (1984). Cerebral asymmetry and the psychophysiology of attention. *International Journal of Psychophysiology: Official Journal of the International Organization of Psychophysiology, 1,* 219–25.

Kamiya, J. (1969). Operant control of the EEG Alpha rhythm and some of its effects on consciousness. In C. T. Tart (Ed.), *Altered states of consciousness.* New York: Wiley.

Kaplan, J. & Zaidel, E. (2001). Error monitoring in the hemispheres: The effect of feedback on lateralized lexical decision. *Cognition, 82,* 157–178.

Kotchoubey, B., Schneider, D., Schleichert, H., Strehl, U., Uhlmann, C., Blankenhorn, V., Froscher, W., & Birbaumer, N. (1996). Self-regulation of slow cortical potentials in epilepsy: A retrial with analysis of influencing factors. *Epilepsy Research, 25,* 269–176.

Liotti, M., Pliszka, S. R., Perez, R., Kothmann, D., Woldorff, M.G. (2005). Abnormal brain activity related to performance monitoring and error detection in children with ADHD. *Cortex, 41*(3), 377–88.

Loo, S. K. (2003). EEG and neurofeedback findings in ADHD. *The ADHD Report, 11*(3), 1–6.

Luck, S. J., Hillyard, S. A., Mouloua, M., Woldorff, M. G., Clark, V. P., Hawkins, H. L. (1994). Effects of spatial cuing on luminance detectability: Psychophysical and electrophysiological evidence for early selection. *Journal of Experimental Psychology Human Perceptual Performance, 20,* 887–904.

Mann, C. A., Lubar, J. F., Zimmerman, A. W., Miller, C. A., Muenchen, R. A. (1992). Quantitative analysis of EEG in boys with attention-deficit-hyperactivity disorder: Controlled study with clinical implications. *Pediatric Neurology, 8*(1), 30–36.

Max, J. E., F. F. Manes, et al. (2005). Prefrontal and executive attention network lesions and the development of attention-deficit/hyperactivity symptomatology. *Journal of the American Academy of Child and Adolescent Psychiatry, 44*(5), 443–50.

McInnes, A., Humphries, T., Hogg-Johnson, S., & Tannock, R. (2003). Listening comprehension and working memory are impaired in attention-deficit hyperactivity disorder irrespective of language impairment. *Journal of Abnormal Child Psychology, 31*(4), 427–443.

Miniussi, C., Rao, A., & Nobre, A. C. (2002). Watching where you look: Modulation of visual processing of foveal stimuli by spatial attention. *Neuropsychologia, 40*, 2448–60.

Monastra, V. J., Lubar, J. F., Linden, M., VanDeusen, P., Green, G., Wing, W., Phillips, A., & Fenger, T. N. (1999). Assessing attention deficit hyperactivity disorder via quantitative electroencephalography: An initial validation study. *Neuropsychology, 13*(3), 424–433.

Monastra, V. J. & Lubar, J. F. (2000). Using quantitative electroencephalography to differentiate attentional from other psychiatric disorders. Presented at the American Psychological Association Convention, Washington, DC.

Monastra, V. J., Monastra, D. M., & George, S. (2002). The effects of stimulant therapy, EEG biofeedback, and parenting style on the primary symptoms of attention-deficit hyperactivity disorder. *Applied Psychophysiology and Biofeedback, 27*(4), 231–249.

Monastra V. J. (2003). Clinical applications of electroencephalographic biofeedback. In M.S. Schwartz & F. Andrasik (Eds.). *Biofeedback: A practitioner's guide* (3rd Ed.). New York: Guilford Press.

Monastra, V. J. (2005). "Electroencephalographic biofeedback (neurotherapy) as a treatment for attention deficit hyperactivity disorder: Rationale and empirical foundation." *Child and Adolescent Psychiatry Clinics of North America, 14*(1), 55–82.

Müller, M. M., Gruber, T., & Keil, A. (2000). Modulation of induced gamma band activity in the human EEG by attention and visual information processing. *International Journal of Psychophysiology: Official Journal of the International Organization of Psychophysiology, 38*, 283–299.

Plotkin, W. B. (1976). On the self-regulation of the occipital Alpha-rhythm: Control strategies, states of consciousness, and the role of physiological feedback. *Journal of Experimental Psychology, General, 105*, 66–99.

Posner, M. I., & Cohen, Y. (1984). Components of visual orienting, In H. Bouma & D. Bouwhuis (Eds.). *Attention and performance X* (pp. 531–556). London: Lawrence Erlbaum Associates Ltd.

Pulvermuller, F., Mohr, B., Schleichert, H., & Veit, R. (2000). Operant conditioning of left-hemispheric slow cortical potentials and its effect on word processing. *Biological Psychology, 53*, 77–215.

Rockstroh, B., Elbert, T., Birbaumer, N., Wolf, P., Duchting-Roth, A., Reker, M., Daum, I., Lutzenberger, W., & Dichgans, J. (1993). Cortical self-regulation in patients with epilepsies. *Epilepsy Research, 14*, 63–72.

Rucklidge, J. J., & Tannock, R. (2002). Neuropsychological profiles of adolescents with ADHD: Effects of reading difficulties and gender. *Journal of Child Psychology and Psychiatry, 43*(8), 988–1003.

Seidman, L. J., Biederman, J., Weber, W., Hatch, M., & Faraone, S. V. (1998). Neuropsychological function in adults with ADHD. *Biological Psychiatry, 44*, 260–268.

Semrud-Clikeman, M., Guy, K., Griffin, J. D., & Hynd, G. W. (2000). Rapid naming deficits in children and adolescents with reading disabilities and attention deficit hyperactivity disorder. *Brain and Language, 75*(1), 70–83.

Senkowski, D., Talsma, D., Herrmann, C. S., & Woldorff, M. G. (2005). Multisensory processing and oscillatory gamma responses: Effects of spatial selective attention. *Experimental brain research. Experimentelle Hirnforschung. Expérimentation cérébrale, 166*, 411–426.

Sheppard, D., Bradshaw, J., & Mattingley, P. L. (1999). Effects of stimulant medication on the lateralization of line bisection judgments of children with attention deficit disorder. *Journal of Neurology, Neurosurgery, and Psychiatry, 66*, 57–63.

Sterman, M. B., R. C. Howe, et al. (1970). Facilitation of spindle-burst sleep by conditioning of electroencephalographic activity while awake. *Science, 167*(921), 1146–1148.

Sterman, M. B. (1977). Sensorimotor EEG operant conditioning: Experimental and clinical effects. *Pavlov Journal of Biological Science, 12,* 63–92.

Valentino, D. A., & Dufresne, R. L. (1991). Attention tasks and EEG power spectra. International Journal of Psychophysiology: Official Journal of the International Organization of Psychophysiology, *11*, 299–301.

Yamaguchi, S., Yamagata, S., & Kobayashi, S. (2000). Cerebral asymmetry of the "top-down" allocation of attention to global and local features. *The Journal of Neuroscience: The Official Journal of the Society for Neuroscience, 20,* RC72.

Zaidel, E. (1995). Interhemispheric transfer in the split brain: Long-term status following complete cerebral commisurotomy. In R. H. Davidson & K. Hugdahl (Eds.), *Brain asymmetry* (pp. 491–532). Cambridge, MA: MIT.

Chapter 8
A Connectionist Perspective on Attentional Effects in Neurodynamics Data

O. David

Abstract The aim of this chapter is to show how the validity of neuropsychological models of attention can be explicitly tested using advanced techniques for the analysis of event-related activity in electroencephalography (EEG) and magnetoencephalography (MEG). These techniques are based on biophysical models of EEG/MEG which afford a neurobiological perspective on event-related-potential (ERP) research and on cognitive neuroscience in general. In particular, they allow one to reinterpret neurodynamical effects of attention in terms of context-dependent changes in neuronal couplings between remote regions embedded in a global network of attention. First, we present the nodes of the network hypothesized and what the relationships are between attention, synchronization, neural coupling, and ERPs. Second, we describe briefly the mathematical details of a recent modeling approach (dynamic causal modeling) to estimate neural couplings from ERPs. Finally, we will show how dynamic causal modeling for ERPs can be used to compare different neuropsychological models using two examples: the mismatch negativity in auditory oddball paradigms and the activation of the ventral visual pathway by emotional attention.

Abbreviations

AAS   Anterior affective system
DCM   Dynamic causal modeling
EEG   Electroencephalography
ERF   Event-related field
ERP   Event-related potential
fMRI   Functional magnetic resonance imaging

O. David
Inserm, U836, Grenoble Institut des Neurosciences, CHU Grenoble, Bât Edmond J Safra, Chemin Fortuné Ferrini, BP 170, 38042 Grenoble CEDEX 9, France, e-mail: odavid@ujf-grenoble.fr

F. Aboitiz and D. Cosmelli (eds.) *From Attention to Goal-Directed Behavior.*   145
© Springer-Verlag Berlin Heidelberg 2009

MEG   Magnetoencephalography
MMN   Mismatch negativity
OFC   Orbitofrontal cortex
STG   Superior temporal gyrus

Introduction

The goal of this chapter is to show how people interested in connectionist models of attention can test their face validity using generative models of neurodynamics data. We will particularly focus on electroencephalography (EEG) and magnetoencephalography (MEG), but our reasoning applies to functional MRI (fMRI) as well. Because generative models are biologically grounded dynamical models, they allow one to reinterpret neurodynamical effects of attention in terms of context-dependent changes in neuronal couplings between remote regions embedded in a global network of attention. To do this, we will begin by following a roadmap that will present first the nodes of the network hypothesized, and then briefly describe the relationships between (1) attention and synchronization, (2) synchronization and neural coupling, (3) neural coupling and attention, (4) neural coupling and event-related potentials (ERPs), and (5) ERPs and attention. Attention can thus be experimentally studied in electrophysiology by focusing on either synchronization or event-related activity. In terms of biophysical modeling, the changes of neural coupling that might exist between different brain regions involved in attentional mechanisms are the common denominator of such neurodynamical measures. We will then focus on a modeling approach to estimate neural couplings from ERPs using a recent technique called "dynamic causal modeling" (DCM) (David, Kiebel, Harrison, Mattout et al., 2006). After a brief description of the mathematical concepts underlying DCM, we will show that DCM for ERP data can be used to test the validity of important concepts in attentional modulation such as bottom-up and top-down effects.

Neural Couplings and the Neurodynamics of Attention

Reviews of neuroimaging studies of human attentional networks have already shown an impressive reproducibility of results (Coull, 1998; Mesulam, 1999; Posner, Sheese, Odludas & Tang, 2006; Wager, Jonides & Reading, 2004). Put briefly, the network responsible for orienting attention to sensory events is more or less independent of the stimulus modality and involves cortical regions (premotor; superior, inferior, and medial parietal; medial prefrontal; anterior cingulate; frontal eye fields) and subcortical structures (pulvinar, superior colliculus, striatum). These regions are usually thought of as the "sources" of attention because they are supposed to exert a top-down effect on sensory-association areas which show activity modulated by attention (Coull, 1998). The sensory-related regions (the "sites" of attentional expression) are defined by the stimulus modality. In visual attention for instance, V4 and inferior temporal areas are thought to be important sites of attentional modulation (Kastner & Pinsk, 2004). Effects of spatial attention have also been observed in early visual areas (Gandhi, Heeger & Boynton, 1999; Martincz, DiRusso, nllo-Vento, Sereno et al., 2001; Posner & Gilbert, 1999; Somers, Dale, Seiffert & Tootell, 1999).

In this chapter, we will assume that the nodes of the attentional network are known, for instance, from MEG/EEG source localization, fMRI, or positron emission tomography (Coull, 1998; Mesulam, 1999; Posner et al., 2006; Wager et al., 2004). We will concentrate on how to identify the links between the different regions of the network.

Attention and Synchronization

In EEG, it has been shown that the gamma (more than 30 Hz) power is associated with orienting of attention (Brovelli, Lachaux, Kahane & Boussaoud, 2005; Posner et al., 2006; Tallon-Baudry, Bertrand, Henaff, Isnard et al., 2005; Taylor, Mandon, Freiwald & Kreiter, 2005) (see Chap. 3 by Lachaux and Ossandón). Fluctuations of EEG power reflect changes of synchronization within local neural networks (Varela, Lachaux, Rodriguez & Martinerie, 2001). Changes in synchronization between distant regions in the gamma band have also been shown (Rose, Sommer & Buchel, 2006; Womelsdorf & Fries, 2007), indicating a crucial role of distributed neural assemblies during attentional processes. Interestingly, the strength of synchronization can be used to predict behavioral response speed (Womelsdorf, Fries, Mitra & Desimone, 2006) or perceptual accuracy (Taylor et al., 2005). However, the functional implications of enhanced synchronization are still unclear.

Synchronization and Neural Coupling

In neuroscience, it is sometimes assumed that synchronization – or oscillatory coupling – between distant regions has a specific functional role (Engel, Fries & Singer, 2001; Varela et al., 2001), such as facilitating selective neuronal communication (Fries, 2005). Here we will avoid speculations about the role of synchronization in neural coding; starting from a physicist's point of view, we will assume that the brain is an open dissipative system composed of interacting components, i.e., neurons.

The Greek root of "synchronization" means "to share the common time." In more formal terms, this can be reformulated as the temporal correlation of different processes. When two physical systems are synchronized together, their common behavior is due to mutual coupling or to a forcing (common input) (Boccaletti, Kurths, Osipov, Valladares et al., 2002). Setting apart the possibility of a common input, if two brain regions show synchronous activity, it necessarily means that there is a sufficiently strong coupling between neurons belonging to those regions. In EEG, this idea can be easily illustrated using neural-mass models (David & Friston, 2003; Freeman, 1978; Valdes, Jimenez, Riera, Biscay et al., 1999; Lopes da Silva, Hoeks, Smits & Zetterberg, 1974; Robinson, Rennie, Wright, Bahramali et al., 2001; Stam, Pijn, Suffczynski & Lopes da Silva, 1999; Van Rotterdam, Lopes da Silva, van den, Viergever et al., 1982; Wendling, Bellanger, Bartolomei & Chauvel, 2000). These models usually comprise cortical macrocolumns, which can be treated as surrogates for cortical areas and, sometimes, thalamic nuclei. They use only one or two state variables to represent the mean activity of a whole neuronal population. These states summarize the behavior of millions of interacting neurons, a procedure that is sometimes referred to loosely as a "mean-field approximation."

Coupling between areas in a neuronal model is known as effective connectivity, i.e., the influence of one system on the other (Friston, 2001). Using a model composed of two areas, one can easily observe the effect of effective connectivity on functional connectivity, i.e., the correlation (here synchronization) between generated signals. Figure 8.1 shows that synchronous oscillations emerge with increasing neuronal coupling between two cortical areas, as described in David, Cosmelli, & Friston (2004) and David, Harrison, & Friston (2005). The connections between regions in Fig. 8.1

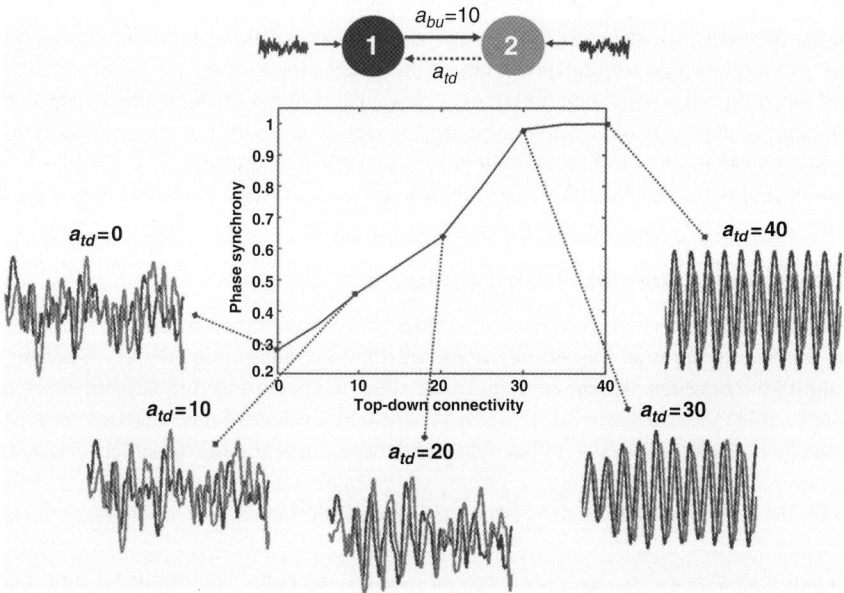

Fig. 8.1 Relationships between effective connectivity (top-down coupling parameter a_{td}) and functional connectivity (phase synchrony). The model is composed of two areas, asymmetrically coupled with forward and backward connections. A propagation delay of 10 ms between areas was assumed. Ongoing activity was simulated using a different stochastic input entering on each area. Synchronous oscillations appear when increasing neuronal coupling between cortical areas

are asymmetrical (forward/bottom-up from 1 to 2, and backward/top-down from 2 to 1; for the details regarding the meaning of these connections, see David et al., 2005). Put briefly, these connections follow rules derived from computational neuroanatomy (Crick & Koch, 1998; Felleman & Van Essen, 1991) which are based upon a tripartitioning of the cortical sheet into supra- and infragranular layers, and granular layer IV. All long-range or extrinsic corticocortical connections are excitatory and are mediated through pyramidal cell axons.

In summary, neuronal coupling and synchrony are probably tightly related to each other in EEG/MEG. In this context, top-down connections (Engel et al., 2001) play a critical role because they are the main determinants for the regulation of oscillatory activity (Fig. 8.1) and for reverberant event-related activity (David et al., 2005; Garrido, Kilner, Kiebel & Friston, 2007a).

Neural Coupling and Attention

So far we have shown that synchronization is related to both attention and neural coupling. This strongly suggests a direct link between attention and changes in neural couplings. Moreover, it hints directly to the connectionist models of attention based on bottom-up and top-down interactions (Coull, 1998; Mesulam, 1999; Pashler, 1998). The most influential neuropsychological theories of attention were proposed in the 1980s (Mesulam, 1981; Posner & Petersen, 1990). The details of those theories differ but they share a common organization, as depicted in Fig. 8.2. Four groups of brain regions can be identified: (1) the so-called sources of attention (prefrontal, parietal, and limbic cortices; subcortical structures such as the thalamus and the superior colliculus) which modulate through top-down connections; (2) the sites of attention, i.e., regions for which the activity is strongly modulated by the attentional context (modality-dependent); (3) possibly the primary sensory regions (modality-dependent); and (4) the arousal system (reticular system), which modulates the activity of the regions involved in attention by bottom-up inputs.

Neuropsychological models thus offer the possibility to address different hypotheses about how the modulation of neural coupling by attention mechanisms can affect measured neurodynamics.

Neural Coupling and Event-Related Potentials

Besides oscillatory activity, another important component of EEG/MEG is the event-related activity. ERPs and event-related fields (ERFs) are obtained by averaging EEG and MEG signals according to some transient stimulus or task (Coles & Rugg, 1995). They show a transient activity that last often less than a few seconds. ERPs and ERFs have been used for decades as putative electrophysiological correlates of perceptual and cognitive operations, such as attention (Hillyard & Kutas, 1983).

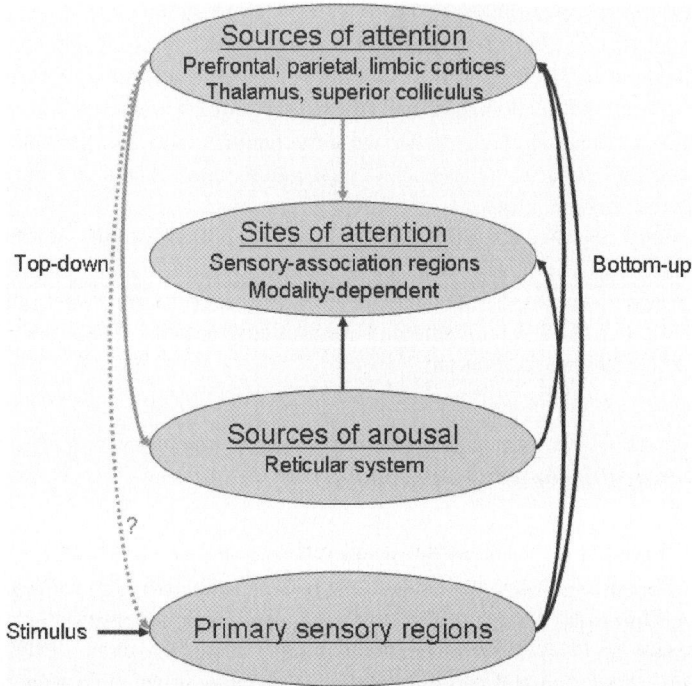

Fig. 8.2 Information flow through the primary sensory system and the attentional system

Like EEG/MEG oscillations, the exact neurobiological mechanisms underlying the generation of ERPs/ERFs are largely unknown. However, it is interesting that the same neural-mass models can generate either oscillatory activity or event-related activity, depending on the shape of the extrinsic input to the modeled network. Neural-mass models are therefore particularly relevant to study the relationships between neural couplings and event-related activity. This was done extensively by David et al. (2005), David, Kilner & Friston (2006), and Rennie, Robinson & Wright (2002). Here we will simply illustrate in Fig. 8.3 how the presence or absence of neural coupling between regions allows or hinders the propagation of event-related activity among successive brain regions. The important point is that neuronal couplings certainly are the cause of many properties of the ERPs/ERFs as well as the bridge between ERP/ERF research and the study of induced and ongoing EEG/MEG activity. In the remainder of the chapter, we will focus on ERPs/ERFs and show how to estimate neural couplings from real data using DCM.

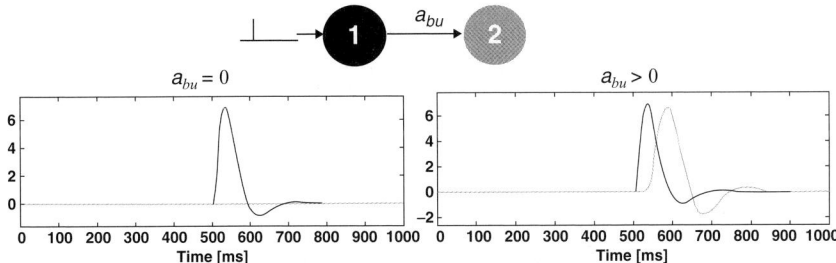

Fig. 8.3 Propagation of event-related activity between two cortical regions in a neural-mass model. Extrinsic neural couplings ensure the propagation of event-related potentials (ERPs) between regions. In general, changes in coupling modify the amplitude and the shape of ERPs. (Adapted from David, Harrison & Friston, 2007)

Attention and Event-Related Potentials

Neuropsychologists have been studying attention for many decades using ERPs. Several ERP components (visual stimuli, P100/N100; auditory stimuli, N100; infrequent "oddball" stimuli, P300/MMN; semantic stimuli, N400; among others) can be modulated by attention, such that their amplitude increases to stimuli which are attended compared with stimuli which are not (Coull, 1998). In visuospatial or auditory modalities, attentional effects are observed in the ERPs as early as 50–100 ms after stimulus (Mangun & Hillyard, 1995), suggesting modulated neuronal activity in extrastriate association areas.

Summary

To sum up, an efficient way to study effective connectivity in networks of attention could be the following: (1) select competing neuropsychological models; (2) translate these models into biophysical models (neural-mass models) in which the connections between different regions can be modulated by the experimental context; (3) estimate the effective connectivity (neural couplings) of these models from real data using DCM for ERPs; and (4) compare the likelihood of the models (model evidence) to identify the most probable neuropsychological models for the data under study. We will illustrate this approach in the remaining sections.

Forward and Inverse Problems

Before presenting the details of DCM, it may be worthwhile to discuss two important modeling concepts at the core of DCM: (1) the "forward problem" and (2) the "inverse problem."

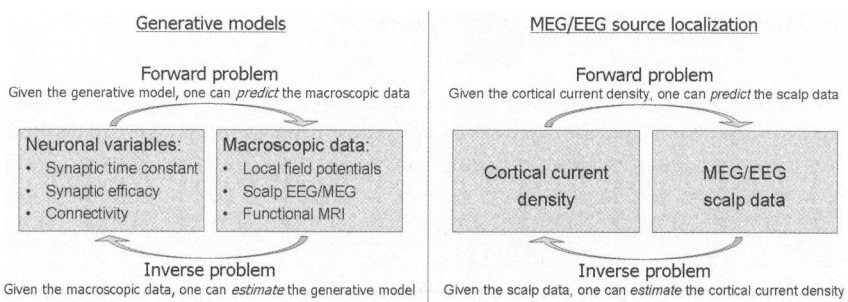

Fig. 8.4 Forward and inverse problems. *Left:* Generative models (dynamic causal modeling, DCM). *Right:* MEG/EEG source localization

DCM is part of the large class of models called "generative," or "biophysical," models of neuroimaging data. The goal of these models (Fig. 8.4) is to propose a way to link in biophysical terms the neuronal variables (e.g., the synaptic time constants, the synaptic efficacies, the ratio of excitation and inhibition, the neuronal connectivity) to the macroscopic data at the brain level (e.g., local field potentials, scalp EEG/MEG, fMRI). In other words, knowing the generative model allows one to predict what the measured data should look like for any perturbation (or stimulation) of the model. Clearly, forward modeling is a difficult issue because one has to make many assumptions to relate millions of neuronal variables to the few data one measures noninvasively in neuroimaging.

Once the forward model has been specified, the second step is to try to estimate the parameters of the model, given measured data. This is called the "inverse problem," and parameter estimation can also be referred to as "model inversion." It is often difficult to get a robust solution to this problem. In DCM, a Bayesian scheme is used to increase the robustness of the estimation of neuronal parameters by adding prior probability distributions on the parameter distributions.

As a reminder, the concepts of forward and inverse problems are relatively common in MEG/EEG source localization (Fig. 8.4). The forward model is the mathematical operator that relates the cortical current density to scalp measurements (Mosher, Leahy & Lewis, 1999). The inverse model is the mathematical inversion of the forward model to estimate the cortical current density from measured scalp data (Baillet, Mosher & Leahy, 2001). DCM for EEG operates in the same mathematical framework as any standard source localization techniques in MEG or EEG.

In DCM, there are thus two major interrelated issues: (1) to design a neuronal model (forward model) with sufficient biological plausibility; (2) to be able to invert the neuronal model from real data to get neuronal estimates on which inferences about brain function are made.

Dynamic Causal Modeling – Theory

The central idea behind DCM (David et al., 2006; Friston, Harrison & Penny, 2003; Kiebel, David & Friston, 2006; Kiebel, Garrido & Friston, 2007) is to treat the brain as a deterministic nonlinear dynamical system that is subject to inputs, and produces outputs. Evoked brain responses are explained by DCM as deterministic responses to some perturbations, i.e., stimuli, in terms of context-dependent coupling which allows for differences in the shape of responses.

DCM for EEG relies on a neuronally plausible model which must be inverted to adjust the data. EEG modeling uses neural-mass models which assume that the data can be approximated as the mean activity of millions of neurons (David & Friston, 2003). DCM for EEG has been developed as a generic tool to analyze evoked potentials obtained at the scalp level for any kind of neuropsychological or cognitive experiment. The generative model of DCM for EEG (David et al., 2005) combines the Jansen model (Jansen & Rit, 1995), a neural-mass model originally developed for explaining visual responses, with simplified rules of corticocortical connectivity (Crick & Koch, 1998) derived from the analysis of connections between the different cortical layers in the visual cortex of monkey. The resultant model (David et al., 2005) is a set of differential equations describing interactions between different inhibitory and excitatory neuronal populations, which can be easily manipulated to embed any hierarchical corticocortical network using forward, backward, and lateral connections, as proposed in computational neuroanatomy (Fig. 8.5). Acknowledging the large cytoarchitectonic variability of the neocortex and other brain structures such as the hippocampus and amygdala, the plausibility of a generic model valid for the whole brain is questionable. However, the crucial point here is that the main purpose of a forward model in the context of DCM is to constrain the generated dynamics in a neuronally plausible way (time constants, propagation delay, directionality of the information transfer, etc.). This is exactly what a generic model can do, with a good tuning between complexity, plausibility, and modularity.

Because DCMs are not restricted to linear or instantaneous systems, they generally depend on a large number of free parameters. These are estimated for DCMs using Bayesian inversion, and inferences about particular connections are made using their posterior or conditional density. The full set of equations for DCM specification and Bayesian parameter estimation can be found in the original papers (David et al., 2005; David et al., 2006; Friston et al., 2003; Kiebel et al., 2006; Penny, Stephan, Mechelli & Friston, 2004). Below we summarize only the most important steps.

A DCM is specified in terms of a state equation and an output equation. The state equation can be written as

$$\dot{x} = f(x, u, \theta), \tag{8.1}$$

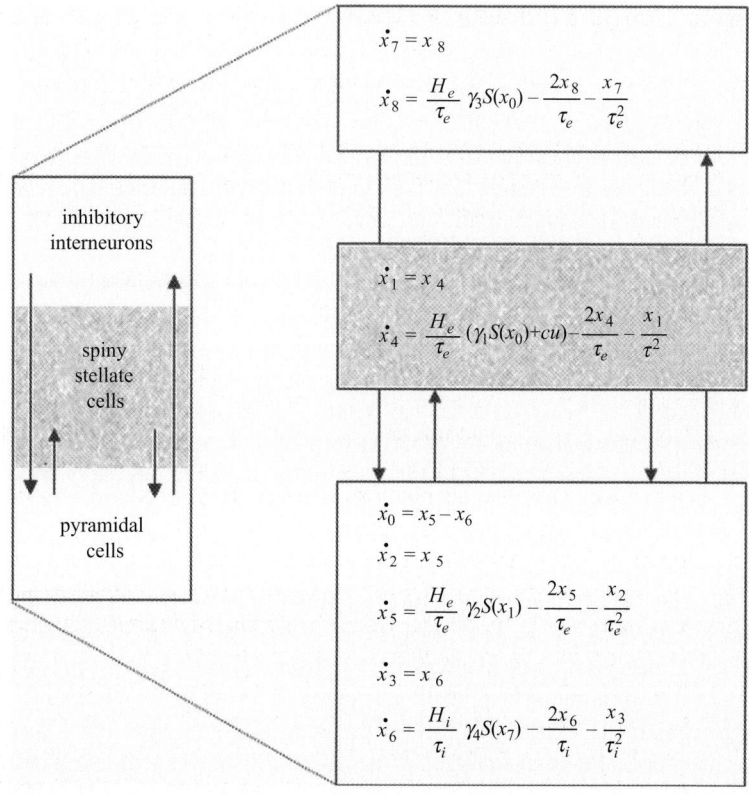

Fig. 8.5 State-space representation of a neural-mass model of a cortical area. Pyramidal cells interact with both excitatory (subscript *e*) and inhibitory (subscript *i*) interneurons. The parameters H_e, H_i, τ_e, and τ_i control the expression of postsynaptic potentials. S is a sigmoid function which transforms the mean membrane potential into a mean firing rate. See David et al. (2005, 2007) for a full description of this model

where x are the neuronal states, u are the extrinsic inputs, and θ are the model parameters (inhibitory and excitatory synaptic time constants and efficacies, intrinsic and extrinsic connectivity, propagation delay). The output equation links the unobserved neuronal states x to the measured data y using a nonlinear instantaneous function g:

$$y = g(x, \theta). \tag{8.2}$$

In DCM for EEG (David et al., 2006; Kiebel et al., 2006), y are a few principal modes of the scalp data and the function g is the head model used for source localization. The head model allows one to match the amplitude of current source densities (mean depolarization of pyramidal cells) according to the source power measured on the scalp. In other words, DCM can be thought of as a source localization scheme

(Baillet et al., 2001) with a dynamical constraint on the reconstructed cortical activity instantiated by the state equation f of the neural-mass model (see also Fig. 8.4).

Operationally, DCM uses a Bayesian scheme for the estimation of model parameters using an expectation-maximization algorithm (Friston, Penny, Phillips, Kiebel et al., 2002). The output of the parameter estimation is posterior probabilities of model parameters $p(\theta|y)$ which are a combination of the likelihood (or confidence in the data) $p(y|\theta)$ and prior expectations about the parameters (for instance, synaptic time constants are expected to be around 5–10 ms) $p(\theta)$:

$$p(\theta|y) \propto p(y|\theta)p(\theta). \tag{8.3}$$

Hyperparameters tune the relative influence of the data and of prior expectations. They are estimated from the data using a restricted maximum likelihood scheme. The most important aspect is that inferences about the model parameters, and particularly about connectivity parameters, can be performed directly from their posterior distribution (under Gaussian assumptions, one estimates and uses the conditional or posterior mean and covariance of the parameters).

The main interest of DCM is to test competing functional hypotheses. For each functional hypothesis, a model m is specified in terms of anatomical connections between regions and of the modulation of some connections by the experimental context. This is equivalent to constructing a specific function f (Eq. 8.1) for each model. After the estimation of parameters of each competing model, the models are compared to find the most likely model, or functional hypothesis. This can be done using Bayesian model selection where the evidence of each model is used to quantify the model plausibility (Penny et al., 2004). The evidence of model m is given by

$$p(y|m) = \int p(y|\theta,m)p(\theta,m)d\theta. \tag{8.4}$$

The evidence can be decomposed into two components: an accuracy term, which quantifies the data fit, and a complexity term, which penalizes models with a large number of parameters. Therefore, the evidence embodies the two conflicting requirements of a good model, namely, that it explains the data and is as simple as possible. The most likely model is the one with the largest evidence. Assuming each data set is independent of the others, one obtains the best model at the group level with n subjects by computing the evidence at the group level by multiplying the marginal likelihoods, or equivalently by adding the log-evidences from each subject (Garrido, Kilner, Kiebel, Stephan et al., 2007b):

$$\ln p(y_1,...,y_n|m) = \sum_{j=1}^{n} \ln p(y_j|m). \tag{8.5}$$

Note that the apparent complexity of the mathematical description of DCM should not be a limitation for psychologists who would like to try a DCM study on their data. Indeed, DCM has been included in the standard release of the software program Statistical Parametric Mapping (SPM), freely available from the Web (http://www.fil.ion.ucl.ac.uk/spm, Functional Imaging Laboratory, University College

London, UK). In Statistical Parametric Mapping, DCM studies can be performed using a graphical user interface. A basic understanding of DCM is only necessary to grasp the limits of the technique and to avoid as much as possible misspecifications of the models.

Dynamic Causal Modeling – Applications

DCM for ERP is a recent technique which has not yet been applied to attention *per se*. In this section, we review the recently published findings of two DCM studies (David et al., 2006; Garrido et al., 2007b; Rudrauf, David, Lachaux, Martinerie et al., 2008) which are nonetheless closely related to attentional mechanisms. Note that DCM for fMRI has been developed using a data set investigating attentional effects on visual perception (Friston et al., 2003). In that data set, the modulation by attention of the effective connectivity between inferior frontal and superior parietal cortices, between V5 and superior parietal cortex, and between V1 and V5 has been demonstrated. Interested readers should refer to Friston et al. (2003) and Penny et al. (2004).

Auditory Oddball

Auditory oddball (infrequent) stimuli typically elicit the so-called mismatch negativity (MMN) (Naatanen, Paavilainen, Rinne & Alho, 2007). The MMN is a negative component of the ERP to any discriminative change in some repetitive aspect of auditory stimulation. It is distributed over frontocentral and central electrodes and peaks at 150–250 ms from stimulus onset. The MMN is based on the presence of a memory trace (approximately 10 s duration) formed by the preceding sound stimuli. It is thought to be generated bilaterally by the auditory and frontal cortices.

DCM for ERP was first applied to an auditory oddball paradigm on a single subject (David et al., 2006). Auditory stimuli, 1,000- or 2,000-Hz tones with 5-ms rise and fall times and 80-ms duration, were presented binaurally for 15 min, every 2 s in a pseudo-random sequence. The 2,000-Hz tones (oddballs) occurred 20% of the time (120 trials) and the 1,000-Hz tones (standards) 80% of the time (480 trials). The subject was instructed to keep a mental record of the number of 2,000-Hz tones. A 128-electrode electroencephalograph was used for the experiment.

Late components, characteristic of rare events, were seen in most frontal electrodes, centered on 250–350 ms after stimulus. As previously reported, early components (i.e., the N100) were almost identical for rare and frequent stimuli (Fig. 8.6). After a conventional source localization analysis, the following DCM was constructed: an extrinsic (thalamic) input entered bilaterally the primary auditory cortex (A1), which was itself connected to ipsilateral orbitofrontal cortex (OFC). In the right hemisphere, an indirect forward pathway was specified from A1 to OFC through the superior

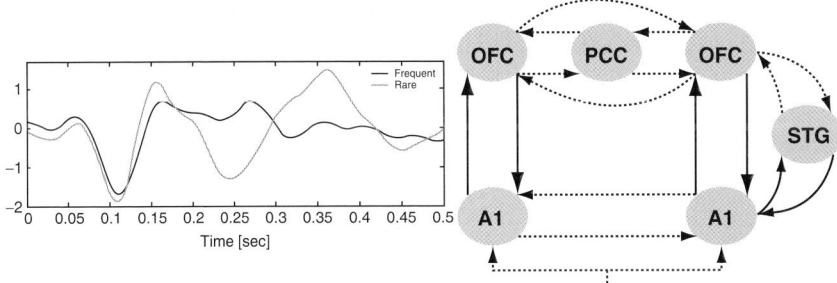

Fig. 8.6 *Left:* ERP for frequent and rare (oddball) auditory events on a central scalp electrode in the subject analyzed. Primary responses (N100) are unchanged whereas a mismatch negativity (MMN) is observed for oddball events. *Right:* Structure of the most probable DCM. *Thick arrows* indicate which connections have been found to be modulated (increase of coupling for rare events as compared to frequent events) to explain the stable N100 and the MMN. See David et al., (2006) for more details. *OFC* orbitofrontal cortex, *PCC* posterior cingulate cortex, *STG* superior temporal gyrus

temporal gyrus (STG). All these connections were reciprocal. At the highest level in the hierarchy, OFC and left posterior cingulate cortex were laterally and reciprocally connected. By testing different possibilities as to which connections showed putative learning-related changes, we showed that the modulation of forward and backward connections between the A1, OFC, and STG was the most likely functional architecture to generate the observed MMN, as compared with hypotheses specifying the modulation of forward connections only, of backward connections only, or of all (including lateral) connections.

Observed effects were confirmed in a subsequent group study (Garrido et al., 2007b; Garrido et al., 2007a), in which a simpler model was tested, including A1, STG, and the inferior frontal gyrus (IFG). Again, DCM demonstrated the best evidence for the simultaneous modulation of forward and backward connections within the network generating the MMN. In summary, DCM analysis of the MMN suggests that a sufficient explanation, given the restricted number of hypotheses tested, for mismatch responses is an increase in forward and backward connections with primary auditory cortex. This results in the appearance of exuberant responses after the N100 in A1 to unexpected stimuli. In computational terms, this could represent a failure to suppress prediction error, relative to learned stimuli, which can be predicted more efficiently (Friston, 2003).

Modulation of the Visual Ventral Stream by Emotional Attention

Emotional salience of stimuli modulates sensory processing, conscious perception, and attention through different sources of top-down control from higher-level regions in parietal and frontal cortex (Anderson, 2005; Vuilleumier, 2005). Alternative models have been proposed to explain the brain mechanisms underlying this influence, invoking feedback to the ventral visual stream from emotion-related structures, i.e., the amygdala, temporal pole, and OFC/ventral medial prefrontal cortex. We will refer to this ensemble as the "anterior affective system" (AAS), for short. These regions, which project back to the visual system, participate in the evaluation of the emotional significance of the visual context and exhibit early responses to visual stimuli which can be modulated by emotional significance (Streit, Dammers, Simsek-Kraues, Brinkmeyer et al., 2003).

The route by which visual information could rapidly reach the structures of the AAS to allow for early modulation by emotion-related information remains an open question. Two main classes of models have been proposed (Vuilleumier, 2005): (1) "two-stage models," in which a rapid preprocessing of visual information occurs sequentially along the hierarchy of the ventral visual stream and in the emotion-related structures, followed by feedback to the visual cortices; and (2) "two-pathway models," in which visual information bypasses the sequential preprocessing and reaches emotion-related structures directly through parallel subcortical pathways.

Using DCM, we tested which type of architecture would best predict real MEG responses in subjects presented with a broad range of unmasked arousing visual stimuli (Rudrauf et al., 2008). Fifteen subjects were presented with 60 pleasant, 60 unpleasant, and 60 neutral movies depicting complex visual scenes, each with a duration of 10 s. MEG signals were recorded using a 151-axial-gradiometer whole-head MEG array CTF Omega system (VSM MedTech, Coquitlam, BC, Canada). The MEG data sampling frequency was 625 Hz, and the acquisition high-pass filter threshold was 0.65 Hz. For each subject, artifact-corrected MEG signals were averaged across trials within each experimental condition (neutral, pleasant, unpleasant stimuli), taking movie onset as the temporal reference. Data processing consisted of three steps: (1) neural network models, explicitly simulating causal interactions between populations of cortical neurons, were first specified; (2) forward modeling of MEG signals (Sarvas, 1987) was used to relate neural activity to scalp signals; and (3) neural parameters were estimated from measured evoked responses. We assessed which model best explained the first 800 ms of MEG data evoked by the stimuli. This time window was chosen as it corresponded to the bulk of the evoked response that served as a basis for the DCM analysis.

The DCMs tested were composed of seven bilateral regions: early visual cortex V1–V2, lateral occipital cortex, ventral occipital cortex, fusiform gyrus, inferotemporal cortex, amygdala/temporal pole, and OFC. These regions, which include the visual ventral stream and the AAS were connected in seven different ways (see Rudrauf et al., 2008 for details). The model comparison strongly and reliably supported a "two-pathway" hypothesis. These models rely on anatomically well established corticocortical long-range fasciculi (Catani et al., 2003), but have never been proposed as a basis for early emotional visual attention. Nevertheless, they could predict real data comparably to more standard models invoking a "retinotectal" subcortical pathway from the retina to the amygdala.

Figure 8.7 shows preliminary results of this study obtained using simplified DCMs in which lateral occipital cortex and ventral occipital cortex were not included. A two-pathway model relying on corticocortical long-range fasciculi (connections

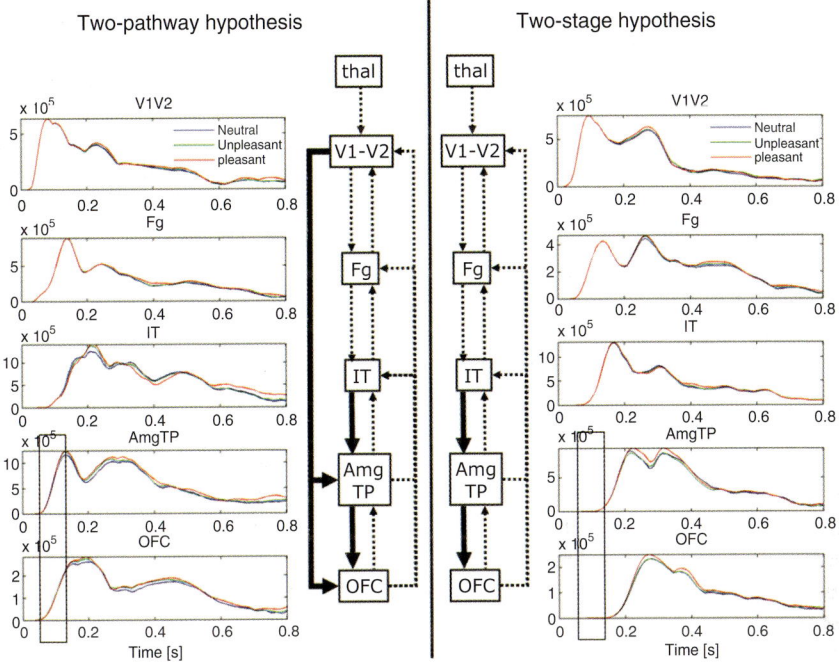

Fig. 8.7 *Left:* Two-pathway model including long-range fasciculi. *Right:* Two-stage model not including long-range fasciculi. Time series are the absolute value of reconstructed MEG activity within each region of the DCM, averaged over 15 subjects. *Thick arrows* indicate the connections a priori modulated by emotional valence of stimuli. There is much more evidence for the two-pathway model than for the two-stage model (not shown). This is partly due to the fact that early activity (between 100 and 200 ms) in the anterior affective system (see the *rectangle*) is not reconstructed by two-stage models. *thal* thalamus, *V1-V2* early visual cortex, *Fg* fusiform gyrus, *IT* inferotemporal cortex, *Amg TP* amygdala/temporal pole

between V1–V2 and the AAS) is compared with a two-stage model. The important point is that a two-stage model does not allow for the activation of the AAS before 170 ms (rectangle in Fig. 8.7), whereas the two-pathway model has no problem in reconstructing this early activity, which was seen in scalp data. This is the main reason why DCM was able to draw a clear-cut distinction between two-pathway models and two-stage models.

Conclusion

In this chapter, we have presented recent advances in the analysis of brain dynamics which are based on the use of biophysical models. Using parameter estimation of the models, one can identify candidate neuronal mechanisms underlying contextual modulation of the EEG or MEG signals, such as during attentional modulation. We have particularly focused on the analysis of event-related activity in EEG and MEG. Note that the same ideas apply to MEG/EEG oscillatory activity (Moran, Kiebel, Stephan, Reilly et al., 2007) and to fMRI (Friston et al., 2003). Signal analysis techniques underlying the use of generative models of brain dynamics are often much more complex than those used for standard statistical analyses and nonlinear dynamical analyses. This is clearly a limitation, but expertise in this field is expanding and there is no doubt that this type of approach is going to take a leading role in cognitive psychology. Why? Because it favors model-driven approaches that open the path to testing specific hypotheses about brain functions. By doing so, DCM-like approaches will probably allow one to significantly refine neuropsychological models. They will also constitute an interesting option for psychopharmacological studies because the estimates of neuronal parameters can provide a better understanding of the specific mechanisms of action different drugs may have on measured neurodynamics.

Acknowledgements We are grateful to Carole Peyrin for critical reading of this chapter. This work was supported by Inserm.

References

Anderson, A. K. (2005). Affective influences on the attentional dynamics supporting awareness. *Journal of Experimental Psychology. General, 134*, 258–281.

Baillet, S., Mosher, J. C., & Leahy, R. M. (2001). Electromagnetic brain mapping. *IEEE Signal Processing Magazine, 18*(6), 14–30.

Boccaletti, S., Kurths, J., Osipov, G., Valladares, D. L., & Zhou, C. S. (2002). The synchronization of chaotic systems. *Physics Reports, 366*, 1–101.

Brovelli, A., Lachaux, J. P., Kahane, P., & Boussaoud, D. (2005). High gamma frequency oscillatory activity dissociates attention from intention in the human premotor cortex. *Neuroimage, 28*, 154–164.

Coles, M. G. H. & Rugg, M. D. (1995). Event-related brain potentials: An introduction. In M. D. Rugg & M. G. H. Coles (Eds.), *Electrophysiology of mind* (pp. 1–26). Oxford: Oxford University Press.

Coull, J. T. (1998). Neural correlates of attention and arousal: Insights from electrophysiology, functional neuroimaging and psychopharmacology. *Progress In Neurobiology, 55*, 343–361.

Crick, F. & Koch, C. (1998). Constraints on cortical and thalamic projections: The no-strong-loops hypothesis. *Nature, 391*, 245–250.

David, O., Cosmelli, D., & Friston, K. J. (2004). Evaluation of different measures of functional connectivity using a neural mass model. *Neuroimage, 21*, 659–673.

David, O. & Friston, K. J. (2003). A neural mass model for MEG/EEG: Coupling and neuronal dynamics. *Neuroimage, 20*, 1743–1755.

David, O., Harrison, L., & Friston, K. J. (2005). Modelling event-related responses in the brain. *Neuroimage, 25*, 756–770.

David, O., Harrison, L., & Friston, K. J. (2007). Neuronal models of EEG and MEG. In K. J. Friston, J. T. Ashburner, S. J. Kiebel, T. E. Nichols, & W. D. Penny (Eds.), *Statistical parametric mapping: The analysis of functional brain images* (1 ed., pp. 414–440). London: Elsevier.

David, O., Kiebel, S. J., Harrison, L. M., Mattout, J., Kilner, J. M., & Friston, K. J. (2006). Dynamic causal modeling of evoked responses in EEG and MEG. *Neuroimage, 30*, 1255–1272.

David, O., Kilner, J. M., & Friston, K. J. (2006). Mechanisms of evoked and induced responses in MEG/EEG. *Neuroimage, 31*, 1580–1591.

Engel, A. K., Fries, P., & Singer, W. (2001). Dynamic predictions: Oscillations and synchrony in top-down processing. *Nature Reviews Neuroscience, 2*, 704–716.

Felleman, D. J., & Van Essen, D. C. (1991). Distributed hierarchical processing in the primate cerebral cortex. *CerebralCortex, 1*, 1–47.

Freeman, W. J. (1978). Models of the dynamics of neural populations. *Electroencephalographyan d Clinical Neurophysiology [Suppl], 34*, 9–18.

Fries, P. (2005). A mechanism for cognitive dynamics: Neuronal communication through neuronal coherence. *Trends in Cognitive Science, 9*, 474–480.

Friston, K. (2003). Learning and inference in the brain. *Neural Networks, 16*, 1325–1352.

Friston, K. J. (2001). Brain function, nonlinear coupling, and neuronal transients. *Neuroscientist, 7*, 406–418.

Friston, K. J., Harrison, L., & Penny, W. (2003). Dynamic causal modelling. *Neuroimage, 19*, 1273–1302.

Friston, K. J., Penny, W., Phillips, C., Kiebel, S., Hinton, G., & Ashburner, J. (2002). Classical and Bayesian inference in neuroimaging: Theory. *Neuroimage, 16*, 465–483.

Gandhi, S. P., Heeger, D. J., & Boynton, G. M. (1999). Spatial attention affects brain activity in human primary visual cortex. *Proceedings of the National Academy of Sciences of the United States of America, 96*, 3314–3319.

Garrido, M. I., Kilner, J. M., Kiebel, S. J., & Friston, K. J. (2007a). Evoked brain responses are generated by feedback loops. *Proceedings of the National Academy of Sciences of the United States of America, 104*, 20961–20966.

Garrido, M. I., Kilner, J. M., Kiebel, S. J., Stephan, K. E., & Friston, K. J. (2007b). Dynamic causal modelling of evoked potentials: A reproducibility study. *Neuroimage, 36*, 571–580.

Hillyard, S. A. & Kutas, M. (1983). Electrophysiology of cognitive processing. *Annual Review of Psychology, 34*, 33–61.

Jansen, B. H. & Rit, V. G. (1995). Electroencephalogram and visual evoked potential generation in a mathematical model of coupled cortical columns. *Biological Cybernetics, 73*, 357–366.

Kastner, S. & Pinsk, M. A. (2004). Visual attention as a multilevel selection process. *Cognitive, Affective and Behavioral Neuroscience, 4*, 483–500.

Kiebel, S. J., David, O., & Friston, K. J. (2006). Dynamic causal modelling of evoked responses in EEG/MEG with lead field parameterization. *Neuroimage, 30*, 1273–1284.

Kiebel, S. J., Garrido, M. I., & Friston, K. J. (2007). Dynamic causal modelling of evoked responses: The role of intrinsic connections. *Neuroimage, 36*, 332–345.

Lopes da Silva, F. H., Hoeks, A., Smits, H., & Zetterberg, L. H. (1974). Model of brain rhythmic activity. The alpha-rhythm of the thalamus. *Kybernetik, 15*, 27–37.

Mangun, G. R. & Hillyard, S. A. (1995). Mechanisms and models of selective attention. In M. D. Rugg & M. G. H. Coles (Eds.), *Electrophysiology of mind: Event-related brain potentials and cognition* (pp. 40–85). Oxford: Oxford University Press.

Martinez, A., DiRusso, F., nIlo-Vento, L., Sereno, M. I., Buxton, R. B., & Hillyard, S. A. (2001). Putting spatial attention on the map: Timing and localization of stimulus selection processes in striate and extrastriate visual areas. *Vision Research, 41*, 1437–1457.

Mesulam, M. M. (1981). A cortical network for directed attention and unilateral neglect. *Annals of Neurology, 10*, 309–325.

Mesulam, M. M. (1999). Spatial attention and neglect: Parietal, frontal and cingulate contributions to the mental representation and attentional targeting of salient extrapersonal events. *Philosophical Transactions of Royal Society of London – Series B: Biological Science, 354*, 1325–1346.

Moran, R. J., Kiebel, S. J., Stephan, K. E., Reilly, R. B., Daunizeau, J., & Friston, K. J. (2007). A neural mass model of spectral responses in electrophysiology. *Neuroimage, 37*, 706–720.

Mosher, J. C., Leahy, R. M., & Lewis, P. S. (1999). EEG and MEG: Forward solutions for inverse methods. *IEEE Transaction on Biomedical Engineering, 46*, 245–259.

Naatanen, R., Paavilainen, P., Rinne, T., & Alho, K. (2007). The mismatch negativity (MMN) in basic research of central auditory processing: A review. *Clinical Neurophysiology, 118*, 2544–2590.

Pashler, H. E. (1998). *The psychology of attention*. Cambridge, MA: MIT.

Penny, W. D., Stephan, K. E., Mechelli, A., & Friston, K. J. (2004). Comparing dynamic causal models. *Neuroimage, 22*, 1157–1172.

Posner, M. I. & Gilbert, C. D. (1999). Attention and primary visual cortex. *Proceedings of the National Academy of Sciences of the United States of America, 96*, 2585–2587.

Posner, M. I. & Petersen, S. E. (1990). The attention system of the human brain. *Annual Review of Neurosciences, 13*, 25–42.

Posner, M. I., Sheese, B. E., Odludas, Y., & Tang, Y. (2006). Analyzing and shaping human attentional networks. *Neural Networks, 19*, 1422–1429.

Rennie, C. J., Robinson, P. A., & Wright, J. J. (2002). Unified neurophysical model of EEG spectra and evoked potentials. *Biological Cybernetics, 86*, 457–471.

Robinson, P. A., Rennie, C. J., Wright, J. J., Bahramali, H., Gordon, E., & Rowe, D. L. (2001). Prediction of electroencephalographic spectra from neurophysiology. *Physics Review E, 63*, 021903.

Rose, M., Sommer, T., & Buchel, C. (2006). Integration of local features to a global percept by neural coupling. *Cerebral Cortex, 16*, 1522–1528.

Rudrauf, D., David, O., Lachaux, J.-P., Martinerie, J., Renault, B., Kovach, C. et al. (2008). Rapid interactions between ventral visual stream and emotion-related structures rely on a two-pathway architecture. *Journal of Neuroscience, 28*, 2793–2803.

Sarvas, J. (1987). Basic mathematical and electromagnetic concepts of the biomagnetic inverse problem. *Physics in Medicine and Biology, 32*, 11–22.

Somers, D. C., Dale, A. M., Seiffert, A. E., & Tootell, R. B. (1999). Functional MRI reveals spatially specific attentional modulation in human primary visual cortex. *Proceedings of the National Academy of Sciences of the United States of America, 96*, 1663–1668.

Stam, C. J., Pijn, J. P., Suffczynski, P., & Lopes da Silva, F. H. (1999). Dynamics of the human alpha rhythm: Evidence for non-linearity? *Clinical Neurophysiology, 110*, 1801–1813.

Streit, M., Dammers, J., Simsek-Kraues, S., Brinkmeyer, J., Wolwer, W., & Ioannides, A. (2003). Time course of regional brain activations during facial emotion recognition in humans. *Neuroscience Letters, 342*, 101–104.

Tallon-Baudry, C., Bertrand, O., Henaff, M. A., Isnard, J., & Fischer, C. (2005). Attention modulates gamma-band oscillations differently in the human lateral occipital cortex and fusiform gyrus. *Cerebral Cortex, 15*, 654–662.

Taylor, K., Mandon, S., Freiwald, W. A., & Kreiter, A. K. (2005). Coherent oscillatory activity in monkey area v4 predicts successful allocation of attention. *Cerebral Cortex, 15*, 1424–1437.

Valdes, P. A., Jimenez, J. C., Riera, J., Biscay, R., & Ozaki, T. (1999). Nonlinear EEG analysis based on a neural mass model. *Biological Cybernetics, 81*, 415–424.

Van Rotterdam, A., Lopes da Silva, F. H., van den, E. J., Viergever, M. A., & Hermans, A. J. (1982). A model of the spatial-temporal characteristics of the alpha rhythm. *Bulletin of Mathematical Biology, 44*, 283–305.

Varela, F., Lachaux, J.-P., Rodriguez, E., & Martinerie, J. (2001). The brainweb: Phase synchronization and large-scale integration. *Nature Reviews Neuroscience, 2*, 229–239.

Vuilleumier, P. (2005). How brains beware: Neural mechanisms of emotional attention. *Trends in Cognitive Science, 9*, 585–594.

Wager, T. D., Jonides, J., & Reading, S. (2004). Neuroimaging studies of shifting attention: A meta-analysis. *Neuroimage, 22*, 1679–1693.

Wendling, F., Bellanger, J. J., Bartolomei, F., & Chauvel, P. (2000). Relevance of nonlinear lumped-parameter models in the analysis of depth- EEG epileptic signals. *Biological Cybernetics, 83*, 367–378.

Womelsdorf, T. & Fries, P. (2007). The role of neuronal synchronization in selective attention. *Current Opinion in Neurobiology, 17*, 154–160.

Womelsdorf, T., Fries, P., Mitra, P. P., & Desimone, R. (2006). Gamma-band synchronization in visual cortex predicts speed of change detection. *Nature, 439*, 733–736.

Part II
From Attention to Behavioral Control

Chapter 9
From Goals to Habits – A View from the Network

J.M. Hurtado

Abstract Imaging and lesion studies support the view that goal-oriented and routine behaviors are supported by different mechanisms. While the execution of routine behaviors can be achieved autonomously by posterior cortical networks, which include sensory and motor areas, the orchestration of complex goal-directed behaviors requires the interaction of prefrontal networks with the posterior system. These interactions are thought to be the neural signature of supervisory attention to action. It is believed that this interaction can serve as a scaffold for the formation of habits and skills. After extensive practice, learned behaviors become resistant to reward devaluation and can execute seamlessly without attentional supervision. Here I review candidate mechanisms at the network and cellular levels that can explain these phenomena.

Abbreviations

BG   Basal ganglia
DA   Dopaminergic
GPe   External segment of globus pallidus
GPi   Internal segment of globus pallidus
MSN   Medium spiny neuron
PFC   Prefrontal cortex
SNr   Substantia nigra pars reticulata
S–R   Stimulus-response
STN   Subthalamic nucleus

J.M. Hurtado
Centro de Investigaciones Médicas, Pontificia Universidad Católica de Chile, Marcoleta 391, Santiago, Chile, e-mail: jmhurtado@inflexa.com

F. Aboitiz and D. Cosmelli (eds.) *From Attention to Goal-Directed Behavior.*   165
© Springer-Verlag Berlin Heidelberg 2009

Introduction

The hallmark of the behavioral flexibility so characteristic of mammals and birds (Butler & Cotterill, 2006) is their ability to overcome the environmental dependency of behavior. Ethologists who study other vertebrate clades have been able to code the entire behavioral repertoire of some species in the form of "ethograms," consisting largely of the specification of stimulus–response (S–R) pairs (Greenberg, 1977). Yet for mammals that approach is only fruitful for innate behaviors. As the mammalian nervous system matures, a set of entirely new competencies arises, such as goal orientation, problem solving, planning, decision making, and other cognitive phenomena that are gathered under the common rubric of "executive processes." These competencies give rise to a complex behavioral landscape that can evolve with relative independence of the immediate environment and that is hard, if ever possible, to capture as a set of S–R rules.

An old idea in neuroscience is that executive processes evolved hand in hand with the expansion of the prefrontal cortices (PFCs), their related thalamic nuclei, and their basal ganglia (BG) territories (Miller & Cohen, 2001). In agreement with this idea, executive capabilities have a developmental course that correlates well with myelination in PFC (Fuster, 2002). Prefrontal regions, together with limbic areas in the temporal lobe and the insula, form a densely interconnected cluster that is relatively removed from the sensory and motor periphery (Young, 1993). This relative isolation from the immediate environment probably contributes to the autonomy of the executive system, yet at the same time sets a barrier on its ability to influence behavior. The reason is that behaviors must be enacted via sensorimotor networks that themselves exhibit a large degree of autonomy.

The autonomy of sensorimotor networks can be appreciated in the resilience of habitual behaviors and in the seamless performance of complex skills; these behaviors have in common that they can be initiated by environmental triggers and proceed to completion automatically, with little or no attentional supervision (Norman & Shallice, 2000). Throughout this chapter I will refer to these simply as "routine" behaviors. Routine behaviors enact various S–R associations and therefore require tight coupling of the network to the immediate environment. They are, nevertheless, autonomous, in the sense that their triggering conditions and their evolution are dictated by the internal network dynamics. This autonomy is often revealed to us in the form of action "slips," complex sequences that we execute without noticing and with oftentimes embarrasing consequences (Reason, 1984).

Behavioral flexibility then seems to depend on the ability of autonomous, environment-independent, processes in prefrontal networks to engage posterior networks that are tightly coupled to the immediate environment, but that nevertheless enjoy a high degree of autonomy. This PFC–posterior engagement is thought to be the neural signature of selective attention (Miller & Cohen, 2001). Behaviors that require overcoming a habit, executing a sequence based on an explicit rule, ignoring irrelevant stimuli, and so on, all require attentional supervision, and they all involve an interaction, suppressive or enhancing, between the two subsystems. In many cases this interaction mediates conflict resolution, as, for example, when learning a new motor skill that conflicts with a previously learned one (Shadmehr & Holcomb, 1999), or in the incongruent condition in a Stroop test, where a rule-based response (name the color of the print) conflicts with a habitual S–R association (name the word).

In this chapter I will review some candidate cellular and network mechanisms in the cortex, thalamus, and BG that may underlie the ability of subsystems to operate autonomously, yet interact reliably. I will focus on interactions that enact goal-oriented behaviors, and their transference, with practice, to becoming habits. The basis for autonomy is the ability of local networks to sustain cell assemblies that are highly resistant to extrinsic perturbations, while at the same time able to undergo rapid and very specific state transitions. A mechanism of interareal pointer assemblies is proposed to mediate highly specific interactions between "top" and "bottom" assemblies. In turn, BG outputs can orient thalamocortical pointers to bias the selection of sensori-motor ensembles. These cortical and subcortical processes can operate in concert during the acquisition phase of skilled behaviors, a process that is both reward- and attention-driven, but then become disengaged once the subject has learned them and they have acquired autonomy.

Cell Assemblies as Autonomous Units

The concept of a cell assembly, originally introduced by Hebb (1949), has been reformulated in different guises (Braitenberg, 1978; Edelman, 1987; Gerstein, Bedenbaugh, and Aertsen 1989; Varela, 1995) and continues to evolve (Freiwald, Kreiter & Singer, 2001; Harris, 2005). A major strength of the concept is that it incorporates connectivity at the outset; assemblies are defined explicitly as sub-graphs in a neural network. In Hebb's formulation, cell assemblies arise from the cortical connectivity matrix by a developmental process of gradual synaptic strength-ening that reflects the history of a subject's interaction with its environment. Repeated activation of cell groups that form recurrent excitatory circuits gives rise to largely *autonomous* corticothalamic ensembles that can "ignite" in the presence of triggering afferents, but that can potentially activate spontaneously in resting condi-tions (Kenet, Bibitchkov, Tsodyks, Grinvald, & Arieli, 2003). This autonomy was key to Hebb's conception, as he considered it a prerequisite for assemblies to medi-ate thought processes that required "holding things in mind" in the presence of envi-ronmental distractors.

Many of the properties Hebb ascribed to cell assemblies followed from the purported reverberatory nature of cortical circuits. Reverberation, a concept originally advanced by Lorente de Nó (1938), is the form of sustained activity thought to arise from recurrent connectivity in recurrent excitatory networks. Recurrent excitation has some additional functional consequences. One is that activation of an assembly's subgraph will tend to drive the activation of the whole; this has been proposed to underlie pattern completion in sensory systems (Gerstein et al., 1989; Lasner & Fransén, 1992). An additional Hebbian insight was that a given cell can partake in many different assemblies, thus enabling cortical networks to utilize its cellular elements in a combinatorial fashion.

The Role of Inhibition

On theoretical grounds it is known that recurrent networks driven solely by excitation lack stability and can reach a state of saturation. Cortical inhibition can circumvent this problem by restricting the spread of excitatory activity and keeping the network elements below saturation. The cortex is equipped with a diverse population of inhibitory interneurons (Markram et al., 2004) that make this possible. Feedforward lateral inhibition may mediate cross-suppression interactions *between* assemblies (Lasner & Fransén, 1992), thus providing a mechanism for competitive selection, whereas feedback inhibition may provide a gain-control mechanism to prevent saturation and adjust the dynamic range within an assembly (Chance, Abbott & Reyes, 2002). Competitive interactions based on mutual suppression have been proposed to mediate response suppression in extrastriate areas, where the response of a cell to a stimulus in its receptive field is reduced when a second stimulus is presented simultaneously outside the receptive field (Desimone, 1998).

Inhibitory networks are also positioned as key players in tuning the temporal dynamics of cell assemblies (Buzsáki, 2006). There is evidence that GABAergic interneurons can fine-tune the timing of action potential generation in pyramidal cells (Pouille & Scanziani, 2001; Wehr & Zador, 2003) and that they contribute to direction selectivity in somatosensory cortex (Wilent & Contreras, 2005). They also participate in the generation of fast oscillations (Whittington, Traub, & Jefferys, 1995) and aid in the oscillatory entrainment of principal cells, both processes that may play a key role in the ignition of cortical assemblies.

Attractors, Stability, and State Transitions

The resurgence of dynamical systems theory in the 1970s placed the cell assembly concept under a new perspective (Hopfield, 1982). Cortical assemblies have, since then, usually been conceived as attractors in a dynamical system. The most basic type of attractor is a point of stable equilibrium, or point attractor. A network can potentially contain several point attractors. Attractors are by definition resistant to small perturbations, but if it is provided with enough noise a network will spontaneously switch between different attractor states. Direct experimental evidence of spontaneous switching between assemblies has been obtained by optical imaging in hippocampal slices, where different cell groups fire together in alternation (Sasaki, Matsuki, & Ikegaya, 2007).

Perhaps the most compelling supporting evidence for cell assemblies behaving as point attractors has been obtained in the cortex of awake animals. Sustained neuronal activity has been observed during the delay periods of a variety of short-term-memory tasks, in prefrontal, inferior temporal, and parietal regions (Fuster, 1973; Goldman-Rakic, 1995; Kim & Shadlen, 1999; Miyashita & Chang, 1988). This activity is usually correlated among a distributed population of neurons (Constantinidis & Goldman-Rakic, 2002; Funahashi & Inoue, 2000; Vaadia et al., 1995), an indication that it is a network phenomenon, and it is believed that it results from reverberation in local assemblies, a process that can be sustained by glutamatergic (NMDA) currents (Wang, 2001). Persistent delay activity also occurs in the mediodorsal thalamic nucleus (Fuster & Alexander, 1973), which connects reciprocally with PFC areas in a topographic manner; this thalamic delay activity is abolished by cooling the PFC (Alexander & Fuster, 1973), an indication that reciprocal corticothalamic connections also participate in reverberation. This suggests that cell assemblies span cortex and thalamus; it is thus sensible to speak of corticothalamic assemblies.

The mechanisms that allow stable sustained activity do not preclude rapid switching in and out of an attractor state when conditions are met. There is a growing consensus that this combination of stability and switchability can be achieved, at a local circuit level, by balancing recurrent excitation with feedback inhibition. Balanced excitation-inhibition is a ubiquitous feature of cortical synaptic activity (Haider, Duque, Hasenstaub, & McCormick, 2006); Liu, 2004; Pouille & Scanziani, 2001; Wehr & Zador, 2003). The hallmark of balance is that excitatory and inhibitory synaptic inputs are tightly correlated, so that increases in background synaptic activity can increase a cell's conductance and at the same time virtually clamp the membrane potential at a depolarized state near or above firing threshold.

An important property of balanced networks is that a stable depolarized state can be initiated by a burst of afferent activity originating from extrinsic locations (Koulakov, Raghavachari, Kepecs & Lisman, 2002; Shu, Hasenstaub, & McCormick, 2003); Tsodyks & Sejnowski, 1995). The depolarized state can also be *terminated* by a similar barrage of afferent activity (Haider et al., 2006; Shu et al., 2003). Of particular interest for our presentation is that these extrinsic perturbations may originate from other cortical areas or as thalamic bursts, themselves provoked by pauses in BG output, among other causes. Rapid switching by such extrinsic perturbations can bias the selection of cell assemblies in local networks that results from intrinsic mechanisms. Balancing mechanisms also serve to tune the stability of high-frequency oscillatory activity (Brunel & Wang, 2003); this is achieved by differentially tuning the frequency components of excitation and inhibition (Hasenstaub et al., 2005).

In addition to local mechanisms, the stability profile of extensive cortical regions can be modified by diffuse ascending neuromodulatory systems (Friston, 2000), which operate via second messenger systems to affect various ion channels and neurotransmitter receptors. In the PFCs, a well-known modulator that affects network stability profiles is dopamine (Cohen, Braver & Brown, 2002). It is now known that dopaminergic (DA) afferents are contributed by almost all midbrain DA nuclei, though in different proportions, with the strongest contribution from the ventral tegmental area (Williams & Goldman-Rakic, 1998). DA axons terminate primarily onto spines in pyramidal cells dendrites, and are thus strategically positioned to influence the gain of glutamatergic synaptic activity (Goldman-Rakic, 1995). DA adjustment of glutamatergic synaptic gain is thought to have dramatic effects on network stability, by shifting the balance of excitation and inhibition (Durstewitz, Seamans & Sejnowski, 2000). DA terminals also contact the soma of various classes of GABAergic interneurons (Sesack, Hawrylak, Melchitzky & Lewis, 1998), and can therefore alter the balance between excitation and inhibition, and hence the firing precision and the relaxation time of cortical reverberation.

Several other ascending systems are positioned to modulate the dynamics of reverberatory circuits and the overall stability profile of cortical networks; a rule that applies to all neuromodulators is that the effects of a given neuromodulator on different cell types are heterogeneous. For example, cholinergic afferents from the basal forebrain have been shown to alter the firing profile of interneurons, by affecting differentially the firing patterns of diverse interneuronal types (Xiang, Huguenard & Prince, 1998). Similar heterogeneous actions on interneurons have been attributed to noradrenergic and serotonergic afferents (Bacci, Huguenard & Prince, 2005)

Continuous Attractors

Assemblies characterized by bistable units can settle in a number of discrete point attractors (Hopfield, 1982), each one corresponding to a particular combination of up–down states of cortical columns (if we consider the cortical column as the basic bistable unit). The network state can, thus, be represented in a state space (also referred to as *phase space*) as a binary vector so that each possible stable state is a point in that space. There are, however, cortical networks that allow *graded* persistent firing, where cells can occupy a virtual continuum of stable firing levels. Firing levels are correlated with stimulus parameters in somatosensory (Romo, Brody, Hernández & Lemus, 1999) and in entorhinal (Egorov, Hamam, Fransén, Hasselmo, & Alonso, 2002) cortices. In this cases a vector representation of the network admits continuous values (Hopfield, 1984), and point attractors can be located near each other, forming quasi-continuous families of stable points. Such "continuous" attractors can be modeled by approximating them as a series of discrete stable states that are close to each other (Brody, Romo & Kepecs, 2003).

The idea of continuous attractors can be extended to situations that do not strictly involve persistent delay period activity, for instance, when cell assemblies fire reliably under some environmental contingency. A classic example would be the assemblies of hippocampal place cells that are revealed by multiple-cell recordings (Wilson & McNaughton, 1993). As a rat forages in a familiar environment, different cell assemblies are ignited for different spatial locations. Some of the cells may be active in more than one location, but the population activity is unique for every location. Every location has a correlative attractor in the network, yet several neighboring attractors are ignited with smooth changes in "stimulus" parameter (i.e., animal's location). When the animal is placed in a different environment, a rearrangement of hippocampal networks takes place; new cell assemblies that are specific to the new environment are formed and they are ignited in a position-dependent manner (Stringer, Rolls, Trappenberg & de Araujo, 2002).

A second example of continuous attractors are the population vectors related to movement direction in primary motor cortex. A cell in primary motor cortex is broadly tuned to movement direction and therefore contributes to a variety of different movements (Georgopoulos, Schwartz & Kettner, 1986). The movement direction achieved is determined by the population vector, that is, the average of the preferred directions of the participant neurons weighted by their firing rates. The primary motor network can thus gravitate to attractors that depend continuously on movement direction. This experimental model also provided an astonishing demonstration of temporal drift of attractors. When animals were trained to reach to a location that was offset from the cue by a certain angle, the population vector rotated during the reaction delay, from pointing to the cued location at the beginning of the trial to the offset target location on movement onset (Georgopoulos, Lurito, Petrides, Schwartz, & Massey, 1989).

The two previous examples do not strictly involve persistent activity, because activity does change depending on environmental parameters. Yet they illustrate an important property that I discuss later: in each case, cell assemblies are *pointers* to locations in extrapersonal space (direction of movement for population vectors or location in the field for hippocampal assemblies). Similarly, cell assemblies may act as pointers *within* the cortex, so an assembly in one area, or cluster of areas, may constitute a pointer to an assembly in a second area. I will propose that similar pointer assemblies, from PFC to sensorimotor areas, mediate the "top-down" control of routine actions.

Pointer Maps, Biasing, and Selective Attention

To trigger behaviorally relevant transitions between different attractors, extrinsic perturbations must carry a specific bias. This problem is best understood in the visual selective attention paradigm in extrastriate areas (Desimone, 1998). When several stimuli with different saliency are simultaneously present in the scene, bottom-up mechanisms (i.e., competitive mutual suppression) left to themselves will move the spotlight of attention to the most salient one. This is because cell assemblies that bind the salient stimulus have the highest stability. If the subject is instructed to attend to a less salient location, a top-down bias signal will *shift* the attentional spotlight away from the salient stimulus (which is now a distractor), and towards the instructed location. The extrinsic bias signal is by no means a random perturbation, but is a highly directed one that pushes the system to a specific state of lower intrinsic stability.

How can we account for the specific *directedness* of biasing signals? An appealing "pointer-map" model has been proposed by Hahnloser et al. (Hahnloser, Douglas, Mahowald & Hepp, 1999, see Hahnloser, Douglas & Hepp, 2002 for an updated version). According to it, recurrent excitation between "bottom" cortical columns is routed through a set of "pointer" neurons in a "top" area. The result of this is that reverberatory buildup of activity in the bottom area is gated by activity in the top area. In this way, different combinations of pointer neurons (that is, a "top" assembly) would selectively activate different assemblies in "bottom" areas. The direction of bias is conferred by a population code in the top area; thus, the mapping is not "point to point" but rather "assembly to assembly."

The same basic biasing mechanism may operate through reciprocal thalamocortical interactions; in this case thalamic "pointers" would gate the formation of cortical assemblies. Since thalamic cells are themselves affected by subcortical afferents, this mechanism may explain the action-selection properties of BG outputs.

Cell Assemblies In the Action Realm

The repertoire of routine behaviors of an animal is arranged in a hierarchy of action schemes (Cooper & Shallice, 2006) that result from the progressive "chunking" of simpler movement sequences into more complex sequences (Sakai, Kitaguchi & Hikosaka, 2003; Verwey, 1994). Basic movement sequences can be triggered by microstimulation in premotor (Graziano, Taylor, Moore, 2002) and parietal (Cooke, Taylor, Moore & Graziano, 2003) areas. The elicited sequences appear purposeful; for example, stimulation in one site causes the mouth to open and the hand to move towards the mouth, as if bringing to the mouth a food item (Graziano et al., 2002). Several sources of evidence suggest that action chunks are enacted by cell assemblies in sensorimotor networks. Single-unit studies have indicated that cells in various motor and parietal areas fire in advance of complex purposeful movement sequences in a sequence-specific manner (Rizzolatti et al., 1988; Tanji & Shima, 1994). This idea is further confirmed by multiple-electrode recordings (Jackson, Gee, Baker & Lemon, 2003; Maynard et al., 1999) suggesting that assemblies in motor cortex are characterized by synchronous firing, where individual cell synchronization is task-specific. Movement-related synchronous firing has also been shown to span several sensorimotor areas (Murthy & Fetz, 1996; Sanes & Donoghue, 1993).

How can schemes that are essentially *sequential* be enacted by cell assemblies? Hebb conjectured that assemblies can ignite in sequence, each one triggering the next, thus giving rise to chains of activation or "phase sequences." Indeed, a type of stimulus-specific sequential activation has been described in sensory systems, notably in the locust antennal lobe (Mazor & Laurent, 2005). In recent years, dynamical systems theory has allowed the formulation of Hebbian phase sequences in a more rigorous mathematical framework (Rabinovich, Huerta, Varona & Afraimovich, 2006). Assemblies that enact discrete movement components can be modeled as saddle-point equilibria in phase space. These saddle-point states are very much like point attractors; they differ, however, in that they allow one unstable trajectory (technically an unstable manifold). This means that they are resilient to perturbations in all but that one particular direction. If different saddle points are connected through their unstable manifolds, a trajectory in phase space is possible that visits a subset of the saddle points in a particular sequence. This corresponds to various cortical assemblies being activated in particular order – a phase sequence. It has been shown that a particular sequence can be specifically triggered by inputs to the system. These inputs may be arranged as pointer maps from extrinsic afferents, such as assemblies from other areas or subcortical (e.g., BG) afferents. This provides a means by which an assembly in one area (say, a "top" area) maps to a *sequence* of assemblies in a different ("bottom") area.

The Anatomical Substrate of Routine Behaviors

Routine behaviors are essentially sequential, and recruit an extensive cortical network that includes primary motor, premotor, secondary motor, somatosensory, dorsal visual, and parietal areas (Gordon, Lee, Flament, Ugurbil, & Ebner, 1998; Harrington et al., 2000). In this type of behavior the dissociation between exclusively motoric, sensory, or cognitive functions falls apart. All areas, including primary motor cortex, exhibit sensory as well as motor-related, and even "cognitive" activity (Carpenter, Georgopoulos, & Pellizzer, 1999). Also, the distinction between planning and execution stages seems to fall apart with routine actions. There is no evidence of *exclusive* areal specialization for planning as opposed to execution of movement; all areas involved participate in different degrees in preparation and initiation of movements (Carpenter et al., 1999; Cisek & Kalaska, 2005; Crutcher & Alexander, 1990; Georgopoulos et al., 1989; Lu & Ashe, 2005). Milner and Goodale (2006) provided substantial evidence for scheme autonomy in their analysis of visuomotor behaviors. For example, a patient with severe visual agnosia can create the precise grip aperture to grab an object while not being aware of the object's shape (Goodale, Milner, Jakobson & Carey, 1991). These properties suggest that the areas in question form a functional unit that can operate autonomously.

This autonomy is partly the consequence of anatomical properties. Analyses of cortical connectivity using graph theory methods suggest cortical areas are connected in a "small-world" fashion: clusters of heavily interconnected areas are somewhat sparsely connected between them (Sporns & Zwi, 2004). The cortical areas involved in autonomous movement generation constitute one such cluster. Network analysis shows that motor areas are embedded within a larger cluster that contains somatosensory and posterior parietal areas (area 7b; Young, 1993). Area 7b provides a major hub whereby the sensorimotor cluster interacts with PFC areas. It is likely that corticothalamic assemblies within a densely interconnected cluster can operate with relative autonomy from top-down PFC influences. In model dynamical systems embedded in small-world networks, robust correlations appear within clusters at low levels of synaptic strength (Morelli, Cerdeira & Zanette, 2005; Zhou & Kurths, 2006). Moreover, information theory analysis of neuroimaging data indicates that autonomous local activity arises in the brain of awake humans (Tononi, McIntosh, Russell & Edelman, 1998).

It is thus plausible that cell assemblies spanning several areas within a functional cluster can "ignite," either spontaneously (i.e., buildup by inter- and intra-areal recurrent excitation) or triggered (e.g., by subcortical afferents) without PFC (i.e., top-down) involvement. This does not preclude that assemblies from different clusters interact and give rise to transient higher-order associations. Indeed, global-scale associations have been described in several studies (Bressler, 1995; Varela,

Lachaux, Rodriguez & Martinerie, 2001). These interactions are likely to mediate behaviors that involve conscious supervision.

Basal Ganglia

The thalamocortical system is anatomically linked to the BG and it is, thus, not surprising that this complex subcortical loop plays a key role in the regulation of cortical function and, ultimately, of behavior. Current ideas on BG function point to a dual role of these structures; they participate in the selection of action plans (Mink, 1996) and also in the learning of habits and skills (Packard & Knowlton, 2002). BG anatomy and physiology is notably complex. In what follows I will focus on some key aspects of the BG network that bear relationship to the formation and selection of thalamocortical cell assemblies. For extensive reviews on anatomy and functional aspects or detailed reviews of anatomy and function, see Bolam, Hanley, Booth & Bevan (2000), Haber (2003), and Wilson (1998).

The following is a list of basic BG properties relevant to our discussion:

1. The BG are arranged in largely parallel, segregated, and topographically organized loops that span cortex, striatum pallidum, and thalamus (Alexander & Crutcher, 1990); thalamocortical connections terminate the loops predominantly in the frontal cortex.
2. Medium spiny neurons (MSN) in the striatum, the BG main input elements, receive converging glutamatergic inputs from functionally related cortical and thalamic regions.
3. MSNs exhibit bistable membrane potentials; switching from hyperpolarized to depolarized states requires a sufficiently large number of coactive cortical afferents (Kasanetz, Riquelme, O'Donnell & Murer, 2006).
4. Midbrain DA afferents contact MSN spines and are positioned to regulate the gain of corticostriatal synapses, and hence the sensitivity and selectivity of the cell to cortical afferents. Additionally, dopamine modulates activity-dependent plasticity in the corticostriatal synapse. Phasic DA afferents are known to signal reward prediction errors (see Chap. 10 by Aboitiz), and therefore they enable synaptic gain and plasticity to be contingent on deviations from reward expectation.
5. MSNs send converging projections onto GABAergic cells in the internal segment of globus pallidus (GPi) and substantia nigra pars reticulata (SNr), that together constitute the BG output nuclei. In spite of being highly convergent, these projections remain parcellated within functional domains.
6. GPi/SNr cells project to the dorsal thalamic nuclei that innervate prefrontal areas (the mediodorsal nucleus), and motor/premotor/supplementary motor and oculomotor areas (the ventroanterior/ventrolateral nuclei).
7. There is a direct projection from prefrontal and sensorimotor cortical areas to the subthalamic nucleus (STN). STN cells are glutamatergic and project to GPi. This results in a net inhibitory pathway that links cortex to thalamus, the so-called hyperdirect pathway. Additionally the STN forms a reciprocal excitatory–inhibitory loop

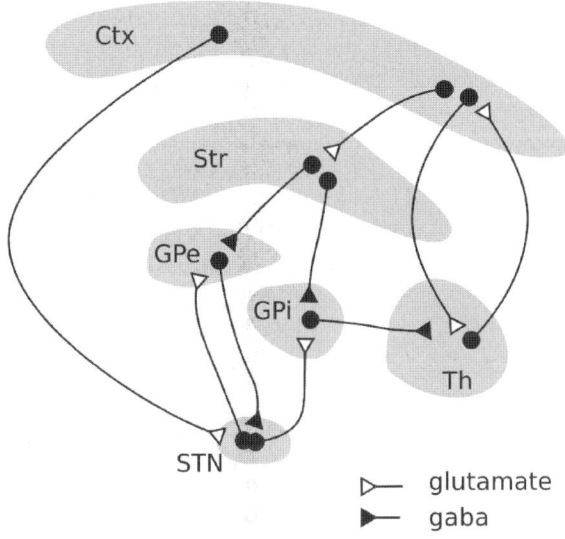

Fig. 9.1 Basal ganglia loops. Only the most prominent neural pathways in the primate basal ganglia are shown. Many connections have been omitted for clarity. *Black dots* represent cell populations, *open triangles* represent glutamatergic terminals, and *filled triangles* represent GABAergic terminals. Substantia nigra pars reticulata is not shown. See the text for details. *Ctx* cortex, *Str* striatum, *GPi* internal segment of globus pallidus, *GPe* external segment of globus pallidus, *STN* subthalamic nucleus, *Th* thalamus.

with the external segment of globus pallidus (GPe). This loop is modulated by striatal projection from MSNs to GPe. See Fig. 9.1 for a summary depiction of the main BG pathways.

Properties 2 and 3 enable MSNs to signal the coactivation of large groups of cells in the corticothalamic system. Synchronous volleys of cortical activity are likely to arrive from within a cluster of interrelated cortical areas and therefore signal the ignition of cell assemblies. Properties 2 and 5 indicate that there is a dramatic reduction in the number of cellular elements along the corticostriatopallidal pathway. This strongly suggests that a redundancy reduction (or "data compression") process takes place along the pathway, yielding an information-rich output through a small number of cells. In agreement with this, it has been found that spiking in BG output cells is largely uncorrelated (Bar-Gad, Havazelet-Heimer, Goldberg, Ruppin, & Bergman, 2000).

In spite of extensive investigative efforts, the effects of dopamine in BG function (property 4) are still controversial. DA afferents not only innervate the striatum but almost every BG nucleus (Cossette, Lévesque & Parent, 1999). The effects are mediated by two receptors families with different intracellular cascades that affect a number of effectors, including various calcium, potassium, and sodium conductances (Moyer, Wolf & Finkel, 2007). DA effects do not follow a simple script. For example, DA modulation of glutamatergic currents through D1 receptors increases NMDA currents and reduces α-amino-3-hydroxy-5-methyl-4-isoxazolepropionate currents, whereas D2 receptor activation causes a decrease in both components. (Cepeda, Buchwald & Levine, 1993). A common hypothesis is that dopamine enhances the signal-to-noise ratio of MSN responses; responses to correlated afferents are enhanced, whereas responses to sparse inputs are reduced (O'Donnell, 2003).

The hyperdirect pathway (property 7) enables cortical activity to affect the BG output nuclei in a direct manner. However, this pathway can be under tight control of striatal outputs, via striatopallidal modulation of the STN–GPe axis. It has been proposed that the onset of synchronized activity in a cortical region generates a fast corollary discharge through the hyperdirect pathway causing inhibition of the thalamic nuclei that feeds the cortical area of origin (Nambu, 2005). BG output activity is therefore determined by the interplay of STN excitatory inputs and striatal inhibition. The resulting GPi activity results in a spatially and temporally complex inhibition of the thalamus that is positioned to bias the selection of corti-cothalamic cell assemblies. In addition to sculpted inhibition, the BG output can exhibit surprising excitatory effects on thalamic cells; it has recently been shown that GABAergic inhibitory postsynaptic potentials of pallidal origin can generate action potentials by a rebound mechanism (Person & Perkel, 2005).

Reward-Driven Stabilization of Routine Behaviors

What is the relationship between action selecting and learning functions of the BG? It is tempting to believe that BG-mediated skill and habit learning will simply follow from repeated selection of a particular sequence. However, a series of instrumental conditioning experiments showed that habits seem to arise when there is no further *covariation* between adjustments in behavior and ensuing rewards, a situation that usually, but not necessarily, results from sequence repetition (Dickinson, 1985). According to these observations, action patterns become habits when rewards achieve a plateau, where further improvements in behavior do not report further increases in reward. Repetition of rewarded behaviors leads to less behavior-reward covariation because it raises the mean reward rate, so further reward increments are smaller relative to the mean. This rule is analogous to the Weber–Fechner law in psychophysics (Schultz, 2006).

What are the neural bases of this phenomenon? Reward-driven habit formation depends critically on plasticity in the sensorimotor portion of the BG (Graybiel, 1998; Packard & Knowlton, 2002). It has been shown that lesions in that region abolish the development of habits (Yin, Knowlton, & Balleine, 2004). In the learn-ing phases of operant conditioning tasks, that region undergoes dramatic plastic changes. Experiments with rats learning to use an auditory cue to navigate a T maze for reward have shown that at the beginning stages of learning, striatal cells fire throughout the trial (Jog, Kubota, Connolly, Hillegaart, & Graybiel, 1999); yet, as the animal learns the reward-predictive value of the cue (which is presented in the middle of the trajectory), striatal responses to the cue progressively vanish. After overtraining, activations only occurs at the start and end points of the maze. This conforms to the idea that striatal activity in acquired habitual behaviors is related to the behavioral "chunks" rather than to sensory cues or movement parameters.

Striatal plasticity is, in turn, dependent on inputs from midbrain DA neurons. DA cells transiently increase their firing rate when a reward-predicting stimulus is presented *unexpectedly* (Schultz & Dickinson, 2000). As a consequence, departures from current S–R "strategies" will result in striatal dopamine release only if they are followed by a reward that is above the expected value. The resulting dopamine-induced plasticity in corticostriatal synapses may then mediate reinforcement of the new reward leading "strategy." On the other hand, if a new behavioral variants fails to raise rewards above the predicted level, no further striatal plasticity takes places and the current S–R behavior remains stable and, therefore, habitual.

Several lines of evidence indicate that cortical motor areas also undergo extensive plastic changes in relation to skill and habit acquisition (Sanes & Donoghue, 2000). These changes are likely to result in the formation of new task-related cell assemblies. For example, in a study where rats had to collect rewards from different locations following an auditory cue, cell populations in motor cortex modified their firing patterns as learning proceeded, so that different groups of cortical cells became related to different S–R pairs. (Cohen & Nicolelis, 2004).

The direction of the plastic changes in cortex and striatum can be partly understood by a recently demonstrated plasticity mechanism at the corticostriatal synapse that is both time-dependent and contingent on DA activity (Pawlak & Kerr, 2008). It was shown that, in the presence of D1 receptor activation, potentiation occurs when an excitatory postsynaptic potential from cortical stimulation follows an MSN action potential, whereas the inverse temporal order results in synaptic depression. This result can explain the progressive suppression of striatal activity with skill learning. In virtue of being a reward-prediction error signal, DA firing would serve to label a simultaneously arriving cortical action potential (say, from a motor command) as an unexpected "reward predictor." As a result of the time-dependent mechanism, an MSN would potentiate its response to the dopamine-labeled motor command, while it will depress responses to motor commands immediately following it. The process would take place recursively: as the DA signal responds to earlier and earlier predictors of reward, the striatal cell would follow course.

These striatal signals can, in turn, bias the selection of cortical assemblies, giving rise to S–R sequences that lead to reward. This biasing process can be mediated by GPi inputs to the thalamus. Thalamic cells are strategically positioned as mediators, since they form reciprocal loops with cortical columns on one hand, and receive BG afferents, on the other. BG outputs can modulate the gain of these corticothalamic loops, and so bias assembly selection in the cortex. This mechanism operates as a form of thalamic pointer map (see the "Pointer Maps, Biasing, and Selective Attention" section) and thus the interactions can be highly specific and dynamic. If the BG bias signal is consistent over time, the selected cortical assemblies can then gain stability by way of intrinsic cortical plasticity mechanisms. A similar effect of BG on cortical assembly stabilization may operate in other cortical regions.

Selective Attention for Frontoposterior Interactions

The mechanisms outlined in the previous section are essentially selectionist and thus they depend critically on the ability of the subject to generate behavioral variability over which selection can operate. Behavioral variability obtained exclusively by random exploration would yield an extremely low learning rate. In contrast, learning of routine actions can be largely accelerated if new behavioral variants are assembled by an additional system that acts as a "scaffold"; for this to work the scaffolding system must contain a map between actions and their outcomes and between those outcomes and their reward value. This would guarantee that the scaffolding system promotes the formation of S–R pairs that match reward delivery.

This is precisely what goal-oriented (or "teleological") behaviors provide. In contrast to routine behaviors (skills or habits) that rely on S–R associations, teleological behaviors are built around associations between actions and their outcomes (Balleine & Dickinson, 1998). If a subject learns that an outcome has been devalued (e.g., the food source has been poisoned), the teleological system will immediately adjust its action schedule to avoid behaviors that lead to that outcome. In contrast, reward devaluation in a purely S–R system will not result in immediate adjustment; instead, the system must undergo a long round of synaptic modification to unlearn the now unrewarding S–R association.

The classic studies by Dickinson (1985) suggested that instrumental learning in mammals is driven by a goal-oriented mechanism; this learning process proceeds until further changes in behavior are not followed by sensible changes in reward, a condition that usually occurs after overtraining. In addition, animal studies have largely confirmed that this teleological/routine behavioral dissociation (Balleine, Delgado & Hikosaka, 2007) has a functional counterpart in the nervous system. For example, lesions in the dorsolateral striatum in rats, the region anatomically linked to sensorimotor cortices, prevents the formation of habits (Yin et al., 2004), whereas lesions in the dorsomedial portion, related to prefrontal regions, prevents the acquisition and expression of goal-oriented (i.e., devaluation-sensitive) behaviors (Yin, Ostlund, Knowlton & Balleine, 2005). This functional dissociation is further supported by studies showing that prefrontal and posterior regions participate differentially in early versus late skill acquisition (Costa, Cohen & Nicolelis, 2004; Sakai et al., 1998). Functional imaging studies support this conclusion; while skill learning recruits prefrontal areas at early stages, this activity is reduced with practice until it becomes restricted mostly to sensorimotor and parietal areas (Kelly & Garavan, 2005; Kübler, Dixon & Garavan, 2006; Shadmehr & Holcomb, 1997). A parallel sequence of changes takes place in the BG territories of prefrontal and posterior networks (Lehéricy et al., 2005). These findings support the hypothesis that a teleological system, involving prefrontal areas and their BG territories, provides a scaffold over which the procedural system, involving sensorimotor/visuomotor/parietal cortices and their BG territories, can build routine actions. The posterior system becomes increasingly autonomous as learning proceeds, until no further support by prefrontal areas is required.

How is the scaffolding interaction between prefrontal and posterior networks enacted? A functional phase sequence in posterior areas could be selected via recurrent excitation with a set of pointers in prefrontal action-outcome assemblies (see the section "Cell Assemblies In the Action Realm"). The existence of action-outcome assemblies have been shown in multiple-electrode recordings in primate PFC (Vaadia et al., 1995). Moreover, prefrontal assembly activation can be sequence-specific (Averbeck, Chafee, Crowe & Georgopoulos, 2002) and be conformed *in toto* prior to task execution.

The prefrontal scaffolding of action sequences in procedural learning appears to be one of a family of processes that involve teleological supervision of motor schemes. The hallmark of this "supervision" is that it requires selective attention and it would be invoked in a variety of tasks that involve some form of planning, decision making, troubleshooting, or the overcoming of habitual responses (Norman & Shallice, 2000).

Attention to action is, thus, probably enacted by heterogeneous mechanisms, operating through corticocortical and corticothalamocortical pathways, that link PFC–posterior networks. The cellular and network features reviewed in previous sections provide a diverse set of candidates to mediate these interactions. Top-down enhancement is likely to be mediated by recruitment of recurrent reverberatory pathways between prefrontal and posterior areas. If streamed through a variant set of pointer assemblies, this excitation can be highly specific (see the section "Pointer Maps, Biasing, and Selective Attention"). BG outputs may participate in this process by sculpting inhibition of thalamic pointer cells, thus adjusting the gain of specific corticothalamic loops (see the section "Basal Ganglia"); in this way, corticothalamocortical interactions can be subject to biasing signals provided by BG afferents into the thalamus. Task inhibition, one of the crucial components of top-down control, can be mediated by bursts of activity into posterior areas, arriving from either cortex or thalamus (see "Attractors, Stability, and State Transitions"). A recent hypothesis, sustained by functional imaging, states that the hyperdirect BG pathway may propagate "stop" signals from cortex to thalamus (Aron & Poldrack, 2006). Striatal adjustment of hyperdirect loop's gain may confer context specificity to these putative "stop" signals.

Conclusions

The neural underpinnings of goal-oriented and habitual behaviors have been extensively studied with imaging techniques and lesion analysis, yet little is known of their mechanisms at the cellular and network level. The mechanisms reviewed here provide a preliminary ground to build testable hypotheses. Behaviors driven by explicit goals are essentially attended and require the interaction between prefrontal (executive) and posterior (procedural, or "routine") networks. The network mechanisms of these interactions are most probably diverse and heterogeneous. Selective enhancement is likely to arise from recurrent corticocortical and corticothalamocortical circuits; specificity may be gained by the dynamic routing of these recurrent connections by pointer assemblies. Selective inhibition is likely achieved by burst discharges of corticocortical or thalamocortical origin; these are streamed through feedforward inhibitory interneurons and cause local reverberation to cease.

The scaffolding hypothesis states that these mechanisms operate in parallel with reward-regulated BG loops: while the goal-setting prefrontal network recruits the sensorimotor network to perform complex behaviors, the latter recruits a loop through the sensorimotor portion of the BG to stabilize particular sequences that predict upcoming rewards. However, we should not conclude from this that there is a functional dissociation between a subcortical reward mechanisms and cortical attention systems. An apparent dissociation in the present formulation occurs because we have limited our presentation to the acquisition of S–R associations, and this analysis begs two important questions: (1) How does a goal-directed system "know" which outcomes will bring reward and (2) how does the reward prediction error property of DA neurons comes about? To answer these two questions it is necessary to examine other portions of frontobasal networks that map potential outcomes to reward values. These functions are thought to be mediated by ventromedial/orbitofrontal cortices and the ventral portions of the striatum (Schultz, Tremblay & Hollerman, 2000; O'Doherty, Dayan, Friston, Critchley, & Dolan, 2003).

References

Alexander, G. E., & Crutcher, M. D. (1990). Functional architecture of basal ganglia circuits: Neural substrates of parallel processing. *Trends in Neuroscience, 13*, 266–271.

Alexander, G. E., & Fuster, J. M. (1973). Effects of cooling prefrontal cortex on cell firing in the nucleus medialis dorsalis. *Brain Research, 61*, 93–105.

Aron, A. R., & Poldrack, R. A. (2006). Cortical and subcortical contributions to Stop signal response inhibition: Role of the subthalamic nucleus. *Journal of Neuroscience, 26*, 2424–2433.

Averbeck, B. B., Chafee, M. V., Crowe, D. A., & Georgopoulos, A. P. (2002). Parallel processing of serial movements in prefrontal cortex. *Proceedings of the National Academy of Sciences of the United States of America, 99*, 13172–13177.

Bacci, A., Huguenard, J. R., & Prince, D. A. (2005). Modulation of neocortical interneurons: Extrinsic influences and exercises in self-control. *Trends in Neuroscience, 28*, 602–610.

Balleine, B. W., & Dickinson, A. (1998). Goal-directed instrumental action: Contingency and incentive learning and their cortical substrates. *Neuropharmacology, 37*, 407–419.

Balleine, B. W., Delgado, M. R., & Hikosaka, O. (2007). The role of the dorsal striatum in reward and decision-making. *Journal of Neuroscience, 27*, 8161–8165.

Bar-Gad, I., Havazelet-Heimer, G., Goldberg, J. A., Ruppin, E., & Bergman, H. (2000). Reinforcement-driven dimensionality reduction – a model for information processing in the basal ganglia. *Journal of Basic and Clinical Physiology and Pharmacology, 11*, 305–320.

Bolam, J. P., Hanley, J. J., Booth, P. A., & Bevan, M. D. (2000). Synaptic organisation of the basal ganglia. *Journal of Anatomy, 196 (Pt 4)*, 527–542.

Braitenberg, V. (1978). Cell assemblies in the cerebral cortex. In R. Heim & G. Palm (Ed.), *Theoretical approaches to complex systems – Lecture notes in biomathematics* (pp. 171–188). New York: Springer.

Bressler, S. L. (1995). Large-scale cortical networks and cognition. *Brain Research. Brain Research Review, 20*, 288–304.

Brody, C. D., Romo, R., & Kepecs, A. (2003). Basic mechanisms for graded persistent activity: Discrete attractors, continuous attractors, and dynamic representations. *Current Opinion in Neurobiology, 13*, 204–211.

Brunel, N., & Wang, X. (2003). What determines the frequency of fast network oscillations with irregular neural discharges? I. Synaptic dynamics and excitation-inhibition balance. *Journal of Neurophysiology, 90*, 415–430.

Butler, A. B., & Cotterill, R. M. J. (2006). Mammalian and avian neuroanatomy and the question of consciousness in birds. *Biology Bulletin, 211*, 106–127.

Buzsáki, G. (2006). *Rhythms of the brain*. New York: Oxford University Press.

Carpenter, A. F., Georgopoulos, A. P., & Pellizzer, G. (1999). Motor cortical encoding of serial order in a context-recall task. *Science, 283*, 1752–1757.

Cepeda, C., Buchwald, N. A., & Levine, M. S. (1993). Neuromodulatory actions of dopamine in the neostriatum are dependent upon the excitatory amino acid receptor subtypes activated. *Proceedings of the National Academy of Sciences of the United States of America, 90*, 9576–9580.

Chance, F. S., Abbott, L. F., & Reyes, A. D. (2002). Gain modulation from background synaptic input. *Neuron, 35*, 773–782.

Cisek, P., & Kalaska, J. F. (2005). Neural correlates of reaching decisions in dorsal premotor cortex: Specification of multiple direction choices and final selection of action. *Neuron, 45*, 801–814.

Cohen, D., & Nicolelis, M. A. L. (2004). Reduction of single-neuron firing uncertainty by cortical ensembles during motor skill learning. *Journal of Neuroscience, 24*, 3574–3582.

Cohen, J. D., Braver, T. S., & Brown, J. W. (2002). Computational perspectives on dopamine function in prefrontal cortex. *Current Opinion in Neurobiology, 12*, 223–229.

Constantinidis, C., & Goldman-Rakic, P. S. (2002). Correlated discharges among putative pyramidal neurons and interneurons in the primate prefrontal cortex. *Journal of Neurophysiology, 88*, 3487–3497.

Cooke, D. F., Taylor, C. S. R., Moore, T., & Graziano, M. S. A. (2003). Complex movements evoked by microstimulation of the ventral intraparietal area. *Proceedings of the National Academy of Sciences of the United States of America, 100*, 6163–6168.

Cooper, R. P., & Shallice, T. (2006). Hierarchical schemas and goals in the control of sequential behavior. *Psychology Review, 113*, 887–916; discussion 917–31.

Cossette, M., Lévesque, M., & Parent, A. (1999). Extrastriatal dopaminergic innervation of human basal ganglia. *Neuroscience Research, 34*, 51–54.

Costa, R. M., Cohen, D., & Nicolelis, M. A. L. (2004). Differential corticostriatal plasticity during fast and slow motor skill learning in mice. *Current Biology, 14*, 1124–1134.

Crutcher, M. D., & Alexander, G. E. (1990). Movement-related neuronal activity selectively coding either direction or muscle pattern in three motor areas of the monkey. *Journal of Neurophysiology, 64*, 151–163.

Desimone, R. (1998). Visual attention mediated by biased competition in extrastriate visual cortex. *Philosophical Transactions of the Royal Society of London - Series B Biological Science, 353*, 1245–1255.

Dickinson, A. (1985). Actions and habits: The development of behavioural autonomy. *Philosophical Transctions of the Royal Society of London - Series B, 308*, 67–78.

Durstewitz, D., Seamans, J. K., & Sejnowski, T. J. (2000). Dopamine-mediated stabilization of delay-period activity in a network model of prefrontal cortex. *Journal of Neurophysiology, 83*, 1733–1750.

Edelman, G. (1987). *Neural darwinism*. New York: Basic Books.

Egorov, A. V., Hamam, B. N., Fransén, E., Hasselmo, M. E., & Alonso, A. A. (2002). Graded persistent activity in entorhinal cortex neurons. *Nature, 420*, 173–178.

Freiwald, W. A., Kreiter, A. K., & Singer, W. (2001). Synchronization and assembly formation in the visual cortex. *Progress in Brain Research, 130*, 111–140.

Friston, K. J. (2000). The labile brain. II. Transients, complexity and selection. *Philosophical Transactions of the Royal Society of London – Series B Biological Science, 355*, 237–252.

Funahashi, S., & Inoue, M. (2000). Neuronal interactions related to working memory processes in the primate prefrontal cortex revealed by cross-correlation analysis. *Cerebral Cortex, 10*, 535–551.

Fuster, J. M. (1973). Unit activity in prefrontal cortex during delayed-response performance: Neuronal correlates of transient memory. *Journal of Neurophysiology, 36*, 61–78.

Fuster, J. M. (2002). Frontal lobe and cognitive development. *Journal of Neurocytology, 31*, 373–385.

Fuster, J. M., & Alexander, G. E. (1973). Firing changes in cells of the nucleus medialis dorsalis associated with delayed response behavior. *Brain Research, 61*, 79–91.

Georgopoulos, A. P., Lurito, J. T., Petrides, M., Schwartz, A. B., & Massey, J. T. (1989). Mental rotation of the neuronal population vector. *Science, 243*, 234–236.

Georgopoulos, A. P., Schwartz, A. B., & Kettner, R. E. (1986). Neuronal population coding of movement direction. *Science, 233*, 1416–1419.

Gerstein, G. L., Bedenbaugh, P., & Aertsen, M. H. (1989). Neuronal assemblies. *IEEE Transactions on Biomedical Engineering, 36*, 4–14.

Goldman-Rakic, P. S. (1995). Cellular basis of working memory. *Neuron, 14*, 477–485.

Goodale, M. A., Milner, A. D., Jakobson, L. S., & Carey, D. P. (1991). A neurological dissociation between perceiving objects and grasping them. *Nature, 349*, 154–156.

Gordon, A. M., Lee, J. H., Flament, D., Ugurbil, K., & Ebner, T. J. (1998). Functional magnetic resonance imaging of motor, sensory, and posterior parietal cortical areas during performance of sequential typing movements. *Experimental Brain Research, 121*, 153–166.

Graybiel, A. M. (1998). The basal ganglia and chunking of action repertoires. *Neurobiology of Learning and Memory, 70*, 119–136.

Graziano, M. S. A., Taylor, C. S. R., & Moore, T. (2002). Complex movements evoked by micro-stimulation of precentral cortex. *Neuron, 34*, 841–851.

Greenberg, N. (1977). An Ethogram of the Blue Spiny Lizard, Sceloporus cyanogenys (Reptilia,Lacertilia, Iguanidae). *Journal of Herpetology, 11*, 177–195.

Haber, S. N. (2003). The primate basal ganglia: Parallel and integrative networks. *Journal of Chemical Neuroanatomy, 26*, 317–330.

Hahnloser, R. H. R., Douglas, R. J., & Hepp, K. (2002). Attentional recruitment of inter-areal recurrent networks for selective gain control. *Neural Computation, 14*, 1669–1689.

Hahnloser, R., Douglas, R. J., Mahowald, M., & Hepp, K. (1999). Feedback interactions between neuronal pointers and maps for attentional processing. *Nature Neuroscience, 2*, 746–752.

Haider, B., Duque, A., Hasenstaub, A. R., & McCormick, D. A. (2006). Neocortical network activity in vivo is generated through a dynamic balance of excitation and inhibition. *Journal of Neuroscience, 26*, 4535–4545.

Harrington, D. L., Rao, S. M., Haaland, K. Y., Bobholz, J. A., Mayer, A. R., Binderx, J. R., & Cox, R. W. (2000). Specialized neural systems underlying representations of sequential movements. *Journal of Cognitive Neuroscience, 12*, 56–77.

Harris, K. D. (2005). Neural signatures of cell assembly organization. *Nature Reviews Neuroscience, 6*, 399–407.

Hasenstaub, A., Shu, Y., Haider, B., Kraushaar, U., Duque, A., & McCormick, D. A. (2005). Inhibitory postsynaptic potentials carry synchronized frequency information in active cortical networks. *Neuron, 47*, 423–435.

Hebb, D. (1949). *The organization of behavior.* New York: Wiley.

Hopfield, J. J. (1982). Neural networks and physical systems with emergent collective computational abilities. *Proceedings of the National Academy of Sciences of the United States of America, 79*, 2554–2558.

Hopfield, J. J. (1984). Neurons with graded response have collective computational properties like those of two-state neurons. *Proceedings of the National Academy of Sciences of the United States of America, 81*, 3088–3092.

Jackson, A., Gee, V. J., Baker, S. N., & Lemon, R. N. (2003). Synchrony between neurons with similar muscle fields in monkey motor cortex. *Neuron, 38*, 115–125.

Jog, M. S., Kubota, Y., Connolly, C. I., Hillegaart, V., & Graybiel, A. M. (1999). Building neural representations of habits. *Science, 286*, 1745–1749.

Kasanetz, F., Riquelme, L. A., O'Donnell, P., & Murer, M. G. (2006). Turning off cortical ensembles stops striatal Up states and elicits phase perturbations in cortical and striatal slow oscillations in rat in vivo. *Journal of Physiology, 577*, 97–113.

Kelly, A. M. C., & Garavan, H. (2005). Human functional neuroimaging of brain changes associated with practice. *Cerebral Cortex, 15*, 1089–1102.

Kenet, T., Bibitchkov, D., Tsodyks, M., Grinvald, A., & Arieli, A. (2003). Spontaneously emerging cortical representations of visual attributes. *Nature, 425*, 954–956.

Kim, J. N., & Shadlen, M. N. (1999). Neural correlates of a decision in the dorsolateral prefrontal cortex of the macaque. *Nature Neuroscience, 2*, 176–185.

Koulakov, A. A., Raghavachari, S., Kepecs, A., & Lisman, J. E. (2002). Model for a robust neural integrator. *Nature Neuroscience, 5*, 775–782.

Kübler, A., Dixon, V., & Garavan, H. (2006). Automaticity and reestablishment of executive control-an fMRI study. *Journal of Cognitive Neuroscience, 18*, 1331–1342.

Lasner, A., & Fransén, E. (1992). Modelling Hebbian cell assemblies comprised of cortical neurons. *Network, 3*, 105–119.

Lehéricy, S., Benali, H., Van de Moortele, P., Pélégrini-Issac, M., Waechter, T., Ugurbil, K., & Doyon, J. (2005). Distinct basal ganglia territories are engaged in early and advanced motor sequence learning. *Proceedings of the National of Academy of Sciences of the United States of America, 102*, 12566–12571.

Liu, G. (2004). Local structural balance and functional interaction of excitatory and inhibitory synapses in hippocampal dendrites. *Nature Neuroscience, 7*, 373–379.

Lorente de Nó, R. (1938). Analysis of the activity of the chains of internuncial neurons. *Journal of Neurophysiology, 1*, 207–244.

Lu, X., & Ashe, J. (2005). Anticipatory activity in primary motor cortex codes memorized movement sequences. *Neuron, 45*, 967–973.

Markram, H., Toledo-Rodriguez, M., Wang, Y., Gupta, A., Silberberg, G., & Wu, C. (2004). Interneurons of the neocortical inhibitory system. *Nature Reviews Neuroscience, 5*, 793–807.

Maynard, E. M., Hatsopoulos, N. G., Ojakangas, C. L., Acuna, B. D., Sanes, J. N., Normann, R. A., & Donoghue, J. P. (1999). Neuronal interactions improve cortical population coding of movement direction. *Journal of Neuroscience, 19*, 8083–8093.

Mazor, O., & Laurent, G. (2005). Transient dynamics versus fixed points in odor representations by locust antennal lobe projection neurons. *Neuron, 48*, 661–673.

Miller, E. K., & Cohen, J. D. (2001). An integrative theory of prefrontal cortex function. *Annual Review of Neuroscience, 24*, 167–202.

Milner, A. D., & Goodale, M. A. (2006). *The visual brain in action*. Oxford: Oxford University Press.

Mink, J. W. (1996). The basal ganglia: Focused selection and inhibition of competing motor programs. *Progress in Neurobiology, 50*, 381–425.

Miyashita, Y., & Chang, H. S. (1988). Neuronal correlate of pictorial short-term memory in the primate temporal cortex. *Nature, 331*, 68–70.

Morelli, L. G., Cerdeira, H. A., & Zanette, D. H. (2005). Frequency clustering of coupled phase oscillators on small-world networks. *The European Physical Journal B – Condensed Matter and Complex Systems, 43*, 243–250.

Moyer, J. T., Wolf, J. A., & Finkel, L. H. (2007). Effects of dopaminergic modulation on the integrative properties of the ventral striatal medium spiny neuron. *Journal of Neurophysiology, 98*, 3731–3748.

Murthy, V. N., & Fetz, E. E. (1996). Oscillatory activity in sensorimotor cortex of awake monkeys: Synchronization of local field potentials and relation to behavior. *Journal of Neurophysiology, 76*, 3949–3967.

Nambu, A. (2005). A new approach to understand the pathophysiology of Parkinson's disease. *Journal of Neurology, 252*(Suppl 4), IV1–IV4.

Norman, D. A., & Shallice, T. (2000). Attention to action: Willed and automatic control of behavior. In M. S. Gazzaniga (Ed.), *Cognitive neuroscience: A reader*. Blackwell Publishing. 376–390.

O'Doherty, J. P., Dayan, P., Friston, K., Critchley, H., & Dolan, R. J. (2003). Temporal difference models and reward-related learning in the human brain. *Neuron, 38*, 329–337.

O'Donnell, P. (2003). Dopamine gating of forebrain neural ensembles. *European Journal of Neuroscience, 17*, 429–435.

Packard, M. G., & Knowlton, B. J. (2002). Learning and memory functions of the Basal Ganglia. *Annual Review of Neuroscience, 25*, 563–593.

Pawlak, V., & Kerr, J. N. D. (2008). Dopamine Receptor Activation Is Required for Corticostriatal Spike-Timing-Dependent Plasticity. *Journal of Neuroscience, 28*, 2435–2446.

Person, A. L., & Perkel, D. J. (2005). Unitary IPSPs drive precise thalamic spiking in a circuit required for learning. *Neuron, 46*, 129–140.

Pouille, F., & Scanziani, M. (2001). Enforcement of temporal fidelity in pyramidal cells by somatic feed-forward inhibition. *Science, 293,* 1159–1163.

Rabinovich, M. I., Huerta, R., Varona, P., & Afraimovich, V. S. (2006). Generation and reshaping of sequences in neural systems. *Biological Cybernetics, 95*, 519–536.

Reason, J. (1984). Lapses of attention in everyday life. In W. Parasuraman & R. Davies (Eds.), *Varieties of attention*. Orlando FL: Academic.515–549.

Rizzolatti, G., Camarda, R., Fogassi, L., Gentilucci, M., Luppino, G., & Matelli, M. (1988). Functional organization of inferior area 6 in the macaque monkey. II. Area F5 and the control of distal movements. *Experimental Brain Research, 71*, 491–507.

Romo, R., Brody, C. D., Hernández, A., & Lemus, L. (1999). Neuronal correlates of parametric working memory in the prefrontal cortex. *Nature, 399*, 470–473.

Sakai, K., Hikosaka, O., Miyauchi, S., Takino, R., Sasaki, Y., & Pütz, B. (1998). Transition of brain activation from frontal to parietal areas in visuomotor sequence learning. *Journal of Neuroscience, 18*, 1827–1840.

Sakai, K., Kitaguchi, K., & Hikosaka, O. (2003). Chunking during human visuomotor sequence learning. *Experimental Brain Research, 152*, 229–242.

Sanes, J. N., & Donoghue, J. P. (1993). Oscillations in local field potentials of the primate motor cortex during voluntary movement. *Proceedings of the National Academy of Sciences of the United Sstates of America, 90*, 4470–4474.

Sanes, J. N., & Donoghue, J. P. (2000). Plasticity and primary motor cortex. *Annual Review in Neuroscience, 23*, 393–415.

Sasaki, T., Matsuki, N., & Ikegaya, Y. (2007). Metastability of active CA3 networks. *Journal of Neuroscience, 27*, 517–528.

Schultz, W. (2006). Behavioral theories and the neurophysiology of reward. *Annual Review in Psychology, 57*, 87–115.

Schultz, W., & Dickinson, A. (2000). Neuronal coding of prediction errors. *Annual Review in Neuroscience, 23*, 473–500.

Schultz, W., Tremblay, L., & Hollerman, J. R. (2000). Reward processing in primate orbitofrontal cortex and basal ganglia. *Cerebral Cortex, 10*, 272–284.

Sesack, S. R., Hawrylak, V. A., Melchitzky, D. S., & Lewis, D. A. (1998). Dopamine innervation of a subclass of local circuit neurons in monkey prefrontal cortex: Ultrastructural analysis of tyrosine hydroxylase and parvalbumin immunoreactive structures. *Cerebral Cortex, 8*, 614–622.

Shadmehr, R., & Holcomb, H. H. (1997). Neural correlates of motor memory consolidation. *Science, 277*, 821–825.

Shadmehr, R., & Holcomb, H. H. (1999). Inhibitory control of competing motor memories. *Experimental Brain Research, 126*, 235–251.

Shu, Y., Hasenstaub, A., & McCormick, D. A. (2003). Turning on and off recurrent balanced cortical activity. *Nature, 423*, 288–293.

Sporns, O., & Zwi, J. D. (2004). The small world of the cerebral cortex. *Neuroinformatics, 2*, 145–162.

Stringer, S. M., Rolls, E. T., Trappenberg, T. P., & de Araujo, I. E. T. (2002). Self-organizing continuous attractor networks and path integration: Two-dimensional models of place cells. *Network, 13*, 429–446.

Tanji, J., & Shima, K. (1994). Role for supplementary motor area cells in planning several movements ahead. *Nature, 371*, 413–416.

Tononi, G., McIntosh, A. R., Russell, D. P., & Edelman, G. M. (1998). Functional clustering: Identifying strongly interactive brain regions in neuroimaging data. *Neuroimage, 7*, 133–149.

Tsodyks, M. V., & Sejnowski, T. (1995). Rapid state switching in balanced neocortical models. *Network, 6*, 111–124.

Vaadia, E., Haalman, I., Abeles, M., Bergman, H., Prut, Y., Slovin, H., & Aertsen, A. (1995). Dynamics of neuronal interactions in monkey cortex in relation to behavioural events. *Nature, 373*, 515–518.

Varela, F. J. (1995). Resonant cell assemblies: A new approach to cognitive functions and neuronal synchrony. *Biological Research, 28*, 81–95.

Varela, F., Lachaux, J. P., Rodriguez, E., & Martinerie, J. (2001). The brainweb: Phase synchronization and large-scale integration. *Nature Reviews Neuroscience, 2*, 229–239.

Verwey, W. B. (1994). Evidence for the development of concurrent processing in a sequential keypressing task. *Acta Psychologica (Amst), 85*, 245–262.

Wang, X. J. (2001). Synaptic reverberation underlying mnemonic persistent activity. *Trends in Neuroscience, 24*, 455–463.

Wehr, M., & Zador, A. M. (2003). Balanced inhibition underlies tuning and sharpens spike timing in auditory cortex. *Nature, 426*, 442–446.

Whittington, M. A., Traub, R. D., & Jefferys, J. G. (1995). Synchronized oscillations in interneuron networks driven by metabotropic glutamate receptor activation. *Nature, 373*, 612–615.

Wilent, W. B., & Contreras, D. (2005). Dynamics of excitation and inhibition underlying stimulus selectivity in rat somatosensory cortex. *Nature Neuroscience, 8*, 1364–1370.

Williams, S. M., & Goldman-Rakic, P. S. (1998). Widespread origin of the primate mesofrontal dopamine system. *Cerebral Cortex, 8*, 321–345.

Wilson, C. (1998). Basal Ganglia. In G. Shepherd (Ed.), *The synaptic organization of the brain* (pp 329–375). New York: Oxford University Press.

Wilson, M. A., & McNaughton, B. L. (1993). Dynamics of the hippocampal ensemble code for space. *Science, 261*, 1055–1058.

Xiang, Z., Huguenard, J. R., & Prince, D. A. (1998). Cholinergic switching within neocortical inhibitory networks. *Science, 281*, 985–988.

Yin, H. H., Knowlton, B. J., & Balleine, B. W. (2004). Lesions of dorsolateral striatum preserve outcome expectancy but disrupt habit formation in instrumental learning. *European Journal of Neuroscience, 19*, 181–189.

Yin, H. H., Ostlund, S. B., Knowlton, B. J., & Balleine, B. W. (2005). The role of the dorsomedial striatum in instrumental conditioning. *European Journal of Neuroscience, 22*, 513–523.

Young, M. P. (1993). The organization of neural systems in the primate cerebral cortex. *Proceedings of the Biological Science, 252*, 13–18.

Zhou, C., & Kurths, J. (2006). Hierarchical synchronization in complex networks with heterogeneous degrees. *Chaos, 16*, 015104.

Chapter 10
Dynamics of a Neuromodulator – I. The Role of Dopaminergic Signaling in Goal-Directed Behavior

F. Aboitiz

Abstract The mesencephalic dopaminergic system plays a key role in the organization of goal-directed behavior, operating in three different domains: activation of the locomotor system via the dorsal corpus striatum, modulating motivational states and predicting rewards via the nucleus accumbens and amygdala, and planning strategies and monitoring errors via the prefrontal cortex. These operations are based on two modes of dopaminergic signaling. Tonic, longer-lasting dopamine release, and short-latency, phasic, stimulus-related dopamine release (70–100 ms poststimulus latency, less than 200-ms duration). Tonic and phasic release have been associated with distinct signaling systems provided by D1-like and D2-like receptors, respectively. The balance between these two signaling modes modulates the vigor, motivation, and flexibility of goal-directed behavior. I finally present a perspective that attempts to unify the multiple functions exerted by dopamine, in the context of evolutionary theory.

Abbreviations

DA   Dopamine
NAc   Nucleus accumbens
PFC   Prefrontal cortex
RRF   Retrorubral fields
SC   Superior colliculus

F. Aboitiz
Departamento de Psiquiatría, Centro de Investigaciones Médicas, Pontificia Universidad Católica de Chile, Marcoleta 391, Santiago, Chile, e-mail: faboitiz@uc.cl

F. Aboitiz and D. Cosmelli (eds.) *From Attention to Goal-Directed Behavior.*
© Springer-Verlag Berlin Heidelberg 2009

SNc   Substantia nigra compacta
SNr   Substantia nigra reticulata
VTA   Ventral tegmental area

Introduction

It is now 50 years since the seminal article by Carlsson et al. (1957) reporting that injection of L-dopa rescued reserpine-treated rabbits from their catatonic state. Using the newly developed technique of spectrophotofluorimetry, the researchers found that L-dopa was metabolized into dopamine (DA), which presumably produced the observed behavioral effects. Although highly controversial in its time, this finding earned Carlsson a shared Nobel prize in 2000, not only for discovering DA as a neurotransmitter but also for establishing that neurotransmission in the brain was chemical rather than electrical (Iversen & Iversen, 2007).

More than any other neurotransmitter, research on DA has had a profound impact in the development of modern neuropsychiatry. Its importance in the palliative treatment of Parkinson's disease, the effects of DA blockers as antipsychotics, the therapeutic effects of stimulants in attention deficit hyperactivity disorder, and the role of the dopaminergic system in addictive disorders testify to its fundamental relevance in behavioral control and in neurological and psychiatric diseases. Subsequent work has unveiled the role of DA as a strong reinforcer in conditioned learning (hence its role in addictive processes; Hyman, Malenka, & Nestler, 2006), but perhaps more importantly as a prediction error signaling system, where the difference between the real and the expected reward determines the error signal and the dopaminergic response, which leads to behavioral plasticity oriented to minimize this difference (Schultz, 2007). Other researchers have pointed out a role of the dopaminergic signal in associating contextual information to unpredicted events, thus facilitating the reinforcement of novel actions (Redgrave & Gurney, 2006). DA also modulates cortical dynamics, and more generally has profound influences on goal-directed behavior, to which cognitive mechanisms appear to be subsidiary mechanisms (see later). Thus, to understand the role of this neurotransmitter in brain dynamics, it is necessary to assess its integrated participation in the different neurobiological and behavioral levels where it acts, both at the cellular and at the systemic levels, and in both subcortical and cortical structures.

In this chapter I will review some key aspects of dopaminergic neurosignaling, with emphasis on the mechanisms of behavioral control during conditioned learning. Furthermore, I will highlight the existence of two different modes of transmission: phasic and tonic signaling, whose appropriate dynamic balance is essential for normal behavior and cognition. Understanding the basics of these processes is fundamental for a thorough discussion of the role of DA in cortical neurodynamics, which will be addressed in Chap. 11. Before delving into these issues, a brief review of the anatomy of the dopaminergic system will be, in my view, of great benefit for the reader.

Anatomy of Mesencephalic Dopaminergic Nuclei

The dopaminergic cell bodies projecting to the telencephalon concentrate in the brainstem (Aboitiz & Montiel, 2001; Bentivoglio & Morelli, 2005; Bjorklund & Dunnett, 2007; Marin, Smeets, & Gonzalez, 1998), making up a complex that has classically been divided into three main groups of nuclei: the ventral tegmental area (VTA; group A10), the substantia nigra compacta (SNc; group A9), and the retrorubral fields (RRF; group A8), which form a dorsocaudal extension of the SNc (Fig. 10.1).

The dopaminergic system is highly conserved among vertebrates, owing to its participation in the activation of the animal during goal-directed (rewarding) behavior. There is a substantial increase of presumably dopaminergic neurons (tyrosine hydroxylase positive) in these nuclei from mouse to humans (from about 25,000 to about 500,000), which has been considered to be due to the increase in the size of the innervated targets in the different species (Bjorklund & Dunnett, 2007). Nevertheless, the brain size difference between a mouse and a human is more than 1,000-fold (http://faculty.washington.edu/chudler/facts.html), suggesting that large-brained species tend to have relatively fewer dopaminergic neurons than small-brained species.

Dopaminergic Projections Target Striatal, Limbic, and Cortical Regions

A more recent description of these nuclei subdivides them into ventral and dorsal tiers, according to their projection sites (Haber & Fudge, 1997). Roughly, the ventral tier (with densely packed, angular calbindin-negative cells, expressing high levels of

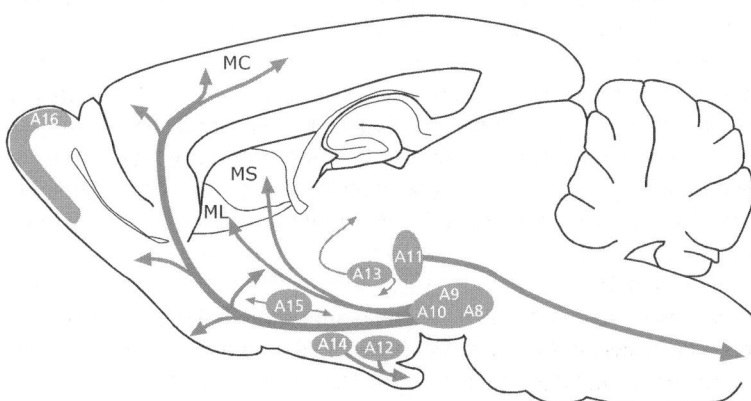

Fig. 10.1 Anatomical localization of the mesencephalic dopaminergic groups that innervate the telencephalon, making up the mesostriatal (*MS*), mesolimbic (*ML*), and mesocortical (*MC*) projection systems. *A10* refers to the ventral tegmental area (VTA); *A9* refers to the substantia nigra compacta (SNc); and *A8* refers to the retrorubral fields (RRF). The scheme represents a rat brain (Bjorklund & Dunnett, 2007)

the DA transporter) relates to the ventral aspect of the VTA and the SNc; while the dorsal tier (with round or fusiform calbindin-positive cells, expressing low levels of the DA transporter) corresponds to the dorsal VTA and SNc plus cells from the RRF cell group. The ventral tier projects mainly to the patch compartment of the dorsal striatum, activating the "direct" striatothalamic pathway which facilitates the onset of motor commands, and inhibiting the "indirect" pathway which suppresses motor activity (Parent & Hazrati, 1995). The dorsal tier innervates the ventral striatum or nucleus accumbens (NAc) and the matrix compartment of the dorsal striatum (Bentivoglio & Morelli, 2005), which have important connections with the limbic system and are important elements of the reward system (Haber & Fudge, 1997); but perhaps more importantly, its activity relates to the prediction error of positive rewards (Waelti, Dickinson, & Schultz, 2001). The projection to the ventral striatum apparently participates in encoding these rewards, while activation of the dorsal striatum specifies the initiation of motor acts oriented to the goal (Breiter, Gollub, Weisskoff, Kennedy, Makris et al., 1997).

The dorsal tier has also been described as projecting to the amygdala (basolateral, lateral nuclei and nucleus of the lateral olfactory tract) and septal nuclei among other regions, and finally to frontal and limbic cortices, in which executive functions such as response inhibition, planning, and working memory take place (Aboitiz & Montiel, 2001; Aboitiz, 2008). Within a relatively conserved framework, there are species differences in the dopaminergic innervation of the frontal cortex (Lewis, Sesack, Levey, & Rosenberg, 1998; S. M. Williams & Goldman-Rakic, 1998), and in the relative proportions of interneuron subtypes in which many dopaminergic terminals synapse (Somogyi, Kisvarday, Freund, & Cowey, 1984; Yanez, Munoz, Contreras, Gonzalez, Rodriguez-Veiga et al., 2005). Perhaps the main species difference in dopaminergic cortical innervation is that in primates, the dopaminergic innervation territory is not limited to the prefrontal cortex (PFC), but covers the entire cortical mantle (Bentivoglio & Morelli, 2005; Lewis et al., 1998). This projection derives from all components of the mesencephalic dopaminergic system, forming a continuous sheet of cells that spans the three main groups (VTA, SNc, and RRF) (Williams & Goldman-Rakic, 1998). An additional difference is the laminar distribution of dopaminergic fibers and receptors, which in primates are abundant in both superficial and deep layers (superficial layer neurons preferentially participate in delay-period activity during working memory tasks; see later) (Friedman & Goldman-Rakic, 1994), while in the rodent they are abundant only in the infragranular layers (Berger, Gaspar, & Verney, 1991; Lewis et al., 1998). Nevertheless, the effects of DA in these species are remarkably similar (Kroner, Krimer, Lewis, & Barrionuevo, 2007).

Although the above scheme has been heuristically useful, other studies have challenged the concept of a neat separation of a ventral tier and a dorsal tier according to their prosencephalic projections. Rather, neurons from these two compartments may project to either of the abovementioned targets. This intermixing, although discrete in rodents, is particularly prominent in primates (Bjorklund & Dunnett, 2007; Francois, Yelnik, Tande, Agid, & Hirsch, 1999; Williams & Goldman-Rakic, 1998). Nevertheless, individual dopaminergic neurons only rarely send projections to more than one of the main subdivisions (mesostriatal, mesolimbic, or mesocortical; Bjorklund & Dunnett, 2007).

Very recently, by using a combined neuroanatomical, immunocytochemical, and electrophysiological approach, Lammel et al. (2008) were able to distinguish different groups of dopaminergic neurons projecting to telencephalic targets. One population of neurons, with cell bodies located in the posterior portion of the VTA, display high firing rates and project to the medial PFC, NAc shell, NAc core, or amygdala; while neurons located in more lateral portions of the VTA (partially overlapping with SNc cells that project to the dorsal striatum) show the classical hyperpolarization-activated cation channel response, a slow firing rate, and project to the lateral shell of the NAc. Finally, a third class consisted of medial PFC-projecting neurons that appeared to lack D2 inhibitory autoreceptors (see later), which permits them to release DA for longer periods without feedback inhibition.

Summarizing, the mesencephalic dopaminergic projection to the prosencephalon includes three main targets: (1) mesostriatal, to the dorsal striatum; (2) mesolimbic, to limbic areas and ventral striatum; and (3) mesocortical, to the cerebral cortex (Fig. 10.1). In addition, there is a dopaminergic projection to globus pallidus (internal and external segments), ventral pallidum, and subthalamic nucleus, which has not yet been included in the above scheme (Smith, Lavoie, Dumas, & Parent, 1989). In this way, the dopaminergic system seems to act threefold in the control of behavior: activating the sensorimotor system (mesostriatal projection), detecting probabilities of reward and monitoring errors in such predictions (mesolimbic component), and participating in the elaboration of strategies for successful behavior and of goal representations to be maintained in working memory (mesocortical component).

The NAc Is Involved in Motivational Processes

The projection to the ventral striatum or NAc (sometimes considered to be part of the so-called extended amygdala; Heimer, Alheid, de Olmos, Groenewegen, Haber et al., 1997), has received special attention in the last few years, owing to its relevance in motivation and addictive behavior, as was demonstrated by early intracranial self-stimulation protocols (Crow, 1972). These findings led to the hypothesis that DA participates in motivation, reward, and learning, encoding information about the salience, predictability, and reward value (positive or negative) of events in the world (Iversen & Iversen, 2007). The NAc is subdivided into a core region and a shell. Phasic DA input to the NAc core has been considered by some to be the main signal encoding reward prediction (Waelti et al., 2001; see the next section), while the shell of this nucleus controls the dopaminergic input to wide areas of the striatum, invigorating behaviors that are coordinated via the NAc core (Iversen & Iversen, 2007). Nevertheless, it has been found that intrashell microinfusion of DA antagonists blocks intracranial self-stimulation, and that in this site (but not in others) there is a striking coincidence between anticipatory DA surges and changes in neuronal firing patterns during self-stimulation behavior (Cheer, Aragona, Heien, Seipel, Carelli et al., 2007). More generally, the dopaminergic input to both NAc and dorsal striatum has been hypothesized to provide a coordinated modulation of both goal-directed behavior and the transition to procedural learning as the animal adapts to the changing world (Iversen, 2006).

There Is Diversity of DA Receptors

DA acts on two main classes of receptors: D1-like (D1 and D5), associated with increases in adenylyl cyclase activity; and D2-like (D2, D3, and D4), which relate to adenylyl cyclase inhibition (Kebabian & Calne, 1979). D1-like and D2 receptors are found in neurons of the corpus striatum (D2 receptors are also found in glutamatergic corticostriatal terminals), while only D2 receptors are found in the SNc. D3 receptors are synthesized by neurons in the NAc. D1, D2, and D4 receptors are found in neurons of frontal cortex, hippocampus, and amygdala, while D5 receptors are synthesized only in the hippocampus (Hall, Farde, Halldin, Hurd, Pauli et al., 1996; Weiner & Molinoff, 1994). Most neuroleptics block the D2 receptor subtype, but several, atypical neuroleptics show affinity for the D3 and D4 subtypes.

Scylla and Charybdis: Phasic Versus Tonic Doperminergic Signaling

Behavioral and physiological approaches to dopaminergic signaling suggest that there are two modes of dopaminergic signaling (Grace, 1995; Grace, Floresco, Goto, & Lodge, 2007; Goto, Otani, & Grace, 2007). On one hand, there is the large but transient (phasic) increase in synaptic DA associated with sensory stimuli, caused by burst firing of dopaminergic neurons. Another mode of transmission consists of tonic, extrasynaptic liberation of DA due to single spiking of dopaminergic neurons or to presynaptic stimulation of dopaminergic terminals by cortical afferents.

Tonic DA – Background Modulator of Excitability

Tonic DA levels represent the background, steady-state levels of extracellular DA. Extracellular DA concentrations change relatively slow, on the order of seconds and minutes, compared with milliseconds in the case of phasic DA release. In contrast with intrasynaptic phasic activity, reuptake by the DA transporter seems to play a minor role in regulating extracellular DA concentrations (see later), and neurotransmitter degradation via the enzyme catechol-O-methyltransferase appears to be the main mechanism of extrasynaptic DA regulation, at least in the cerebral cortex

(Grace et al., 2007; Yavich, Forsberg, Karayiorgou, Gogos, & Mannisto, 2007). Basal striatal DA concentrations are in the 5–10-nM range, which is sufficient to tonically activate the high-affinity D2 and D4 receptors (Richfield, Penney, & Young, 1989). This mode of signaling has been claimed to depend on presynaptic activation of dopaminergic terminals by glutamatergic afferents (in the corpus striatum; Grace, 1991), and on relatively periodical single-spike dopaminergic neuron activity (2–10 Hz; Grace et al., 2007).

A variety of behaviors, including movement, stress, and sex, induce slow DA concentration increases that can be measured by microdialysis procedures (Schultz, 2007). Stable, tonic DA concentrations enable a large variety of behavioral functions controlled by the striatum and cortex, functions that become impaired after DA depletion in Parkinson's disease patients (Schultz, 2007).

Phasic DA Liberation – Orienting Response to Salient Stimuli

Phasic DA release is caused by burst activity of dopaminergic neurons, resulting in a sudden release of DA within the synaptic cleft (reaching low millimolar ranges). Liberated DA is rapidly inactivated by the DA transporter, which recycles the neurotransmitter to the presynaptic terminal (Grace et al., 2007). Thus, this effect is quite powerful but largely intrasynaptic. This response can be detected by single-cell electrophysiological techniques (Grace, 1990) or by voltammetry methods using microelectrodes that measure the currents related to oxidation and reduction of DA (Schultz, 2007). Likewise, imaging studies in humans using raclopride (a selective D2 antagonist) in the NAc probably reflect the action of phasic DA bursting (Grace et al., 2007; Sesack, Aoki, & Pickel, 1994). Phasic dopaminergic signaling is known to induce long-term potentiation in the striatum, hippocampus, and PFC (Frey, Matthies, Reymann, & Matthies, 1991; Gurden, Tassin, & Jay, 1999; Huang, Simpson, Kellendonk, & Kandel, 2004; Reynolds, Hyland, & Wickens, 2001).

Sensory-Triggered Responses In the SNc

Phasic DA release is likely related to glutamatergic input (Floresco & Grace, 2003; Floresco, West, Ash, Moore, & Grace, 2003; Lodge & Grace, 2006a). Potential sources for this input include the superior colliculus (SC), PFC, preoptic hypothalamus, and pedunculopontine tegmentum (Dommett, Coizet, Blaha, Martindale, Lefebvre et al., 2005; Grace et al., 2007; Redgrave & Gurney, 2006). Several lines of evidence point to the deep layers of the SC as the main source of short-latency sensory input into the SNc (70–100-ms poststimulus latency, less than 200-ms duration), usually producing a strong burst response in dopaminergic neurons (Coizet, Comoli, Westby, & Redgrave, 2003; Comoli et al., 2003; Dommett et al., 2005; Redgrave & Gurney, 2006). In turn, dopaminergic neurons from the SNc

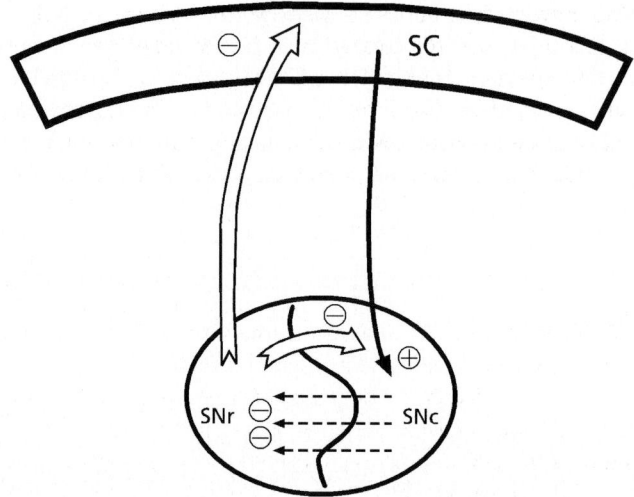

Fig. 10.2 Phasic dopamine (DA) release in the SNc facilitates orienting behavior. The superior colliculus (*SC*) responds to salient stimuli generating a short-latency input to the SNc; the latter releases DA into the GABAergic substantia nigra reticulata (*SNr*), thus blocking its inhibitory influence onto the SC and the SNc and allowing the generation of an orienting saccadic movement (Redgrave & Gurney, 2006)

release DA from a profuse dendritic plexus extending into the neighboring substantia nigra reticulata (SNr), the latter exerting a continuous inhibitory influence over both the SNc and the SC. In this manner, phasic DA release is able to modulate the GABAergic control of the SNr over these structures (Bjorklund & Dunnett, 2007; Redgrave & Gurney, 2006). Thus, collicular input on the SNc activates dopaminergic neurons, which produce inhibition of the GABAergic neurons in the SNr, liberating both the SNc and the SC from a sustained inhibition by the SNr (Fig. 10.2). In the SC, this disinhibition allows the excitation required to trigger a saccadic response toward the stimulus (Redgrave & Gurney, 2006).

Sensory-evoked phasic DA release has been claimed to have an important role in attention and preattention mechanisms. Visually, SC neurons tend to respond to spatially localized changes in luminosity that signal movement or appearance or disappearance of objects in the visual field, while being relatively insensitive to object-specific characteristics (Sparks & Jay, 1986; Stein & Meredith, 1993; Wurtz & Albano, 1980). Thus, the sensory information triggering phasic DA release depends on features such as salience and predetermined configurations, rather than more elaborate recognition processes. Phasic dopaminergic activation to salient stimuli is enhanced if the stimulus is unexpected or novel (novel but inconspicuous stimuli are, however, ineffective; Horvitz, 2000; Ljungberg, Apicella, & Schultz, 1992). However, other strongly attention inducing stimuli such as conditioned aversive events or conditioned inhibitors are ineffective in triggering dopaminergic discharges (Mirenowicz & Schultz, 1996; Tobler, Dickinson, & Schultz, 2003). This has been claimed to support the concept that the dopaminergic response does not purely reflect attentional mechanisms but reflects the reward properties of stimuli. In general, phasic dopaminergic activation may reflect a combined sensitivity to rewarding and physically salient events (Schultz, 2007), or may be related to stimuli that are associated with reward, either through experience or by innate mechanisms.

Balance Between Tonic and Phasic Activity In the VTA

About half of the mesencephalic dopaminergic neurons are not spontaneously active but are in a hyperpolarized state owing to their constant GABAergic input from the ventral pallidum (Grace & Bunney, 1985). Inactivation of the latter by the NAc permits the sporadical single-spike activity of dopaminergic neurons that contributes to maintaining the tonic DA levels in different brain components (Floresco & Phillips, 2001; Floresco et al., 2003). In turn, the NAc is activated by the ventral subiculum of the hippocampus (supplying contextual information), by the PFC (facilitating behavioral flexibility), and by the amygdala (involved in emotional responses; Fig. 10.3). Thus, all these systems may potentially contribute to increase the tonic DA levels in the brain (Grace et al., 2007).

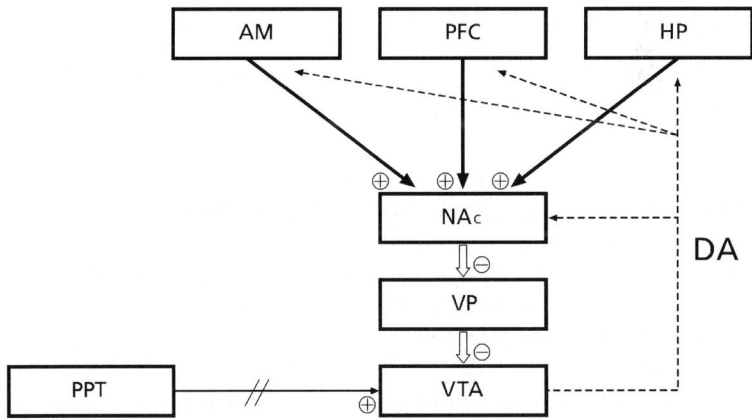

Fig. 10.3 Control of tonic and phasic DA release in the VTA. This structure is normally inhibited by GABAergic input from the ventral pallidum (*VP*). Inhibition of the latter by the nucleus accumbens (*NAc*) releases the suppressive effect of the VP onto the VTA, permitting spontaneous DA discharge in some neurons. This process is facilitated by excitatory telencephalic inputs to the NAc (*HP*, hippocampus; *AM*, amygdala; *PFC*, prefrontal cortex). Input from the pedunculopontine tegmentum (*PPT*) facilitates phasic release (Grace et al., 2007)

Burst firing in VTA dopaminergic neurons has been found to be produced by stimulation of the pedunculopontine tegmentum (Mena-Segovia, Bolam, & Magill, 2004). However, in the VTA, glutamate by itself is incapable of triggering burst dopaminergic neuron firing, suggesting the presence of a gate enabling these neurons to respond to the excitatory input (Wang & French, 1993). This putative gate has been recently identified to be an input from the laterodorsal tegmentum (Lodge & Grace, 2006b). Some evidence suggests that this gating may be partly mediated by the nicotinic acetylcholine receptor (Mameli-Engvall et al., 2006). Additional studies have determined that burst DA firing in the VTA only takes place in neurons that are spontaneously (tonically) active. In other words, only dopaminergic neurons whose sustained GABAergic inhibition has been released and are spontaneously firing are susceptible to generate phasic bursting after stimulation; that is, phasic activity occurs in a subset of the tonically active dopaminergic neurons (Floresco et al., 2003; Fig. 10.2). In fact, a recent report indicates that hipppocampal stimulation can effectively gate VTA burst firing by decreasing ventral pallidal inhibition (Lodge & Grace, 2006a).

DA and Behavior – Reward Prediction and Contextual Associations

The reduced behavioral responses to motivating stimuli in experimental animals when interfering with dopaminergic transmission has been a key factor in relating DA to reward. Furthermore, studies consistently showed that activation of the midbrain dopaminergic system and/or of its projection to the NAc is essential for intracranial self-stimulation protocols (Schultz, 2007; see also Chap. 9). However, reward is not produced by the activation of specific receptors, but is rather defined by its action on behavior. Rewards "generate approach and consummatory behavior, constitute outcomes of the preceding stimuli and actions, serve as positively reinforcing signals and produce reward predictions through associative conditioning" (Schultz, 2007).

The relation of DA to reward is not simple. Treatment with DA receptor blockers or artificial GABA infusion into the VTA causes animals to be unable to link sequences of actions to obtain a food reward, although they will eat normally if moved close to the food source (Berridge & Robinson, 1998). This suggests that DA does not make up internal evaluations of reward, but rather specifies the ability to act based on those valuations. In other words, DA may mediate the binding between the hedonic evaluation of the stimuli and the assignment of these values to other objects or behaviors (Montague, Hyman, & Cohen, 2004). A more recent report showed that infusion of DA in the ventral medial PFC (but not in dorsal medial PFC) promotes behavioral responses according to outcome value instead of antecedent stimuli, thus activating goal-expectancy behavior rather than behaviorally fixed instrumental behavior (Hitchcott, Quinn & Taylos, 2007).

Dopaminergic Neurons as Prediction Error Detectors

Dopaminergic neurons show phasic activation with reward and reward-predicting stimuli (Schultz, 1998), and become transiently depressed with stimuli predicting nonreward (Tobler et al., 2003). More precisely, the dopaminergic response to reward does not occur unconditionally but has been claimed to be more related to the prediction error of reward, being positive in cases of a positive prediction error (unexpected rewards), and negative in cases of negative errors (failure to receive expected rewards), so

DA response = reward occurred – reward predicted.

Prediction errors are claimed to be fundamental for learning, as the error signal might trigger synaptic modifications leading to subsequent behavioral changes and changes in predictions, reiterating this process until rewards match the expectations and synaptic transmission becomes unchanging and stable. Dopaminergic activation produced by unexpected rewards propagates backward during conditioned learning, that is, activation produced by the reward tends to progressively decrease, and activation produced by the predicting stimulus tends to progressively increase, as the predicting stimulus becomes fully associated with the reward (Schultz, 2007; Schultz, Dayan, & Montague, 1997; Fig. 10.4). In other words, while the predicting stimulus has not been associated with reward, it elicits little or no dopaminergic activation, while the unexpected reward elicits a strong activation. As the predicting stimulus becomes associated with reward, it begins to elicit dopaminergic activity while the reward itself diminishes its dopaminergic activation capacity (as it becomes each time more predictable).

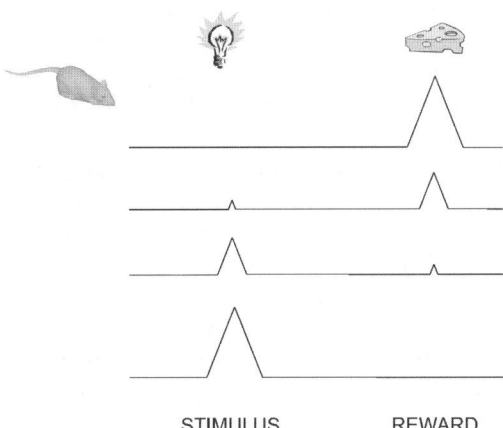

STIMULUS   REWARD

Fig. 10.4 Backward propagation of DA release during learning. In the naive animal, DA is released when the unexpected reward is presented. As the animal associates a predicting stimulus with the reward, the dopaminergic response is displaced to the predicting stimulus (Schultz, 2007)

Prediction errors occur not only with the expected reward value of the stimulus (a stimulus may produce more or less reward than expected), but also with the timing of stimulus delivery (Nakahara, Itoh, Kawagoe, Takikawa, & Hikosaka, 2004). Delaying an expected reward by some 500 ms produces a dopaminergic depression at the expected time and a positive dopaminergic response if it is delivered later, while anticipating the reward produces a larger than normal dopaminergic response.

In humans, a recent report relates abnormal midbrain activation in relation to prediction errors in first-episode psychotic patients, who showed attenuated responses to prediction error in reward trials but augmented responses to prediction error in neutral trials (Murray, Corlett, Clark, Pessiglione, Blackwell et al., 2008). This finding highlights the role of reward-prediction mechanisms in neuropsychiatric conditions and in normal behavioral control.

Dopaminergic signaling also codes for reward uncertainty (i.e., the degree of uncertainty of a given reward within a given distribution of probable outcomes; Fiorillo, Tobler, & Schultz, 2003; Schultz et al., 1997). Uncertainty reduces the reward value of stimuli in risk-averse individuals and increases it in risk-seekers, in compliance with the expected-utility theory of microeconomics (von Neumann & Morgenstern, 1944). Uncertainty-related dopaminergic activation is different from, and uncorrelated to, the reward-related dopaminergic responses in several aspects, and consists of a tonic increase in the single-cell activity during the interval between the reward-predicting stimulus and the reward. This signal evokes low DA concentrations which are suitable to activate the high-affinity D2 receptors, while the reward signals induce stimulation of the high-affinity, intrasynaptic D1 receptors (Schultz, 2007). This kind of dopaminergic response deserves much further study owing to its behavioral and neurodynamical implications.

Dopaminergic neurons also respond to aversive stimuli, which have been described to elicit a short-latency (less than 100 ms) phasic dopaminergic suppression (Ungless, Magill, & Bolam, 2004), or relatively slow and long-lasting depressions (several seconds) (Mirenowicz & Schultz, 1996). Slow and long-lasting activation of putative dopaminergic neurons with aversive stimuli has been reported, in the form of initial activations or rebound responses after depressions (Chiodo, Caggiula, Antelman, either & Lineberry, 1979; Hommer & Bunney, 1980). However, some evidence suggests that some of these neurons might not be dopaminergic (Ungless et al., 2004).

Phasic DA Release Also Provides Contextual Associations During Learning

In the interpretation given earlier, phasic DA release works as a reward-prediction device, selecting behaviors that maximize future rewards (Montague et al., 2004; Schultz & Dickinson, 2000; Tobler et al., 2003; Waelti et al., 2001). However, according to Redgrave & Gurney (2006), in real-life conditions the reward value of many unexpected events is unknown at the time that phasic DA release takes place. For these authors, this type of signaling relates to preattentive, presaccadic processing in which there is not much information about the appetitive or aversive reinforcement consequences. They claim that, perhaps more than predicting the occurrence of reward, phasic DA release has a role in the reselection of actions that triggered an unpredicted event. In other words, every time a salient, unexpected stimulus is produced, phasic dopaminergic signaling in the corpus striatum, amygdala, and PFC permits association of the sensory, motor, and contextual situations immediately previous to this event, so that the animal may develop a "causative theory" of the events that led to this unpredicted stimulus and will become able to generate them in the future (Redgrave & Gurney, 2006). If this stimulus is subsequently associated with positive or negative reinforcement, the animal will know what to do in order to approach or avoid this situation, respectively.

Discussion – A Unifying View of DA Action Through Evolution

I have reviewed some of the evidence pointing to DA as a fundamental modulator of many aspects of goal-directed behavior. It is in fact surprising how many survival-dependent behaviors, from reward prediction to cognitive events, are regulated by DA. Is there a common thread underlying these functions? In a way, cognitive processes may be considered as subsidiary elements of goal-directed conducts; the former make much less sense in the absence of objectives to be fulfilled. Thus, the evolution of cognition may be viewed as a consequence of the elaboration of goal-oriented conducts, where the organism takes advantage of increasingly more subtle cues to better predict the occurrence of reward or punishment. As predicting cues become more sophisticated, in many instances they are more separated in time and space from the rewarding objective, forcing the organism to develop elaborate strategies to achieve or avoid the rewarding or punishing events.

DA works at every level of this process, from reward signaling to cognitive processing. Initially, dopaminergic systems probably served a reward-response role (as observed in invertebrates; Panksepp & Huber, 2004); this permitted the organisms to make contextual associations of rewarding experiences (Redgrave & Gurney, 2006). These associations rapidly evolved as predictive devices, which were necessarily accompanied with the elaboration of goal-directed behaviors and cognitive processing. Dopaminergic neurotransmission was possibly co-opted to regulate all these newly developing functions, thus broadening its domain of regulatory functions.

However, dopaminergic functions may be more primitive than reward signaling. D1-like and D2-like DA receptors have been found to have separate evolutionary origins, D1-like pairing closely with β-adrenergic receptors, and both together representing a sister group of the family of α1 adrenergic receptors. This whole set is in turn the sister group of a clade consisting of two families, the D2-like receptors and the α2 adrenergic receptors (Xhaard, Rantanen, Nyronen, & Johnson, 2006; Fig.10.5). Thus, D2-like receptors, and their functions, possibly originated

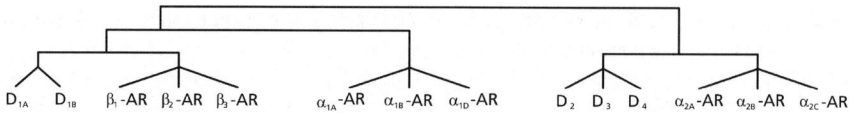

Fig. 10.5 Evolutionary relations between adrenergic receptors and D1 and D2 DA receptors (Xhaard et al., 2006)

earlier than the D1-like elements, involved in phasic signaling at least in vertebrates. One possible ancestral function of D2-like receptors and α2 adrenergic receptors was to invigorate behavior via inhibition of inhibitory processes or other mechanisms. Rewarding stimuli may have produced some increases in catecholamine release, associated with the behavioral responses to these inputs, which produced collateral activation of β-adrenergic receptors or the like. It is proposed that, by gene duplication, the D1-like family of DA receptors evolved as a response device for large DA surges related to reward.

Thus, an evolutionary framework provides a unifying perspective for the diverse actions of DA in behavior, and of its multiple physiological functions. Perhaps one main point to be drawn from this review is that the complex regulation of cortical neurodynamics by DA has deep roots in behavioral control mechanisms, originating from an ancestral system involved in behavioral activation.

Acknowledgements Research referred to in this work has been partly funded by FONDECYT grant no. 1050721 and by the Millennium Center for Integrative Neuroscience. I also thank Marcela Henríquez for helpful comments.

References

Aboitiz, F. (2008). Evolution of the amygdalar complex. In F. Aboitiz & A. Dagnino-Subiabre (Eds.), *The neurobiology of stress: An evolutionary approach* (pp. 1–30). Kerala, India: Research Signpost.

Aboitiz, F., & Montiel, J. (2001). Anatomy of "mesencephalic" dopaminergic cell groups in the central nervous system. In J. Segura (Ed.), *Role of reactive catecholamine species in neurodegeneration and apoptosis of dopaminergic neurons* (pp. 1–19). New York: F. P. Graham.

Bentivoglio, M., & Morelli, M. (2005). The organisation and circuits of mesencephalic dopaminergic neurons and the distribution of dopamine receptors in the brain. In S. B. Dunnett, M. Bentivoglio, A. Björklund, & T. Hökfelt (Eds.), *Handbook of chemical neuroanatomy* (pp. 1–107). Amsterdam: Elsevier.

Berger, B., Gaspar, P., & Verney, C. (1991). Dopaminergic innervation of the cerebral cortex: Unexpected differences between rodents and primates. *Trends in Neuroscience, 14*(1), 21–27.

Berridge, K. C., & Robinson, T. E. (1998). What is the role of dopamine in reward: Hedonic impact, reward learning, or incentive salience? *Brain Research. Brain Research Review, 28*(3), 309–369.

Bjorklund, A., & Dunnett, S. B. (2007). Dopamine neuron systems in the brain: An update. *Trends in Neuroscience, 30*(5), 194–202.

Breiter, H. C., Gollub, R. L., Weisskoff, R. M., Kennedy, D. N., Makris, N., Berke, J. D., et al. (1997). Acute effects of cocaine on human brain activity and emotion. *Neuron, 19*(3), 591–611.

Carlsson, A., Lindqvist, M., & Magnusson, T. (1957). 3,4-Dihydroxyphenylalanine and 5-hydrox-ytryptophan as reserpine antagonists. *Nature, 180*(4596), 1200.

Coizet, V., Comoli, E., Westby, G. W., & Redgrave, P. (2003). Phasic activation of substantia nigra and the ventral tegmental area by chemical stimulation of the superior colliculus: An electro-physiological investigation in the rat. *European Journal of Neuroscience, 17*(1), 28–40.

Comoli, E., Coizet, V., Boyes, J., Bolam, J. P., Canteras, N. S., Quirk, R. H., et al. (2003). A direct projection from superior colliculus to substantia nigra for detecting salient visual events. *Nature Neuroscience, 6*(9), 974–980.

Crow, T. J. (1972). Catecholamine-containing neurones and electrical self-stimulation. 1. A review of some data. *Psychological Medicine, 2*(4), 414–421.

Cheer, J. F., Aragona, B. J., Heien, M. L., Seipel, A. T., Carelli, R. M., & Wightman, R. M. (2007). Coordinated accumbal dopamine release and neural activity drive goal-directed behavior. *Neuron, 54*(2), 237–244.

Chiodo, L. A., Caggiula, A. R., Antelman, S. M., & Lineberry, C. G. (1979). Reciprocal influences of activating and immobilizing stimuli on the activity of nigrostriatal dopamine neurons. *Brain Research, 176*(2), 385–390.

Dommett, E., Coizet, V., Blaha, C. D., Martindale, J., Lefebvre, V., Walton, N., et al. (2005). How visual stimuli activate dopaminergic neurons at short latency. *Science, 307*(5714), 1476–1479.

Fiorillo, C. D., Tobler, P. N., & Schultz, W. (2003). Discrete coding of reward probability and uncertainty by dopamine neurons. *Science, 299*(5614), 1898–1902.

Floresco, S. B., & Grace, A. A. (2003). Gating of hippocampal-evoked activity in prefrontal corti-cal neurons by inputs from the mediodorsal thalamus and ventral tegmental area. *Journal of Neuroscience, 23*(9), 3930–3943.

Floresco, S. B., & Phillips, A. G. (2001). Delay-dependent modulation of memory retrieval by infusion of a dopamine D1 agonist into the rat medial prefrontal cortex. *Behaviour Neuroscience, 115*(4), 934–939.

Floresco, S. B., West, A. R., Ash, B., Moore, H., & Grace, A. A. (2003). Afferent modulation of dopamine neuron firing differentially regulates tonic and phasic dopamine transmission. *Nature Neuroscience, 6*(9), 968–973.

Francois, C., Yelnik, J., Tande, D., Agid, Y., & Hirsch, E. C. (1999). Dopaminergic cell group A8 in the monkey: Anatomical organization and projections to the striatum. *Journal of Comparative Neurology, 414*(3), 334–347.

Frey, U., Matthies, H., Reymann, K. G., & Matthies, H. (1991). The effect of dopaminergic D1 receptor blockade during tetanization on the expression of long-term potentiation in the rat CA1 region in vitro. *Neuroscience Letters, 129*(1), 111–114.

Friedman, H. R., & Goldman-Rakic, P. S. (1994). Coactivation of prefrontal cortex and inferior parietal cortex in working memory tasks revealed by 2DG functional mapping in the rhesus monkey. *Journal of Neuroscience, 14*(5 Pt 1), 2775–2788.

Goto, Y., Otani, S., & Grace, A. A. (2007). The Yin and Yang of dopamine release: A new perspective. *Neuropharmacology, 53*(5), 583–587.

Grace, A. A. (1990). Evidence for the functional compartmentalization of spike generating regions of rat midbrain dopamine neurons recorded in vitro. *Brain Research, 524*(1), 31–41.

Grace, A. A. (1991). Phasic versus tonic dopamine release and the modulation of dopamine system responsivity: A hypothesis for the etiology of schizophrenia. *Neuroscience, 41*(1), 1–24.

Grace, A. A. (1995). The tonic/phasic model of dopamine system regulation: Its relevance for understanding how stimulant abuse can alter basal ganglia function. *Drug and Alcohol Dependence, 37*(2), 111–129.

Grace, A. A., & Bunney, B. S. (1985). Low doses of apomorphine elicit two opposing influences on dopamine cell electrophysiology. *Brain Research, 333*(2), 285–298.

Grace, A. A., Floresco, S. B., Goto, Y., & Lodge, D. J. (2007). Regulation of firing of dopaminer-gic neurons and control of goal-directed behaviors. *Trends in Neuroscience, 30*(5), 220–227.

Gurden, H., Tassin, J. P., & Jay, T. M. (1999). Integrity of the mesocortical dopaminergic system is necessary for complete expression of in vivo hippocampal-prefrontal cortex long-term potentiation. *Neuroscience, 94*(4), 1019–1027.

Haber, S. N., & Fudge, J. L. (1997). The primate substantia nigra and VTA: Integrative circuitry and function. *Critical Reviews in Neurobiology, 11*(4), 323–342.

Hall, H., Farde, L., Halldin, C., Hurd, Y. L., Pauli, S., & Sedvall, G. (1996). Autoradiographic localization of extrastriatal D2-dopamine receptors in the human brain using [125I]epidepride. *Synapse, 23*(2), 115–123.

Heimer, L., Alheid, G. F., de Olmos, J. S., Groenewegen, H. J., Haber, S. N., Harlan, R. E., et al. (1997). The accumbens: Beyond the core-shell dichotomy. *Journal of Neuropsychiatry and Clinical Neurosciences, 9*(3), 354–381.

Hitchcott, P. K., Quinn, J. J., & Taylor, J. R. (2007). Bidirectional modulation of goal-directed actions by prefrontal cortical dopamine. *Cerebral Cortex, 17*, 2820–2827.

Horvitz, J. C. (2000). Mesolimbocortical and nigrostriatal dopamine responses to salient non-reward events. *Neuroscience, 96*(4), 651–656.

Huang, Y. Y., Simpson, E., Kellendonk, C., & Kandel, E. R. (2004). Genetic evidence for the bidirectional modulation of synaptic plasticity in the prefrontal cortex by D1 receptors. *Proceedings of the Nationall Academy of Sciences of the United States of America, 101*(9), 3236–3241.

Hyman, S. E., Malenka, R. C., & Nestler, E. J. (2006). Neural Mechanisms of Addiction: The Role of Reward-Related Learning and Memory. *Annual Reviews in Neuroscience, 29*, 565–598.

Iversen, L. (2006). Neurotransmitter transporters and their impact on the development of psychopharmacology. *British Journal of Pharmacology, 147*(Suppl 1), S82–S88.

Iversen, S. D., & Iversen, L. L. (2007). Dopamine: 50 years in perspective. *Trends in Neuroscience, 30*(5), 188–193.

Kebabian, J. W., & Calne, D. B. (1979). Multiple receptors for dopamine. *Nature, 277*(5692), 93–96.

Kroner, S., Krimer, L. S., Lewis, D. A., & Barrionuevo, G. (2007). Dopamine increases inhibition in the monkey dorsolateral prefrontal cortex through cell type-specific modulation of interneurons. *Cerebral Cortex, 17*(5), 1020–1032.

Lammel, S., Hetzel, A., Häckel, O., Jones, I., Liss, B., & Roeper, J. (2008) Unique properties of mesoprefrontal neurons within a dual mesocorticolimbic dopamine system. *Neuron, 57*, 760–773.

Lewis, D. A., Sesack, S. R., Levey, A. I., & Rosenberg, D. R. (1998). Dopamine axons in primate prefrontal cortex: Specificity of distribution, synaptic targets, and development. *Advances in Pharmacology, 42*, 703–706.

Ljungberg, T., Apicella, P., & Schultz, W. (1992). Responses of monkey dopamine neurons during learning of behavioral reactions. *Journal of Neurophysiology, 67*(1), 145–163.

Lodge, D. J., & Grace, A. A. (2006a). The hippocampus modulates dopamine neuron responsivity by regulating the intensity of phasic neuron activation. *Neuropsychopharmacology, 31*(7), 1356–1361.

Lodge, D. J., & Grace, A. A. (2006b). The laterodorsal tegmentum is essential for burst firing of ventral tegmental area dopamine neurons. *Proceedings of the National Academy of Sciences of United States of America, 103*(13), 5167–5172.

Mameli-Engvall, M., Evrard, A., Pons, S., Maskos, U., Svensson, T. H., Changeux, J. P., et al. (2006). Hierarchical control of dopamine neuron-firing patterns by nicotinic receptors. *Neuron, 50*(6), 911–921.

Marin, O., Smeets, W. J., & Gonzalez, A. (1998). Evolution of the basal ganglia in tetrapods: A new perspective based on recent studies in amphibians. *Trends in Neuroscience, 21*(11), 487–494.

Mena-Segovia, J., Bolam, J. P., & Magill, P. J. (2004). Pedunculopontine nucleus and basal ganglia: Distant relatives or part of the same family? *Trends in Neurosci, 27*(10), 585–588.

Mirenowicz, J., & Schultz, W. (1996). Preferential activation of midbrain dopamine neurons by appetitive rather than aversive stimuli. *Nature, 379*(6564), 449–451.

Montague, P. R., Hyman, S. E., & Cohen, J. D. (2004). Computational roles for dopamine in behavioural control. *Nature, 431*(7010), 760–767.

Murray, G. K., Corlett, P. R., Clark, L., Pessiglione, M., Blackwell, A. D., Honey, G., et al. (2008). Substantia nigra/ventral tegmental reward prediction error disruption in psychosis. *Molecular Psychiatry, 13*, 267–276.

Nakahara, H., Itoh, H., Kawagoe, R., Takikawa, Y., & Hikosaka, O. (2004). Dopamine neurons can represent context-dependent prediction error. *Neuron, 41*(2), 269–280.

Panksepp, J. B., & Huber, R. (2004). Ethological analyses of crayfish behavior: A new invertebrate system for measuring the rewarding properties of psychostimulants. *Behavioral Brain Research, 153*(1), 171–180.

Parent, A., & Hazrati, L. N. (1995). Functional anatomy of the basal ganglia. I. The cortico-basal ganglia-thalamo-cortical loop. *Brain Research. Brain Research Reviews, 20*(1), 91–127.

Redgrave, P., & Gurney, K. (2006). The short-latency dopamine signal: A role in discovering novel actions? *Nature Reviews Neuroscience, 7*(12), 967–975.

Reynolds, J. N., Hyland, B. I., & Wickens, J. R. (2001). A cellular mechanism of reward-related learning. *Nature, 413*(6851), 67–70.

Richfield, E. K., Penney, J. B., & Young, A. B. (1989). Anatomical and affinity state comparisons between dopamine D1 and D2 receptors in the rat central nervous system. *Neuroscience, 30*(3), 767–777.

Schultz, W. (1998). Predictive reward signal of dopamine neurons. *Journal of Neurophysiology, 80*(1), 1–27.

Schultz, W. (2007). Behavioral dopamine signals. *Trends in Neuroscience, 30*(5), 203–210.

Schultz, W., Dayan, P., & Montague, P. R. (1997). A neural substrate of prediction and reward. *Science, 275*(5306), 1593–1599.

Schultz, W., & Dickinson, A. (2000). Neuronal coding of prediction errors. *Annual Review of Neuroscience, 23*, 473–500.

Sesack, S. R., Aoki, C., & Pickel, V. M. (1994). Ultrastructural localization of D2 receptor-like immunoreactivity in midbrain dopamine neurons and their striatal targets. *Journal of Neuroscience, 14*(1), 88–106.

Smith, Y., Lavoie, B., Dumas, J., & Parent, A. (1989). Evidence for a distinct nigropallidal dopaminergic projection in the squirrel monkey. *Brain Research, 482*(2), 381–386.

Somogyi, P., Kisvarday, Z. F., Freund, T. F., & Cowey, A. (1984). Characterization by Golgi impregnation of neurons that accumulate 3H-GABA in the visual cortex of monkey. *Experimental Brain Research, 53*(2), 295–303.

Sparks, D. L., & Jay, M. F. (1986). The functional organization of the primate superior colliculus: A motor perspective. *Progress in Brain Research, 64*, 235–241.

Stein, B., & Meredith, M. (1993). *The merging of the senses.* Cambridge, MA: MIT.

Tobler, P. N., Dickinson, A., & Schultz, W. (2003). Coding of predicted reward omission by dopamine neurons in a conditioned inhibition paradigm. *Journal of Neuroscience, 23*(32), 10402–10410.

Ungless, M. A., Magill, P. J., & Bolam, J. P. (2004). Uniform inhibition of dopamine neurons in the ventral tegmental area by aversive stimuli. *Science, 303*(5666), 2040–2042.

von Neumann, J., & Morgenstern, O. (1944). *The theory of games and economic behavior.* New Jersey: Princeton University Press.

Waelti, P., Dickinson, A., & Schultz, W. (2001). Dopamine responses comply with basic assumptions of formal learning theory. *Nature, 412*(6842), 43–48.

Wang, T., & French, E. D. (1993). L-glutamate excitation of A10 dopamine neurons is preferentially mediated by activation of NMDA receptors: Extra- and intracellular electrophysiological studies in brain slices. *Brain Research, 627*(2), 299–306.

Weiner, N., & Molinoff, P. B. (1994). Catecholamines. In G. J. Siegel, B. W. Agranoff, R. W. Albers, & P. B. Molinoff (Eds.), *Nasic Neurochemistry. Molecular, cellular and medical aspects* (5th edn., pp. 261–282). New York: Raven Press.

Williams, S. M., & Goldman-Rakic, P. S. (1998). Widespread origin of the primate mesofrontal dopamine system. *Cerebral Cortex, 8*(4), 321–345.

Wurtz, R. H., & Albano, J. E. (1980). Visual-motor function of the primate superior colliculus. *Annual Reviews in Neuroscience, 3*, 189–226.

Xhaard, H., Rantanen, V. V., Nyronen, T., & Johnson, M. S. (2006). Molecular evolution of adrenoceptors and dopamine receptors: Implications for the binding of catecholamines. *Journal of Medicinal Chemistry, 49*(5), 1706–1719.

Yanez, I. B., Munoz, A., Contreras, J., Gonzalez, J., Rodriguez-Veiga, E., & DeFelipe, J. (2005). Double bouquet cell in the human cerebral cortex and a comparison with other mammals. *Journal of Comparative Neurology, 486*(4), 344–360.

Yavich, L., Forsberg, M. M., Karayiorgou, M., Gogos, J. A., & Männistö, P. T. (2007). Site-specific role of catechol-O-methyltransferase in dopamine overflow within prefrontal cortex and dorsal striatum. *Journal of Neuroscience, 27*(38), 10196–10209.

Chapter 11
Dynamics of a Neuromodulator – II.
Dopaminergic Balance and Cognition

F. Aboitiz

Abstract Dopamine (DA) strongly modulates the activity of neuronal ensembles in the prefrontal cortex. Activity of D1-like receptors, produced by phasic DA release, produces a net depressing effect by stimulating inhibitory interneurons, while on the other hand it enhances the activity levels of pyramidal cells that are strongly active. This produces an increase in the signal-to-noise ratio and contributes to the maintenance of behaviorally relevant circuits such as attentional systems and working memory ensembles. On the other hand, at much lower concentrations, D2-receptor-like activity produces a general disinhibition of cortical networks, favoring the maintenance of multiple representations and contributing to updating representations according to contextual changes. The balance between these two DA signaling systems is crucial, as failure to maintain the behavioral goal results in distractibility, while failure to update it with new sensory evidence results in perseverance. Thus, the mesencephalic DA system is a strong modulator of forebrain neurodynamics, both cortical and subcortical. A disbalance in the DA signaling systems might be one of the physiopathological mechanisms underlying neuropsychiatric disorders such as schizophrenia and attention deficit–hyperactivity disorder.

Abbreviations

ADHD   Attention deficit–hyperactivity disorder
BLA   Basolateral amygdala
cAMP   Cyclic AMP
COMT   Catechol-O-methyltransferase
DA   Dopamine

F. Aboitiz
Departamento de Psiquiatría, Centro de Investigaciones Médicas, Pontificia Universidad
Católica de Chile, Marcoleta 391, Santiago, Chile, e-mail: faboitiz@uc.cl

F. Aboitiz and D. Cosmelli (eds.) *From Attention to Goal-Directed Behavior.*   205
© Springer-Verlag Berlin Heidelberg 2009

DISC1   Disrupted in schizophrenia 1
NAc   Nucleus accumbens
NE   Norepinephrine
PFC   Prefrontal cortex
VTA   Ventral tegmental area

Introduction

A striking characteristic of the prefrontal cortex (PFC) is its extensive dopaminergic innervation. Being the structure perhaps most relevant for behavioral organization and the control of goal-directed behavior, much research has being devoted to unveiling the role of this neurotransmitter as a modulator of cortical neurodynamics. Many early and recent studies have highlighted the role of dopamine (DA) in cognitive function, especially working memory and attentional mechanisms (Goldman-Rakic, 1995a). DA has been shown to have important but highly complex effects on the dynamics of PFC networks, biasing activity in different directions depending on the circumstances (Seamans & Yang, 2004; Wang, Tegnér, Constantinidis & Goldman-Rakic, 2004). Furthermore, activation of different types of DA receptors produces antagonic effects, a situation that establishes a shifting balance between distinct forms of DA signaling, which is required for appropriate behavior and cognitive control under different circumstances. Disturbances in this balance are believed to relate to several neuropathological states.

   This chapter addresses the role of different mechanisms of DA signaling in the control of cortical neurodynamics and cognitive functions (especially working memory and attention), highlighting the different modes of signaling that were discussed in Chap. 10. Furthermore, the relation between DA dysfunctions and neuropsychiatric disease is discussed at the end of the chapter.

DA Modulation of PFC Activity

DA receptor activity has been proposed to have both facilitatory and inhibitory effects of PFC neurons, but the overall picture of the effects of DA in the PFC is confusing (Muly, Szigeti, & Goldman-Rakic, 1998; Yang, Seamans, & Gorelova, 1996). DA axons innervate both pyramidal neurons and GABAergic interneurons (Sesack, Bressler, & Lewis, 1995; Verney, Alvarez, Geffard, & Berger, 1990; Williams & Goldman-Rakic, 1993); although at least in macaques the innervation of pyramidal cells seems to be more abundant (Krimer, Jacob & Goldman-Rakic, 1997). Both types of neurons express multiple types of DA receptors (Bergson, Mrzljak, Smiley, Pappy, Levenson, et al., 1995; Khan, Gutiérrez, Martin, Peñafiel, Rivera, et al., 1998; Kroner, Krimer, Lewis, & Barrionuevo, 2007; Lidow, Wang,

Fig. 11.1 Distribution of dopamine (*DA*) receptors in prefrontal cortical cells. D2 receptors tend to localize in fast-spiking, parvalbumin-positive interneurons, while D1 receptors tend to distribute both in interneurons and in dendritic spines of pyramidal neurons

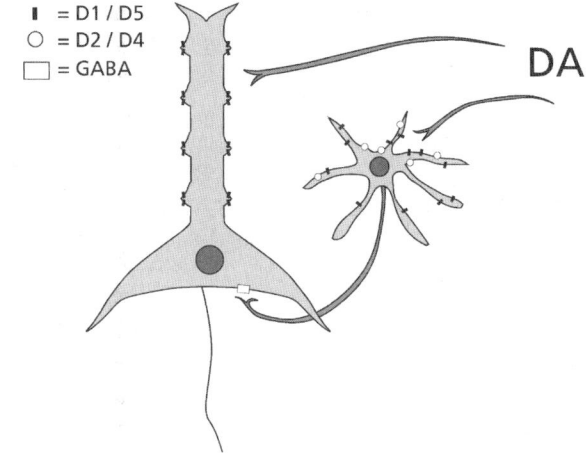

Cao, & Goldman-Rakic, 1998; Mrzljak, Bergson, Pappy, Huff, Levenson, et al., 1996; Muly, Szigeti, & Goldman-Rakic, 1998). Nevertheless, in both rat and monkey, expression of D1 receptors on pyramidal cells is substantially greater than expression of D2-like receptors (Gaspar, Bloch, & Le Moine, 1995), whereas both types of receptors have been localized in interneurons (Khan et al., 1998; Mrzljak et al., 1996; Muly et al., 1998; Paspalas & Goldman-Rakic, 2005; Sesack, Bressler, & Lewis, 1995; Vincent, Pabreza, & Benes, 1995; Vysokanov, Flores-Hernandez, & Surmeier, 1998; Wedzony, Chocyk, Mackowiak, Fijal, & Czyrak, 2000; Fig. 11.1).

Activation of D1 And D2 Receptors Produces Antagonic Effects

D1 receptor stimulation has been found to produce activation of prefrontal pyramidal neurons, as observed in intrinsic neuronal currents (Gorelova & Yang, 2000; Henze, Gonzalez-Burgos, Urban, Lewis, & Barrionuevo, 2000; Seamans & Yang, 2004; Tseng & O'Donnell, 2004). Activation of these receptors has been reported to facilitate NMDA-mediated postsynaptic currents, but it is also found to inhibit non-NMDA-mediated responses evoked by local stimulation and facilitation of GABAergic activity (Gao, Krimer, & Goldman-Rakic, 2001; Gorelova, Seamans, & Yang, 2002; Seamans, Durstewitz, Christie, Stevens, & Sejnowski, 2001; Seamans, Gorelova, Durstewitz, & Yang, 2001; Trantham-Davidson, Neely, Lavin, & Seamans, 2004; Wang & O'Donnell, 2001; Yan & Surmeier, 1997). Other studies indicate that D1 signaling may have effects on the hyperpolarization-activated cation current I_h, which participates in dendritic integration (Luthi & McCormick, 1998). However, another study reported that in striatal neurons, the I_h current is blocked by D2 but not by D1 receptor activity (Deng, Zhang, & Xu, 2007).

The most robust effect of D1 activation in PFC seems to be on inhibitory interneurons. In the monkey, DA axons make specific contacts with parvalbumin-expressing, fast-spiking interneurons and other interneurons (but not with calretinin-expressing interneurons), increasing their excitability via D1 receptors (Gao, Wang, & Goldman-Rakic, 2003; Gorelova et al., 2002; Kroner et al., 2007; Sesack et al., 1995; Sesack, Hawrylak, Guido, & Levey, 1998; Zhou & Hablitz, 1999). Calretinin-positive interneurons include double bouquet and neurogliaform cells which innervate distal dendrites of pyramidal cells, thus regulating the vertical integration of their synaptic inputs (Freund, 2003; Yanez, Munoz, Contreras, Gonzalez, Rodriguez-Veiga, et al., 2005). On the other hand, the DA-innervated, parvalbumin-expressing, fast-spiking interneurons include chandelier or basket-type neurons that make strong inhibition in the soma or initial axonal segment of pyramidal cells (Bjorklund & Dunnett, 2007; Freund, 2003). These interneurons have been postulated to work as pacemakers of rhythmic activity, acting as a network that synchronizes oscillatory activity of pyramidal neurons at different frequencies (Freund, 2003). The excitation of fast-spiking interneurons by DA may enable them to fire through sustained periods of synaptic activity and to maintain a balance between excitation and inhibition during these periods (Kroner et al., 2007).

The functional role of D2-like receptors is perhaps less clear. D2 or D4 receptor stimulation has been described to reduce GABA-mediated inhibitory currents and NMDA-mediated synaptic responses (Seamans & Yang, 2004; Tseng & O'Donnell, 2004), which points to a reduction in GABAergic transmission due to presynaptic mechanisms or postsynaptic receptor downregulation (Trantham-Davidson et al., 2004).

As can be seen, "it is difficult to come up with a coherent picture of what DA is doing in the PFC based simply on a summary of … individual effects" (Seamans & Yang, 2004). These authors published an extensive review of DA action in the PFC, featuring 18 key properties of DA transmission. One thing that becomes clear from this work is that the effects of DA are not straightforward: they depend on concentration levels, target neurons, basal activity, receptor activation, and signaling cascades. Perhaps as a very broad summary, D1 activation may increase NMDA currents, but at the same time it decreases glutamate release and increases interneuron excitability and inhibitory postsynaptic potentials in pyramidal cells. On the other hand, D2 activation leads to a reduction in inhibitory postsynaptic potentials, and reduction of NMDA currents (Seamans & Yang, 2004) (Fig. 11.2). Note that NMDA receptors are linked to processes of synaptic plasticity rather than neurotransmission per se. Thus DA, acting on D1 receptors, exerts a predominantly inhibitory effect on excitability of single pyramidal cells (specifically, spontaneous spike firing suppression), which is largely mediated by GABA release. This effect is strongly modulated by tonic D2 activity levels, which tend to increase excitability and occlude the effects of inhibitory inputs (Seamans & Yang, 2004). This antagonism may provide a processing tone that serves to regulate the gain within prefrontal networks, where activity of D2 receptors may curtail the prolonged effects of D1 receptor activation (Seamans & Yang, 2004; Kroner et al., 2007). Higher than normal DA concentrations as occur

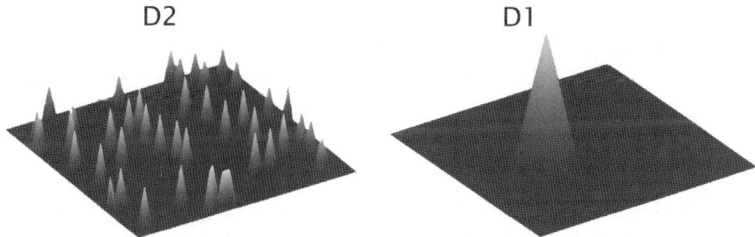

Fig. 11.2 Cortical landscapes of activity under the influences of D2 receptor activity (tonic DA liberation) and D1-receptor activity (phasic DA liberation). The D2 state promotes the activation of multiple representations, while the D1 state produces a generalized inhibition, and allows the activation of very few networks (Seamans and Yang, 2004)

during stress or as a result of pharmacological intervention could shift the balance towards disproportionately activating inhibitory interneurons, contributing to inhibition of PFC activity (Kroner et al., 2007).

Although tonic DA is thought to provide crucial control over higher cognitive function (Akil et al., 1999; Schultz, 1998, 2002), the precise mechanisms of action are still debated. In general, it is considered that the background DA levels regulate the sensitivity of PFC neurons to event-related inputs, including phasic DA activity (Williams & Goldman-Rakic, 1995). Thus, tonic DA downmodulates phasic DA release via stimulation of the highly sensitive D2 and D4 receptors present on DA terminals (Grace, 1991, 1995). In addition, novelty-related increases in tonic DA in the hippocampus facilitate long-term potentiation and memory encoding (Lisman & Grace, 2005). Others have proposed that tonic DA levels play a key role in the regulation of PFC synaptic plasticity, by converting long-term synaptic depression to potentiation (Matsuda, Marzo, & Otani, 2006). These authors observed that weak tetanic stimulation of parallel fibers in layers 1 and 2, which otherwise induces long-term depression in layer 5 pyramidal neurons, is able to induce long-term potentiation in these cells if the preparation is primed with even low DA concentrations ($3\,\mu M$), an effect that occurs via the combined action of D1 and D2 receptors. In this context, pathological conversions of long-term potentiation to long-term depression due to insufficient DA tone may occur in some psychiatric diseases such as schizophrenia, where there is also NMDA-receptor dysfunction (Sokolov, 1998).

DA Strongly Interacts with Other Neuromodulators

DA is not the only neuromodulator in the PFC, there being strong norepinephrinergic, serotoninergic, and cholinergic influences, which interact with the actions of DA. In learning tasks, DA activity has been related to set formation, i.e., learning to attend to those features and motor responses that are relevant for the task

(Redgrave & Gurney, 2006; Robbins & Roberts, 2007; Robbins, 2005). On the other hand, disruption of serotonine transmission specifically impairs reversal learning, while norepinephrine (NE) depletion impairs set shifting and acetylcholine depletion leads to deficits in serial reversal learning, implying that these neurotransmitters operate synergistically in different stages of frontal executive operations (Robbins & Roberts, 2007; Robbins, 2005).

In particular, NE transmission relates to arousal state and, like DA, has phasic and tonic modes of signaling. Phasic NE firing in the locus coeruleus relates to a subject's interest in a stimulus or task, while tonic NE may range from silent firing (REM sleep), slight (non-REM sleep), or high levels during stress (Foote, Aston-Jones, & Bloom, 1980). The actions of both NE and DA signaling have been proposed to generate changes in I_h channels, via changes in cyclic AMP (cAMP) concentrations, but in different dendritic spines of superficial-layer pyramidal neurons (Arnsten, 2007). These two signaling systems are proposed to cooperate in increasing the specificity of PFC neurons during spatial working memory tasks. Activation of the Gi-related α2A NE receptors is proposed to suppress I_h channels in spines receiving inputs from neurons with similar spatial properties, thus increasing firing in the preferred spatial direction. On the other hand, D1 receptor activation excites I_h channels on spines with inputs with dissimilar spatial properties, consequently decreasing firing in nonpreferred directions (Arnsten, 2007; Wang et al., 2007). However, evidence also suggests opposing influences of cortical NE and DA in the stress response (Pascucci, Ventura, Latagliata, Cabib and Stefano, 2007; see "Attentional Mechanisms May Be Involved in DA Regulation of Working Memory").

DA in Working Memory and Attention

The original discovery that DA depletion in the sulcus principalis of the PFC impaired performance in spatial delayed tasks (Brozoski, Brown, Rosvold, & Goldman, 1979) stimulated the study of the dopaminergic innervation of the PFC and its role in cognitive function, especially working memory and attention. In working memory tasks, PFC neurons show task-related firing specific to each period of the task: there are cells that respond to the appearance of the stimulus (cue), others that activate during the response period, and still others that fire sustainedly during the delay period when the sensory representation is maintained online in the absence of environmental stimuli (Fuster, 1995; Schultz, Apicella, Ljungberg, Romo, & Scarnati, 1993). The latter are specific for the key attribute of the stimulus, and their activity depends on DA signaling. Dopaminergic cells projecting to the PFC fire during the cue period in spatial working memory tasks (Schultz et al., 1993). Furthermore, PFC neurons are thought to be organized in a mnemonic map organized in cortical columns representing space discretely. The reverberatory intracolumnar and intercolumnar connectivity, modulated by GABAergic interneurons and by dopaminergic terminals, is considered to shape the receptive fields in working memory tasks, in terms of stimulus specificity, location, and timing of response (Goldman-Rakic, 1995b; Kritzer & Goldman-Rakic, 1995; Schwartz & Goldman-Rakic, 1984).

The Inverted-U-Shaped Relation of DA and Working Memory

Agonists of the D1 receptor in the PFC potentiate neural activity during working memory and attentional processes, by increasing delay- and response-related activity relative to background activity (Sawaguchi, Matsumura, & Kubota, 1988, 1990). There is an inverted-U (or bell-shaped) relationship between the levels of DA and working memory performance, in which intermediate DA levels are optimal for behavioral performance but higher doses impair it and induce perseveration (Goldman-Rakic, 1996; Goldman-Rakic, Lidow, Smiley, & Williams, 1992; Sawaguchi & Goldman-Rakic, 1991; Williams & Goldman-Rakic, 1995). This is consistent with the Yerkes–Dodson principle of optimal levels of arousal or stress for behavioral performance. This concentration dependency of the effects of DA may result at least in part from DA effects on interneurons (Muly et al., 1998). Furthermore, in PFC neurons of behaving monkeys, low-level D1 stimulation of delay-specific neurons enhances spatial tuning of memory cells, whereas higher levels decrease their excitability and direction sensitivity (Vijayraghavan, Wang, Birnbaum, Williams, & Arnsten, 2007). This study also demonstrated that this effect was mediated by activation of the Gs signaling cascade coupled to adenylyl cyclase. It was hypothesized that, at lower DA doses, suppression of firing in the nonpreferred directions involves D1-receptor-mediated enhancement of GABAergic tuning mechanisms, while at higher doses, the suppression of firing in the preferred direction was due to the opening of I_h channels by way of cAMP. This finding is consistent with the observation that high levels of D1 receptor blockade completely inhibit cell firing, whereas low levels increase neuronal firing (Williams & Goldman-Rakic, 1995). Thus, a low dose of D1 receptor antagonist allowed PFC neurons to unmask their receptive field and memory properties via GABAergic inhibition. At higher doses, inhibition is too high, and cell firing becomes substantially less specific and decreased.

Interestingly, the effects of D1 stimulation on the PFC are also dependent on the demands of a cognitive task. Infusion of a D1 receptor agonist in rats can improve or impair performance depending on the task's signal-to noise demands (Chudasama & Robbins, 2004). The study by Vijayraghavan et al. (2007) indicates that D1 receptor stimulation can be helpful if the task demands precise spatial tuning and the subject must suppress irrelevant information. This is consistent with imaging studies in normal humans indicating that stimulant medication reduces the area of PFC activated during the task and increases efficiency of the blood oxygen level dependent response in functional MRI assessments (Mattay et al., 2003; Mehta et al., 2000).

Although, as mentioned, most studies have pointed to a role of D1 receptors in working memory, the less evident D2 receptors have been found to participate in these tasks by modulating motor-related neural activity in memory-guided saccades (Wang, Vijayraghavan, & Goldman-Rakic, 2004).

Attentional Mechanisms May Be Involved in DA Regulation of Working Memory

One interpretation of the aforementioned findings is that the occupation of D1 receptors prevents distraction during the delay period of working memory tasks (Robbins, 2005). In this line, a D1 agonist was found to improve working memory at relatively long delays, but impairs it at shorter delays (Floresco & Phillips, 2001). Furthermore, D1 agonists were found to have an effect in an attentional task that resembles tests of continuous attentional performance in humans (the five-choice serial reaction task; Granon et al., 2000). In this study, rats with low baseline levels of performance in this task significantly improved their perform-ance after administration of D1 receptor agonists, while animals with high baseline performances showed deficits after treatment with the agonist. This suggests that the high-performance group was closer to the optimal level of performance than the low-performance group; administration of the agonist implied deviation from this optimal point in high performers and approach to the optimal point in low performers. This observation may have important clinical consequences consider-ing individual variability in the responses to stimulants (Robbins, 2005).

DA depletion with 6-hydroxydopamine in the frontal cortex of marmosets significantly impairs performance in a test that resembles the Wisconsin Card Sorting Test (this test requires transference of attention from one attribute of the stimulus to another when the key attribute of the stimulus is changed during the task). Moreover, lesioned animals were more distracted than controls, although their visual discrimination abilities remained intact (Crofts et al., 2001). In the same experiment, it was observed that if the 6-hydroxydopamine lesion was produced during acquisition of discrimination learning, animals were impaired in maintaining attention to one of the perceptual dimensions. The general interpretation of these findings is that D1 receptor activation contributes to stabilizing representations, and its malfunction implies distractibility due to lability of the underlying neural net-works (Robbins, 2005).

Methylphenidate, which blocks the DA transporter (and the noradrenaline transporter as well) improves spatial working memory performance in normal volunteers, which is associated with reductions of regional blood flow in frontopa-rietal circuits associated with the task (Mattay et al., 2003; Mehta et al., 2000). Interestingly, the beneficial effects of this drug were related to the basal working memory scores in each subject: individuals with low scores showed more improve-ment than subjects with high scores (Mehta et al., 2000). This is consistent with the findings of Granon et al. (2000) mentioned above. In the study of Mattay et al.

(2003), subjects homozygous for the native Val allele of the catechol-*O*-methyl-transferase (*COMT*) gene (which has been associated with poorer working memory and attentive mechanisms) also exhibited the greatest benefits of methylphenidate administration, when compared with subjects carrying the hypofunctional Met allele, which increases DA availability (especially extracellular). Additional studies have confirmed this inverted-U-shaped response in relation to *COMT* polymorphisms (Goldberg et al., 2003). In the same line, schizophrenic patients homozygous for the Met *COMT* polymorphism showed more cognitive stability but less cognitive flexibility than patients homozygous for the Val allele (heterozygotes performed at intermediate levels; Nolan, Bilder, Lachman, & Volavka, 2004). More recently, using in vivo voltammetry, Yavich et al. (2007) showed that in *COMT*-deficient mice, removal of DA was slower than in control animals in the PFC but not in the striatum, indicating that poor *COMT* activity is related to an increase in tonic DA levels in the PFC.

Overall, this evidence indicates that prefrontal DA contributes to the balance between the maintenance of representations on one hand, and switching attention when the circumstances require it. It has been proposed that the dynamics of phasic versus tonic DA signaling, acting via D1 and D2 receptors, respectively, maintains this delicate balance between focusing and periodical updating of contextual information to generate appropriate behavior (Aboitiz, López, López-Calderón, & Carrasco, 2006; Robbins, 2005). According to Seamans & Yang (2004), moderate levels of D1 receptor activation increase inhibition in the PFC, but at the same time enhance persistent activity driven by particularly strong inputs. Thus, circuits that are active during working memory tasks become enhanced and stabilized, while circuits with low activity become suppressed, increasing the strong circuits' salience and the signal-to-noise ratio. However, if D1 activation is too high, inhibition predominates, resulting in disruption of the working memory ensembles (this might explain the inverted-U-shaped curve of performance in response to D1 stimulation). On the other hand, when D2 effects predominate there is a net reduction in inhibition, with multiple inputs generating several but low-salience working memory circuits, allowing multiple representations and attentional shifts.

The Balance between Limbic and Cortical Control over Behavior

DA also exerts a strong influence on limbic structures, and especially on the balance between these structures and cortical activity during the execution of behavior. I will briefly review some effects of DA signaling in the interaction between limbic components such as the hippocampus and amygdala, and cortical activity.

Limbic Versus Cortical Drive In the Nucleus Accumbens

Phasic DA signaling triggers activity in only a subgroup of neurons in the nucleus accumbens (NAc), and may have an important role in early stages of reward-associative learning by modifying the synaptic strength of specific limbic inputs to the NAc (Grace et al., 2007; see also Fig. 10.3 in Chap. 10). On the other hand, tonic changes in extracellular levels depend on the activity of a large number of DA neurons, thus modulating the activity of a large number of NAc neurons and their inputs. Thus, the interaction between tonic and phasic DA signaling is believed to modulate input selection in the NAc and behavioral response selection in goal-directed behavior (Bratcher, Farmer-Dougan, Dougan, Heidenreich, & Garris, 2005; Floresco, Todd, & Grace, 2001). In this context, the balance between phasic and tonic DA signaling modulates the relative weight of limbic versus prefrontal inputs to the NAc. Single neurons in the NAc receive convergent inputs from the PFC and limbic structures such as hippocampal ventral subiculum and basolateral amygdala (French & Totterdell, 2002, 2003). Furthermore, D2 receptor stimulation selectively attenu-ates the medial prefrontal input to this structure, and DA antagonists increase the prefrontal influence over the NAc (Goto & Grace, 2005a; O'Donnell & Grace, 1994; West & Grace, 2002). In the same line, increasing tonic DA levels by inactivation of the ventral pallidum produces an attenuation of the prefrontal drive on the NAc, but does not affect hippocampal input to this structure; and activation of the ventral pal-lidum decreases tonic D2 suppression of the PFC on the NAc. Thus, increasing tonic DA levels results in an attenuation of the prefrontal influence on the NAc via stimula-tion of D2 receptors (Grace et al., 2007). On the other hand, phasic DA, acting largely via D1 receptors, increases the response of NAc neurons to hippocampal (ventral subiculum) inputs and does not affect prefrontal afferents (Goto & Grace, 2005b).

Therefore, increases in DA input to the NAc produce a concomitant decrease in prefrontal input (via tonic activation of D2 receptors) and an increase in hippocam-pal input (via phasic activation of D1 receptors). This shift has important behavioral consequences, as rats with severed hippocampal input and D1 receptor inactivation showed impairments both in acquisition of a discrimination skill and in their capac-ity to switch strategies during the task, which is consistent with the concept of a role of phasic DA transmission in reward-associated learning. On the other hand, inac-tivation of the medial PFC and administration of a D2 receptor antagonist in the NAc resulted in an spared capacity to acquire the discrimination skill, but rats were severely impaired in changing strategies, showing intense perseveration (Goto & Grace, 2005b).

This evidence indicates that the medial PFC is in constant competition with other inputs such as the hippocampus to control NAc activity. In pathological conditions, this balance is shifted in one direction or the other. For example, in rats sensitized to cocaine, induction of long-term potentiation via subicular stimulation is strongly potentiated, while prefrontal input is diminished. Again, this results in spared acquisition of discriminating tasks but a strong perseverance when strategy changes are needed. Similar observations have been made in L-dopa-medicated parkinsonian patients (Cools, Lewis, Clark, Barker, & Robbins, 2007).

D1/D2 Receptors Modulate Amygdalar Input to the PFC

In the PFC, DA projections from the mesencephalic ventral tegmental area (VTA) overlap with inputs from the basolateral amygdala (BLA) (Floresco & Tse, 2007), pointing to a modulatory effect of DA over BLA inputs to the PFC. A recent study reported that VTA stimulation or local application of DA reduces both BLA-evoked inhibitory and excitatory effects in distinct populations of medial PFC neurons (Floresco & Tse, 2007). Furthermore, administration of D2 and D4 receptor agonists reduces BLA-evoked inhibition by decreasing GABA release, while activation of D1 receptors attenuates BLA-evoked firing of prefrontal neurons. It is known that blockade of D4 receptors impairs both acquisition and extinction of a conditioned fear response (Pfeiffer & Fendt, 2006). Thus, the balance between D2-like-receptor and D1 receptor activity may be crucial in determining the effects of DA on BLA input to the PFC. Disbalances in these regulatory systems may be related to certain neuropsychiatric disorders. For example, increased BLA activity together with decreased VTA activity might result in a pathological hypoactivation of the medial PFC, as observed in patients with depression (Siegle, Thompson, Carter, Steinhauer, & Thase, 2007). On the other hand, in cases of substance abuse there would be a hypersensitive mesocortical DA system, leading to increased medial PFC activation (Childress et al., 1999).

Clinical Discussion

The crucial modulatory role of DA in cognition, emotion, and behavior implies that disbalances in this system might be associated with important neuropsychiatric conditions. At least four of these disorders have been related to a disregulation of this system, although the evidence is not equally strong for all cases. In Parkinson's disease, there is general agreement that a depletion of DA is at the base of this condition. Two other conditions, schizophrenia and attention deficit–hyperactivity disorder (ADHD), have been linked to DA mainly on the basis of the therapeutic effects of DA blockers or stimulants, respectively. Furthermore, catecholamines are strongly involved in the stress response, which may be at the basis of several neuropsychiatric and other clinical disorders. Nevertheless, the above discussion does not strictly show that depletion of DA is at the base of this disease but rather that modulation of DA signaling may compensate for the deficits produced, which might be related to other systems.

Parkinson's Disease – DA Depletion

Parkinson's disease has long been characterized by the depletion of DA levels in the basal ganglia (Birkmayer & Hornykiewicz, 1961; Ehringer & Hornykiewicz,

1960), which is currently treated by L-dopa and related dopaminergic agonists. Although there has been some controversy about the genetic bases for this disease, recent reports have pointed to at least six genes involved in familial forms (Gosal, Ross, & Toft, 2006).

Administration of DA precursors to Parkinson's disease patients has been claimed to restore tonic DA concentrations in these patients, enabling behavioral processes such as eye movements, walking, and arm movements, which are not associated with phasic DA signaling (DeLong, 1983). Tonic DA receptor stimulation apparently invigorates a large number of behavioral processes that depend on striatal and cortical networks (Niv, Daw, Joel, & Dayan, 2007).

Cognitive deficits have been reported in the early stages of Parkinson's disease, and show some improvement after L-dopa treatment (Lange et al., 1992). However, some behavioral functions, such as reversal learning (likely mediated by ventral prefrontal circuitry), evidence a significant detriment after medication. Thus, while L-dopa improves functioning in the more damaged striatal and dorsolateral prefrontal areas in early Parkinson's disease, it may overdose some regions of the PFC (particularly orbitofrontal regions), leading to some of the cognitive deficits associated with treatment (Robbins, 2005). This suggests that normally the basal DA tone in dorsolateral PFC is lower than that of the orbitofrontal cortex.

Stress – Increase in Tonic Catecholamines

Stress is known to increase catecholamines in the PFC, likely corresponding to an augmentation in tonic firing (Deutch, Clark, & Roth, 1990). Both DA and NE increases result in rise of the levels of intracellular cAMP and possibly gate I_h channel opening (Wang et al., 2007). This is proposed to interrupt corticocortical networks that are necessary for persistent activity, inducing network collapse by disconnecting both preferred and nonpreferred inputs (Arnsten, 2007), which consequently generates disfunctions in cognitive processes such as attention and working memory.

A recent report indicates that rats experiencing restraint stress initially show a short-lasting increase of NE release in the medial PFC, and of DA release in the NAc; followed by a sustained increase of DA in the medial PFC and a decrease of DA in the NAc (Pascucci et al., 2007). Furthermore, NE depletion in the PFC abolished the initial increase of NE and DA in the PFC, and DA depletion in the NAc eliminated the subsequent increase in DA in the PFC, suggesting antagonic effects of NE and DA in the response to stress.

Schizophrenia – Hypersensitivity to DA?

The discovery that antipsychotics exert their action by blocking DA receptors in the brain (Carlsson & Lindqvist, 1963; van Rossum, 1966) led to the "DA hypothesis" of schizophrenia, which proposed that this condition was caused by an excess of DA signaling in the cerebral cortex (Matthysse, 1973; Snyder, 1974; see also Chap. 15 by Gaspar et al.). In parallel, Niemegeers & Janssen (1965) found that drugs such as haloperidol could block the stimulant actions of amphetamine in experimental animals, by facilitating DA release (Moore & Gudelsky, 1977). In addition, amphetamine was able to induce schizophrenia-like psychotic symptoms (at least the "first rank" symptoms) in human volunteers, and exacerbated psychotic symptoms in schizophrenic patients (Angrist & Sudilovsky, 1978).

Nevertheless, this hypothesis was controversial as no evidence could be obtained for increased DA receptors (especially of the D2 type) in first-episode patients or in unmedicated patients (Iversen & Iversen, 2007). However, some recent results add new support to the DA hypothesis, indicating an increased dopaminergic response to low doses of amphetamine in schizophrenic patients; another study reported that there is an increased occupancy of D2 receptors in the brains of schizophrenics, which could mask the differences in receptor numbers with normal subjects when using specific DA receptor ligands (Abi-Dargham et al., 2000; Iversen, 2006).

In agreement with this evidence, Grace (1991, 1995) proposed that, in the schizophrenic, a pathological impairment in PFC activity caused a decrease in the tonic, extracellular DA levels in the striatum and NAc, leading to a compensatory upregulation of DA responsivity. This would cause abnormally large phasic DA responses in the compensated regions, giving rise to several of the positive symptoms of this disease. This hypothesis was claimed to account for the increased levels of D2 DA receptors in the NAc of these patients in the absence of a hyperdopaminergic state. Moreover, we mentioned earlier that the Met polymorphism in *COMT* causes in humans an impairment in DA inactivation in the PFC, a consequently deficient cognitive function, and an increased risk of developing schizophrenia (Tunbridge, Harrison, & Weinberger, 2006). One interpretation is that at least the positive, more florid symptoms of schizophrenia can be caused by an increase of DA release and/or a sensitization to DA signaling in these patients (Iversen & Iversen, 2007; Seeman et al., 2005). Recently, a hypoactive form of the gene disrupted in schizophrenia 1 (*DISC1*) has been found to cosegregate with schizophrenia in some families (Ishizuka, Paek, Kamiya, & Sawa, 2006). *DISC1* normally controls catecholamine-induced cAMP levels during the stress response, implying that the hypofunctioning allele might be related to an exaggerated response to adverse conditions, providing an increased susceptibility to the disease (Arnsten, 2007).

On the other hand, a different line of evidence suggests a disbalance of glutamatergic neurotransmission as the base of this condition (Patil et al., 2007). One possibility is that the two systems, glutamatergic and dopaminergic, are interdependent, so a disbalance in one of them implies a consequent alteration in the other, either as a compensatory effect or as a concatenated dysregulation. Lisman, Coyle, Green, Javitt, Benes, et al. (2008) proposed that NMDA receptor malfunction in schizophrenia generates an overexcitation of inhibitory interneurons, leading to an overactivation of pyramidal neurons in the cerebral cortex, especially in the hippocampus. Hippocampal overactivity would then induce a hyperdopaminergic state by releasing the inhibition that the NAc exerts over the VTA via the ventral pallidum (see Chap. 10). Summarizing, rather than a single factor, perhaps there are many interrelated variables that become altered in schizophrenia, including GABAergic, dopaminergic, and NMDA neurotransmission. Therefore, perhaps a systems-level approach is the most promising research strategy in relation to this and other mental diseases.

ADHD – Phasic/Tonic DA Dysregulation?

ADHD, both in children and in adults, is currently treated by DA-based psychostimulants (amphetamines or methylphenidate), although novel treatments with NE agonists are also effective (Biederman, 2005; Castellanos et al., 1997; Iversen, 2006).

ADHD has a relative high heritability rate (approximately 80%) (Biederman, 2005; Taylor, 1998); nevertheless, more detailed genetic studies have not been able to identify genes with high penetrance in the phenotype (See Chap. 14 by Carrasco et al.). The proposed candidates are usually, but not exclusively, associated with dopaminergic transmission. The 7 variable number tandem repeat repetition allele in exon 3 of the gene coding the DA receptor D4 (D4-7R) has frequently been associated with ADHD (Barr et al., 2000; Carrasco et al., 2006; Holmes et al., 2002; Manor et al., 2002; Muglia, Jain, Macciardi, & Kennedy, 2000). This allele apparently codifies a subsensitive form of the postsynaptic receptor D4 (Missale, Nash, Robinson, Jaber, & Caron, 1998; Swanson, Kinsbourne, Nigg, Lanphear, Stefanatos, et al., 2007) and is found predominantly in the frontal cortex (Diamond, 2007; Meador-Woodruff et al., 1996). The D4-7R receptor has also been associated with novelty-seeking temperament and with addictive conducts, both conditions frequently coexisting with ADHD (Ding et al., 2002). This allele has a relatively recent origin (about 40,000 years ago), and it has been suggested that it was subject to intense selection favoring migratory behavior in the period in which Europe and America were subject to human colonization (Ding et al., 2002).

There are other candidate polymorphisms for ADHD, such as alleles of the D2 and D5 dopaminergic receptors, the gene coding for COMT, and the DA transporter DAT1 (Carrasco et al., 2006; Comings et al., 1996; Hawi et al., 2003; Kuntsi & Stevenson, 2000), which is found in high concentrations in the corpus striatum but in low concentrations in the PFC (Diamond, 2007). Serotoninergic- and noradrenergic-related genes have also been proposed (Quist et al., 2003; Smith et al., 2003).

Considering the model of DA action by Seamans & Yang (2004), in ADHD there might be a disbalance in favor of the so-called D2 state, where tonic DA predominates and multiple representations may be activated at the expense of focused activity as in working memory tasks ("D1 state," see Fig. 11.2). Medication with stimulants might then work by boosting phasic DA release and stabilizing inhibitory mechanisms, while potentiating only a few circuits of working memory. Nevertheless, other evidence suggests that like in schizophrenia, impulsive behavior (a characteristic symptom in some ADHD patients) is caused by an abnormally large phasic DA response (Grace, 1991). It is thus possible that more than an increase in D2-like receptor activity, in ADHD there is a shifting disbalance between tonic and phasic signaling. DA neurons show a sustained increase in activity during the interval between the reward-predicting stimulus and the reward itself (both events associated with phasic release), whose synaptic effects are likely mediated by D2 receptors (see Chap. 10). It is likely that this increased tonic activity that occurs between phasic signals permits both motivation and the goal representation to be kept active while the reward is still to come. Deficits in reward-associative mechanisms, leading to delay aversion (in which there is an impairment in associating behavior with its consequences), have been considered to represent a core symptom of ADHD (Sagvolden, Aase, Zeiner & Berger, 1998), and might result from the disbalance between phasic (stimulus-related) activity and tonic activation during the interval between the predicting stimulus and the occurrence of reward. Thus, some ADHD subjects might be characterized by the presence of highly motivated, but very short-lasting states produced by phasic DA surges that are unable to maintain the tonic firing increase between stimulus and reward. This might be due to a chronic, abnormally high tonic DA activity that masks the tonic increase associated with the reward-predicting stimulus. In ADHD there are two main symptomatic extremes, one being purely inattentive and the other predominance of hyperactivity and impulsivity. I suggest that this variability reflects different emphases on tonic versus phasic signaling across subjects.

Furthermore, the interpretation of an increased "D2 state" in ADHD is in apparent discordance with findings indicating that the D4-7R allele, linked to ADHD, codifies a hypofunctional receptor (Missale et al., 1998). A malfunctioning D4 receptor would weaken the action of the "D2 state," thus favoring general inhibition and concentration. One possibility is that in subjects with the D4-7R allele, there are compensatory mechanisms generating a higher-than normal tonic transmission and an exacerbation of the "D2 state." Alternatively, recent findings indicate that despite being associated with ADHD, the D4-7R allele may confer protection to ADHD subjects in terms of cognitive performance and brain structure (Swanson et al., 2007). This observation might be in agreement with a weakening of tonic DA signaling in relation to a hypofunctional D4-7R receptor. Further studies are needed to further evaluate these possibilities.

Concluding Statement

DA is a strong behavioral and cognitive modulator, on whose appropriate dynamic balance the integrity of goal-directed behavior depends. DA acts at different levels, motor activation, motivation, and behavioral organization to achieve goals in the mid or long term. The participation of this neurotransmitter (and perhaps the contribution of noradrenergic systems) in this complex multilevel behavioral sequence is possibly a result of its evolutionary history and its early origins as a reward-related signal (see Chap. 10). Dysfunctions in this system lead to highly disturbing behavioral, emotional, and cognitive deficits underlying several forms of neuropsychiatric diseases. Cognitive functions, mediated by complex cortical dynamics, appear to be subsidiary elements of complex goal-directed behavior, especially as organisms evolve increasingly subtler mechanisms to associate rewards with predicting stimuli. In this context, cognitive dysfunction in neuropsychiatric disease might be better conceived as one aspect of a more general deficit in the organization of goal-directed behavior.

Acknowledgements Research referred to in this work was partly funded by FONDECYT grant no. 1050721 and by the Millennium Center for Integrative Neuroscience. I also thank Ximena Carrasco for helpful comments.

References

Abi-Dargham, A., Rodenhiser, J., Printz, D., Zea-Ponce, Y., Gil, R., Kegeles, L. S., et al. (2000). Increased baseline occupancy of D2 receptors by dopamine in schizophrenia. *Proceedings of the National Academy of Sciences of the United States of America, 97*(14), 8104–8109.

Aboitiz, F., López, V., López-Calderón, J., & Carrasco, X. (2006). Beyond endophenotypes: An interdisciplinary approach to attentional deficit-hyperactivity disorder. In M. Vanchevsky (Ed.), *Focus in cognitive psychology research* (pp. 183–205). New York: Nova Science Publishers.

Akil, M., Pierri, J. N., Whitehead, R. E., Edgar, C. L., Mohila, C., Sampson, A. R., et al. (1999). Lamina-specific alterations in the dopamine innervation of the prefrontal cortex in schizophrenic subjects. *Ameican Journal of Psychiatry, 156*(10), 1580–1589.

Angrist, B., & Sudilovsky, A. (1978). Central nervous system stimulants: Historical aspects and clinical effects. In L. L. Iversen (Ed.), *Handbook of psychopharmacology* (Vol. 11, pp. 99–165). New York: Plenum Press.

Arnsten, A. F. T. (2007). Catecholamine and second messenger influences on prefrontal cortical networks of "representational knowledge": A rational bridge between genetics and the symptoms of mental illness. *Cerebral Cortex, 17*, i6–i15.

Barr, C. L., Wigg, K. G., Bloom, S., Schachar, R., Tannock, R., Roberts, W., et al. (2000). Further evidence from haplotype analysis for linkage of the dopamine D4 receptor gene and attention-deficit hyperactivity disorder. *American Journal of Medical Genetics, 96*(3), 262–267.

Bergson, C., Mrzljak, L., Smiley, J. F., Pappy, M., Levenson, R., & Goldman-Rakic, P. S. (1995). Regional, cellular, and subcellular variations in the distribution of D1 and D5 dopamine receptors in primate brain. *Journal of Neuroscience, 15*(12), 7821–7836.

Biederman, J. (2005). Attention-deficit/hyperactivity disorder: A selective overview. *Biological Psychiatry, 57*(11), 1215–1220.

Birkmayer, W., & Hornykiewicz, O. (1961). [The L-3,4-dioxyphenylalanine (DOPA)-effect in Parkinson-akinesia.]. *Wiener Klinische Wochenschrift, 73*, 787–788.

Bjorklund, A., & Dunnett, S. B. (2007). Dopamine neuron systems in the brain: An update. *Trends in Neuroscience, 30*(5), 194–202.

Bratcher, N. A., Farmer-Dougan, V., Dougan, J. D., Heidenreich, B. A., & Garris, P. A. (2005). The role of dopamine in reinforcement: Changes in reinforcement sensitivity induced by D1-type, D2-type, and nonselective dopamine receptor agonists. *Journal of Experimental Analysis of Behavior, 84*(3), 371–399.

Brozoski, T. J., Brown, R. M., Rosvold, H. E., & Goldman, P. S. (1979). Cognitive deficit caused by regional depletion of dopamine in prefrontal cortex of rhesus monkey. *Science, 205*(4409), 929–932.

Carlsson, A., & Lindqvist, M. (1963). Effect of chlorpromazine or haloperidol on formation of 3methoxytyramine and normetanephrine in mouse brain. *Acta pharmacologica et toxicologica, 20,* 140–144.

Carrasco, X., Rothhammer, P., Moraga, M., Henriquez, H., Chakraborty, R., Aboitiz, F., et al. (2006). Genotypic interaction between DRD4 and DAT1 loci is a high risk factor for attention-deficit/hyperactivity disorder in Chilean families. *American Journal of Medical Genetics. Part B, Neuropsychiatric Genetics , 141*(1), 51–54.

Castellanos, F. X., Giedd, J. N., Elia, J., Marsh, W. L., Ritchie, G. F., Hamburger, S. D., et al. (1997). Controlled stimulant treatment of ADHD and comorbid Tourette's syndrome: Effects of stimulant and dose. *Journal of the American Academy of Child & Adolescent Psychiatry, 36*(5), 589–596.

Comings, D. E., Wu, S., Chiu, C., Ring, R. H., Gade, R., Ahn, C., et al. (1996). Polygenic inheritance of Tourette syndrome, stuttering, attention deficit hyperactivity, conduct, and oppositional defiant disorder: The additive and subtractive effect of the three dopaminergic genes–DRD2, D beta H, and DAT1. *American Journal of Medical Genetics, 67*(3), 264–288.

Cools, R., Lewis, S. J., Clark, L., Barker, R. A., & Robbins, T. W. (2007). L-DOPA disrupts activity in the nucleus accumbens during reversal learning in Parkinson's disease. *Neuropsychopharmacology, 32*(1), 180–189.

Crofts, H. S., Dalley, J. W., Collins, P., Van Denderen, J. C., Everitt, B. J., Robbins, T. W., et al. (2001). Differential effects of 6-OHDA lesions of the frontal cortex and caudate nucleus on the ability to acquire an attentional set. *Cerebral Cortex, 11*(11), 1015–1026.

Childress, A. R., Mozley, P. D., McElgin, W., Fitzgerald, J., Reivich, M., & O'Brien, C. P. (1999). Limbic activation during cue-induced cocaine craving. *Ameican Journal of Psychiatry, 156*(1), 11–18.

Chudasama, Y., & Robbins, T. W. (2004). Dopaminergic modulation of visual attention and working memory in the rodent prefrontal cortex. *Neuropsychopharmacology, 29*(9), 1628–1636.

DeLong, M. R. (1983). The neurophysiologic basis of abnormal movements in basal ganglia disorders. *Neurobehavioral Toxicology and Teratology, 5*(6), 611–616.

Deng, P., Zhang, Y., & Xu, Z. C. (2007). Involvement of I(h) in dopamine modulation of tonic firing in striatal cholinergic interneurons. *Journal of Neuroscience, 27*(12), 3148–3156.

Deutch, A. Y., Clark, W. A., & Roth, R. H. (1990). Prefrontal cortical dopamine depletion enhances the responsiveness of mesolimbic dopamine neurons to stress. *Brain Research, 521*(1–2), 311–315.

Diamond, A. (2007). Consequences of variations in genes that affect dopamine in prefrontal cortex. *Cerebral Cortex, 17,* i161–i170.

Ding, Y. C., Chi, H. C., Grady, D. L., Morishima, A., Kidd, J. R., Kidd, K. K., et al. (2002). Evidence of positive selection acting at the human dopamine receptor D4 gene locus. *Proceedings of the National Academy of Sciences of the United States of America, 99*(1), 309–314.

Ehringer, H., & Hornykiewicz, O. (1960). [Distribution of noradrenaline and dopamine (3-hydroxytyramine) in the human brain and their behavior in diseases of the extrapyramidal system.]. *Klinische Wochenschrift, 38,* 1236–1239.

Floresco, S. B., & Phillips, A. G. (2001). Delay-dependent modulation of memory retrieval by infusion of a dopamine D1 agonist into the rat medial prefrontal cortex. *Behavioral Neuroscience, 115*(4), 934–939.

Floresco, S. B., Todd, C. L., & Grace, A. A. (2001). Glutamatergic afferents from the hippocampus to the nucleus accumbens regulate activity of ventral tegmental area dopamine neurons. *Journal of Neuroscience, 21*(13), 4915–4922.

Floresco, S. B., & Tse, M. T. (2007). Dopaminergic regulation of inhibitory and excitatory transmission in the basolateral amygdala-prefrontal cortical pathway. *Journal of Neuroscience, 27*(8), 2045–2057.

Foote, S. L., Aston-Jones, G., & Bloom, F. E. (1980). Impulse activity of locus coeruleus neurons in awake rats and monkeys is a function of sensory stimulation and arousal. *Proceedings of the National Academy of Sciences of the United States of America, 77*(5), 3033–3037.

French, S. J., & Totterdell, S. (2002). Hippocampal and prefrontal cortical inputs monosynaptically converge with individual projection neurons of the nucleus accumbens. *Journal of Comparative Neurology, 446*(2), 151–165.

French, S. J., & Totterdell, S. (2003). Individual nucleus accumbens-projection neurons receive both basolateral amygdala and ventral subicular afferents in rats. *Neuroscience, 119*(1), 19–31.

Freund, T. F. (2003). Interneuron Diversity series: Rhythm and mood in perisomatic inhibition. *Trends in Neuroscience, 26*(9), 489–495.

Fuster, J. M. (1995). *Memory in the cerebral cortex*. Cambridge, MA: MIT Press.

Gao, W. J., Krimer, L. S., & Goldman-Rakic, P. S. (2001). Presynaptic regulation of recurrent excitation by D1 receptors in prefrontal circuits. *Proceedings of the National Academy of Sciences of the United States of America, 98*(1), 295–300.

Gao, W. J., Wang, Y., & Goldman-Rakic, P. S. (2003). Dopamine modulation of perisomatic and peridendritic inhibition in prefrontal cortex. *Journal of Neuroscience, 23*(5), 1622–1630.

Gaspar, P., Bloch, B., & Le Moine, C. (1995). D1 and D2 receptor gene expression in the rat frontal cortex: Cellular localization in different classes of efferent neurons. *European Journal of Neuroscience, 7*(5), 1050–1063.

Goldberg, T. E., Egan, M. F., Gscheidle, T., Coppola, R., Weickert, T., Kolachana, B. S., et al. (2003). Executive subprocesses in working memory: Relationship to catechol-O-methyltransferase Val158Met genotype and schizophrenia. *Archives of General Psychiatry, 60*(9), 889–896.

Goldman-Rakic, P. S. (1995a). Anatomical and functional circuits in prefrontal cortex of nonhuman primates. Relevance to epilepsy. *Advances in Neurology, 66*, 51–63; discussion 63–55.

Goldman-Rakic, P. S. (1995b). Architecture of the prefrontal cortex and the central executive. *Annals of the New York Academy of Sciences, 769*, 71–83.

Goldman-Rakic, P. S. (1996). Regional and cellular fractionation of working memory. *Proceedings of the National Academy of Sciences of the United States of America, 93*(24), 13473–13480.

Goldman-Rakic, P. S., Lidow, M. S., Smiley, J. F., & Williams, M. S. (1992). The anatomy of dopamine in monkey and human prefrontal cortex. *Journal of Neural Transmission Suppl, 36*, 163–177.

Gorelova, N., Seamans, J. K., & Yang, C. R. (2002). Mechanisms of dopamine activation of fast-spiking interneurons that exert inhibition in rat prefrontal cortex. *Journal of Neurophysiology, 88*(6), 3150–3166.

Gorelova, N. A., & Yang, C. R. (2000). Dopamine D1/D5 receptor activation modulates a persistent sodium current in rat prefrontal cortical neurons in vitro. *Journal of Neurophysiology, 84*(1), 75–87.

Gosal, D., Ross, O. A., & Toft, M. (2006). Parkinson's disease: The genetics of a heterogeneous disorder. *European Journal of Neurology, 13*(6), 616–627.

Goto, Y., & Grace, A. A. (2005a). Dopamine-dependent interactions between limbic and prefrontal cortical plasticity in the nucleus accumbens: Disruption by cocaine sensitization. *Neuron, 47*(2), 255–266.

Goto, Y., & Grace, A. A. (2005b). Dopaminergic modulation of limbic and cortical drive of nucleus accumbens in goal-directed behavior. *Nature Neuroscience, 8*(6), 805–812.

Grace, A. A. (1991). Phasic versus tonic dopamine release and the modulation of dopamine system responsivity: A hypothesis for the etiology of schizophrenia. *Neuroscience, 41*(1), 1–24.

Grace, A. A. (1995). The tonic/phasic model of dopamine system regulation: Its relevance for understanding how stimulant abuse can alter basal ganglia function. *Drug and Alcohol Dependence, 37*(2), 111–129.

Grace, A. A., Floresco, S. B., Goto, Y., & Lodge, D. J. (2007). Regulation of firing of dopaminergic neurons and control of goal-directed behaviors. *Trends in Neuroscience, 30*(5), 220–227.

Granon, S., Passetti, F., Thomas, K. L., Dalley, J. W., Everitt, B. J., & Robbins, T. W. (2000). Enhanced and impaired attentional performance after infusion of D1 dopaminergic receptor agents into rat prefrontal cortex. *Journal of Neuroscience, 20*(3), 1208–1215.

Hawi, Z., Lowe, N., Kirley, A., Gruenhage, F., Nothen, M., Greenwood, T., et al. (2003). Linkage disequilibrium mapping at DAT1, DRD5 and DBH narrows the search for ADHD susceptibility alleles at these loci. *Molecular Psychiatry, 8*(3), 299–308.

Henze, D. A., Gonzalez-Burgos, G. R., Urban, N. N., Lewis, D. A., & Barrionuevo, G. (2000). Dopamine increases excitability of pyramidal neurons in primate prefrontal cortex. *Journal of Neurophysiology, 84*(6), 2799–2809.

Holmes, J., Payton, A., Barrett, J., Harrington, R., McGuffin, P., Owen, M., et al. (2002). Association of DRD4 in children with ADHD and comorbid conduct problems. *American Journal of Medical Genetics, 114*(2), 150–153.

Ishizuka, K., Paek, M., Kamiya, A., & Sawa, A. (2006). A review of Disrupted-In-Schizophrenia-1 (DISC1): Neurodevelopment, cognition, and mental conditions. *Biological Psychiatry, 59*(12), 1189–1197.

Iversen, L. (2006). Neurotransmitter transporters and their impact on the development of psychopharmacology. *British Journal of Pharmacology, 147*(Suppl 1), S82–88.

Iversen, S. D., & Iversen, L. L. (2007). Dopamine: 50 years in perspective. *Trends in Neuroscience, 30*(5), 188–193.

Khan, Z. U., Gutierrez, A., Martin, R., Penafiel, A., Rivera, A., & De La Calle, A. (1998). Differential regional and cellular distribution of dopamine D2-like receptors: An immunocytochemical study of subtype-specific antibodies in rat and human brain. *Journal of Comparative Neurology, 402*(3), 353–371.

Krimer, L. S., Jakab, R. L., & Goldman-Rakic, P. S. (1997). Quantitative three-dimensional analysis of the catecholaminergic innervation of identified neurons in the macaque prefrontal cortex. *Journal of Neuroscience, 17*(19), 7450–7461.

Kritzer, M. F., & Goldman-Rakic, P. S. (1995). Intrinsic circuit organization of the major layers and sublayers of the dorsolateral prefrontal cortex in the rhesus monkey. *Journal of Comparative Neurology, 359*(1), 131–143.

Kroner, S., Krimer, L. S., Lewis, D. A., & Barrionuevo, G. (2007). Dopamine increases inhibition in the monkey dorsolateral prefrontal cortex through cell type-specific modulation of interneurons. *Cerebral Cortex, 17*(5), 1020–1032.

Kuntsi, J., & Stevenson, J. (2000). Hyperactivity in children: A focus on genetic research and psychological theories. *Clinical Child Psychology and Psychiatry, 3*(1), 1–23.

Lange, K. W., Robbins, T. W., Mardsen, C. D., James, M., Owen, A. M., & Paul, G. M. (1992). L-Dopa withdrawal in Parkinson's disease selectively impairs cognitive performance in tests sensitive to frontal lobe dysfunction. *Psychopharmacology, 107*(2–3), 394–404.

Lidow, M. S., Wang, F., Cao, Y., & Goldman-Rakic, P. S. (1998). Layer V neurons bear the majority of mRNAs encoding the five distinct dopamine receptor subtypes in the primate prefrontal cortex. *Synapse, 28*(1), 10–20.

Lisman, J. E., & Grace, A. A. (2005). The hippocampal-VTA loop: Controlling the entry of information into long-term memory. *Neuron, 46*(5), 703–713.

Lisman, J. E., Coyle, J. T., Green, R. W., Javitt, D. C., Benes, F. M., Heckers, S., & Grace, A. A. (2008). Circuit-based framework for understanding neurotransmitter and risk gene interactions in schizophrenia. *Trends in Neuroscience, 31*(5), 234–242.

Luthi, A., & McCormick, D. A. (1998). H-current: Properties of a neuronal and network pacemaker. *Neuron, 21*(1), 9–12.

Manor, I., Tyano, S., Eisenberg, J., Bachner-Melman, R., Kotler, M., & Ebstein, R. P. (2002). The short DRD4 repeats confer risk to attention deficit hyperactivity disorder in a family-based design and impair performance on a continuous performance test (TOVA). *Molecular Psychiatry, 7*(7), 790–794.

Matsuda, Y., Marzo, A., & Otani, S. (2006). The presence of background dopamine signal converts long-term synaptic depression to potentiation in rat prefrontal cortex. *Journal of Neuroscience, 26*(18), 4803–4810.

Mattay, V. S., Goldberg, T. E., Fera, F., Hariri, A. R., Tessitore, A., Egan, M. F., et al. (2003). Catechol O-methyltransferase val158-met genotype and individual variation in the brain response to amphetamine. *Proceedings of the National Academy of Sciences of the United States of America, 100*(10), 6186–6191.

Matthysse, S. (1973). Antipsychotic drug actions: A clue to the neuropathology of schizophrenia? *Federation Proceedings, 32*(2), 200–205.

Meador-Woodruff, J. H., Damask, S. P., Wang, J., Haroutunian, V., Davis, K. L., & Watson, S. J. (1996). Dopamine receptor mRNA expression in human striatum and neocortex. *Neuropsychopharmacology, 15*(1), 17–29.

Mehta, M. A., Owen, A. M., Sahakian, B. J., Mavaddat, N., Pickard, J. D., & Robbins, T. W. (2000). Methylphenidate enhances working memory by modulating discrete frontal and parietal lobe regions in the human brain. *Journal of Neuroscience, 20*(6), RC65.

Missale, C., Nash, S. R., Robinson, S. W., Jaber, M., & Caron, M. G. (1998). Dopamine receptors: From structure to function. *Physiological Reviews, 78*(1), 189–225.

Moore, K. E., & Gudelsky, G. A. (1977). Drug actions on dopamine turnover in the median eminence. *Advances in Biochemical Psychopharmacology, 16*, 227–235.

Mrzljak, L., Bergson, C., Pappy, M., Huff, R., Levenson, R., & Goldman-Rakic, P. S. (1996). Localization of dopamine D4 receptors in GABAergic neurons of the primate brain. *Nature, 381*(6579), 245–248.

Muglia, P., Jain, U., Macciardi, F., & Kennedy, J. L. (2000). Adult attention deficit hyperactivity disorder and the dopamine D4 receptor gene. *American Journal of Medical Genetics, 96*(3), 273–277.

Muly, E. C., 3rd, Szigeti, K., & Goldman-Rakic, P. S. (1998). D1 receptor in interneurons of macaque prefrontal cortex: Distribution and subcellular localization. *Journal of Neuroscience, 18*(24), 10553–10565.

Niemegeers, C. J., & Janssen, P. A. (1965). A comparative study of the inhibitory effects of haloperidol and trifluperidol on learned shock-avoidance behavioural habits and on apomorphine-induced emesis in mongrel dogs and in beagles. *Psychopharmacologia, 8*(4), 263–270.

Niv, Y., Daw, N. D., Joel, D., & Dayan, P. (2007). Tonic dopamine: Opportunity costs and the control of response vigor. *Psychopharmacology (Berl), 191*(3), 507–520.

Nolan, K. A., Bilder, R. M., Lachman, H. M., & Volavka, J. (2004). Catechol O-methyltransferase Val158Met polymorphism in schizophrenia: Differential effects of Val and Met alleles on cognitive stability and flexibility. *Ameican Journal of Psychiatry, 161*(2), 359–361.

O'Donnell, P., & Grace, A. A. (1994). Tonic D2-mediated attenuation of cortical excitation in nucleus accumbens neurons recorded in vitro. *Brain Research, 634*(1), 105–112.

Pascucci, T., Ventura, R., Latagliata, E. C., Cabib, S., & Puglisi-Allegra, S. (2007). The medial prefrontal cortex determines the accumbens dopamine response to stress through the opposing influences of norepinephrine and dopamine. *Cerebral Cortex, 17*, 2796–2804.

Paspalas, C. D., & Goldman-Rakic, P. S. (2005). Presynaptic D1 dopamine receptors in primate prefrontal cortex: Target-specific expression in the glutamatergic synapse. *Journal of Neuroscience, 25*(5), 1260–1267.

Patil, S. T., Zhang, L., Martenyi, F., Lowe, S. L., Jackson, K. A., Andreev, B. V., Avedisova, A. S., Bardenstein, L. M., Gurovich, I. Y., Morozova, M. A., Mosolov, S. N., Neznanov, N. G., Reznik, A. M., Smulevich, A. B., Tochilov, V. A., Johnson, B. G., Monn, J. A., & Schoepp, D. A. (2007). Activation of mGlu2/3 receptors as a new approach to treat schizophrenia: A randomized Phase 2 clinical trial. *Nature Medicine, 13*(9), 1102–1107.

Pfeiffer, U. J., & Fendt, M. (2006). Prefrontal dopamine D4 receptors are involved in encoding fear extinction. *Neuroreport, 17*(8), 847–850.

Quist, J. F., Barr, C. L., Schachar, R., Roberts, W., Malone, M., Tannock, R., et al. (2003). The serotonin 5-HT1B receptor gene and attention deficit hyperactivity disorder. *Molecular Psychiatry, 8*(1), 98–102.

Redgrave, P., & Gurney, K. (2006). The short-latency dopamine signal: A role in discovering novel actions? *Nature Reviews Neuroscience, 7*(12), 967–975.

Robbins, T., & Roberts, A. (2007). Differential regulation of fronto-executive function by the monoamines and acetylcholine. *Cerebral Cortex, 17*, i151–i160.

Robbins, T. W. (2005). Chemistry of the mind: Neurochemical modulation of prefrontal cortical function. *Journal of Comprative Neurology, 493*(1), 140–146.

Sagvolden, T., Aase, H., Zeiner, P., & Berger, D. F. (1998). Altered reinforcement mechanisms in Attention-Deficit/Hyperactivity Disorder. *Behavioral Brain Research, 94*, 61–71.

Sawaguchi, T., & Goldman-Rakic, P. S. (1991). D1 dopamine receptors in prefrontal cortex: Involvement in working memory. *Science, 251*(4996), 947–950.

Sawaguchi, T., Matsumura, M., & Kubota, K. (1988). Dopamine enhances the neuronal activity of spatial short-term memory task in the primate prefrontal cortex. *Neuroscience Research, 5*(5), 465–473.

Sawaguchi, T., Matsumura, M., & Kubota, K. (1990). Effects of dopamine antagonists on neuronal activity related to a delayed response task in monkey prefrontal cortex. *Journal of Neurophysiology, 63*(6), 1401–1412.

Schultz, W. (1998). Predictive reward signal of dopamine neurons. *Journal of Neurophysiology, 80*(1), 1–27.

Schultz, W. (2002). Getting formal with dopamine and reward. *Neuron, 36*(2), 241–263.

Schultz, W., Apicella, P., Ljungberg, T., Romo, R., & Scarnati, E. (1993). Reward-related activity in the monkey striatum and substantia nigra. *Progress Brain Research, 99*, 227–235.

Schwartz, M. L., & Goldman-Rakic, P. S. (1984). Callosal and intrahemispheric connectivity of the prefrontal association cortex in rhesus monkey: Relation between intraparietal and principal sulcal cortex. *Journal of Comparative Neurology, 226*(3), 403–420.

Seamans, J. K., Durstewitz, D., Christie, B. R., Stevens, C. F., & Sejnowski, T. J. (2001). Dopamine D1/D5 receptor modulation of excitatory synaptic inputs to layer V prefrontal cortex neurons. *Proceedings of the National Academy of Sciences of the United States of America, 98*(1), 301–306.

Seamans, J. K., Gorelova, N., Durstewitz, D., & Yang, C. R. (2001). Bidirectional dopamine modulation of GABAergic inhibition in prefrontal cortical pyramidal neurons. *Journal of Neuroscience, 21*(10), 3628–3638.

Seamans, J. K., & Yang, C. R. (2004). The principal features and mechanisms of dopamine modulation in the prefrontal cortex. *Progress in Neurobiology, 74*(1), 1–58.

Seeman, P., Weinshenker, D., Quirion, R., Srivastava, L. K., Bhardwaj, S. K., Grandy, D. K., et al. (2005). Dopamine supersensitivity correlates with D2High states, implying many paths to psychosis. *Proceedings of the National Academy of Sciences of the United States of America, 102*(9), 3513–3518.

Sesack, S. R., Bressler, C. N., & Lewis, D. A. (1995). Ultrastructural associations between dopamine terminals and local circuit neurons in the monkey prefrontal cortex: A study of calretinin-immunoreactive cells. *Neuroscience Letters, 200*(1), 9–12.

Sesack, S. R., Hawrylak, V. A., Guido, M. A., & Levey, A. I. (1998). Cellular and subcellular localization of the dopamine transporter in rat cortex. *Advances in Pharmacology, 42*, 171–174.

Siegle, G. J., Thompson, W., Carter, C. S., Steinhauer, S. R., & Thase, M. E. (2007). Increased amygdala and decreased dorsolateral prefrontal BOLD responses in unipolar depression: Related and independent features. *Biological Psychiatry, 61*(2), 198–209.

Smith, K. M., Daly, M., Fischer, M., Yiannoutsos, C. T., Bauer, L., Barkley, R., et al. (2003). Association of the dopamine beta hydroxylase gene with attention deficit hyperactivity disorder: Genetic analysis of the Milwaukee longitudinal study. *American Journal of Medical Genetics. Part B, Neuropsychiatric Genetics, 119*(1), 77–85.

Snyder, S. H. (1974). Proceedings: Drugs, neurotransmitters, and psychosis. *Psychopharmacology Bulletin, 10*(4), 4–5.

Sokolov, B. P. (1998). Expression of NMDAR1, GluR1, GluR7, and KA1 glutamate receptor mRNAs is decreased in frontal cortex of "neuroleptic-free" schizophrenics: Evidence on reversible up-regulation by typical neuroleptics. *Jouranl of Neurochemistry, 71*(6), 2454–2464.

Swanson, J. M., Kinsbourne, M., Nigg, J., Lanphear, B., Stefanatos, G. A., Volkow, N., et al. (2007). Etiologic subtypes of attention-deficit/hyperactivity disorder: Brain imaging, molecular genetic and environmental factors and the dopamine hypothesis. *Neuropsychology Review, 17*(1), 39–59.

Taylor, E. (1998). Clinical foundations of hyperactivity research. *Behavioral Brain Research, 94*(1), 11–24.

Trantham-Davidson, H., Neely, L. C., Lavin, A., & Seamans, J. K. (2004). Mechanisms underlying differential D1 versus D2 dopamine receptor regulation of inhibition in prefrontal cortex. *Journal of Neuroscience, 24*(47), 10652–10659.

Tseng, K. Y., & O'Donnell, P. (2004). Dopamine-glutamate interactions controlling prefrontal cortical pyramidal cell excitability involve multiple signaling mechanisms. *Journal of Neuroscience, 24*(22), 5131–5139.

Tunbridge, E. M., Harrison, P. J., & Weinberger, D. R. (2006). Catechol-o-methyltransferase, cognition, and psychosis: Val158Met and beyond. *Biological Psychiatry, 60*(2), 141–151.

van Rossum, J. M. (1966). The significance of dopamine-receptor blockade for the mechanism of action of neuroleptic drugs. *Archives Internationales de Pharmacodynamie et de Thérapie, 160*(2), 492–494.

Verney, C., Alvarez, C., Geffard, M., & Berger, B. (1990). Ultrastructural double-labelling study of dopamine terminals and GABA-containing neurons in rat anteromedial cerebral cortex. *European Journal of Neuroscience, 2*(11), 960–972.

Vijayraghavan, S., Wang, M., Birnbaum, S. G., Williams, G. V., & Arnsten, A. F. (2007). Inverted-U dopamine D1 receptor actions on prefrontal neurons engaged in working memory. *Nature Neuroscience, 10*(3), 376–384.

Vincent, S. L., Pabreza, L., & Benes, F. M. (1995). Postnatal maturation of GABA-immunoreactive neurons of rat medial prefrontal cortex. *Journal of Comparative Neurology, 355*(1), 81–92.

Vysokanov, A., Flores-Hernandez, J., & Surmeier, D. J. (1998). mRNAs for clozapine-sensitive receptors co-localize in rat prefrontal cortex neurons. *Neuroscience Letters, 258*(3), 179–182.

Wang, J., & O'Donnell, P. (2001). D(1) dopamine receptors potentiate nmda-mediated excitability increase in layer V prefrontal cortical pyramidal neurons. *Cerebral Cortex, 11*(5), 452–462.

Wang, M., Ramos, B. P., Paspalas, C. D., Shu, Y., Simen, A., Duque, A., et al. (2007). Alpha2A-adrenoceptors strengthen working memory networks by inhibiting cAMP-HCN channel signaling in prefrontal cortex. *Cell, 129*(2), 397–410.

Wang, X. J., Tegnér, J., Constantinidis, C., & Goldman-Rakic, P. S. (2004). Division of labor among distinct subtypes of inhibitory neurons in a cortical microcircuit of working memory. *Proceedings of the National Academy of Sciences of the United States of America, 101*(5), 1368–1373.

Wang, M., Vijayraghavan, S., & Goldman-Rakic, P. S. (2004). Selective D2 receptor actions on the functional circuitry of working memory. *Science, 303*(5659), 853–856.

Wedzony, K., Chocyk, A., Mackowiak, M., Fijal, K., & Czyrak, A. (2000). Cortical localization of dopamine D4 receptors in the rat brain–immunocytochemical study. *Journal of Physiology and Pharmacology, 51*(2), 205–221.

West, A. R., & Grace, A. A. (2002). Opposite influences of endogenous dopamine D1 and D2 receptor activation on activity states and electrophysiological properties of striatal neurons: Studies combining in vivo intracellular recordings and reverse microdialysis. *Journal of Neuroscience, 22*(1), 294–304.

Williams, G. V., & Goldman-Rakic, P. S. (1995). Modulation of memory fields by dopamine D1 receptors in prefrontal cortex. *Nature, 376*(6541), 572–575.

Williams, S. M., & Goldman-Rakic, P. S. (1993). Characterization of the dopaminergic innervation of the primate frontal cortex using a dopamine-specific antibody. *Cerebral Cortex, 3*(3), 199–222.

Yan, Z., & Surmeier, D. J. (1997). D5 dopamine receptors enhance Zn2+-sensitive GABA(A) currents in striatal cholinergic interneurons through a PKA/PP1 cascade. *Neuron, 19*(5), 1115–1126.

Yanez, I. B., Munoz, A., Contreras, J., Gonzalez, J., Rodriguez-Veiga, E., & DeFelipe, J. (2005). Double bouquet cell in the human cerebral cortex and a comparison with other mammals. *Journal of Comparative Neurology, 486*(4), 344–360.

Yang, C. R., Seamans, J. K., & Gorelova, N. (1996). Electrophysiological and morphological properties of layers V-VI principal pyramidal cells in rat prefrontal cortex in vitro. *Journal of Neuroscience, 16*(5), 1904–1921.

Yavich, L., Forsberg, M. M., Karayiorgou, M., Gogos, J. A., & Männistö, P. T. (2007). Site-specific role of catechol-O-methyltransferase in dopamine overflow within prefrontal cortex and dorsal striatum. *Journal of Neuroscience, 27*(38), 10196–10209.

Zhou, F. M., & Hablitz, J. J. (1999). Dopamine modulation of membrane and synaptic properties of interneurons in rat cerebral cortex. *Journal of Neurophysiology, 81*(3), 967–976.

Part III
Clinical and Developmental Issues

Chapter 12
Prefrontal Cortex and Control of Behavior – Evidence from Neuropsychological Studies

A. Slachevsky, P. Reyes, G. Rojas, and J.R. Silva

Abstract "Few subjects in neurology have been associated with as much and paradox as the behavioral afflictions of the prefrontal cortex" (Marsel Mesulam, 2002).

The case of Phineas Gage, the first well-described patient with a prefrontal lesion, revealed that this sort of lesion could cause severe trouble in everyday life and a profound disturbance of personality. Disturbances of patients with prefrontal lesions are a consequence of disruptions in the main role of the prefrontal cortex: executive control, i.e., a cognitive function underlying the human faculty to act or think not only in reaction to external events but also to internal goals and states. The neuropsychological study of patients with prefrontal lesions suggests the existence of three dimensions of executive control subserved by different prefrontal regions: emotional, motivational, and cognitive. Indeed, damage to the prefrontal cortex may lead to a set of symptoms collectively known as "dysexecutive syndrome," characterized by changes in those dimensions. In this chapter, we discuss how the neuropsychological method, i.e., the study of brain–behavior relationship in patients with brain lesions, contributes to a better understanding of the role of the prefrontal cortex in behavior control.

Introduction – Anatomy

Knowledge of neuroanatomy is necessary to the understanding of clinical disorders associated with frontal lobe damage and of frontal lobe functions. For this reason, we will present a brief description of frontal lobe anatomy (Damasio & Anderson, 2003). The frontal lobe is the largest sector of the telencephalon and the one with

A. Slachevsky

Instituto de Ciencias Biomédicas, Facultad de Medicina, Universidad de Chile,
Avda. Independencia 1027, Santiago, Chile, e-mail: aslachevsky@adsl.tie.cl

F. Aboitiz and D. Cosmelli (eds.) *From Attention to Goal-Directed Behavior.*
© Springer-Verlag Berlin Heidelberg 2009

the clearest borders. Traditionally, we consider three main regions in the frontal lobe: the dorsolateral, the mesial, and the inferior or orbital sectors. Another way to divide the frontal lobe is to describe (Damasio, 1996):

1. A motor sector (mostly the precentral gyrus, Brodmann area 4)
2. A premotor sector (the most anterior segment of the inferior part half of the precentral gyrus, the posterior third of the superior and middle frontal gyri, and the pars opercularis of the inferior frontal gyrus, Brodmann areas 44, 6, and 8)
3. A prefrontal sector that encompasses all the remaining aspect of the frontal lobe (Brodmann areas 45, 46, 9, and 10 on the lateral surface, 24 and 32 on the mesial surface, and 12, 11, 13, 25, and 47 on the orbital sector)

The prefrontal cortex, the association cortex of the frontal lobe, is the cortex that receives projections from the mediodorsal nucleus of the thalamus (Rolls, 2002). Macroscopically, three major aspects may be distinguished: mesial, dorsolateral, and orbital (the major part of the orbital cortex is referred to as "ventromedial prefrontal cortex" and the remaining more lateral portion is referred to as "ventrolateral prefrontal cortex"). Several divisions of prefrontal cortex have been proposed. On the basis of divisions of the mediodorsal nucleus of the thalamus, the prefrontal cortex may be divided into three main regions:

1. The orbitofrontal or inferior (ventral) cortex: this region is situated in the orbital (ventral) surface of the prefrontal cortex and receives projections from the magnocellular, medial, part of the mediodorsal nucleus
2. The dorsolateral prefrontal cortex, which receives projections from the parvocellular, lateral, part of the mediodorsal nucleus
3. The frontal eye field, which receives projections from the pars paralamellaris (most lateral) part of the mediodorsal nucleus

On the basis of an evolutionary interpretation of cortical architectonics, Pandya and Yeterian (1996) have divided the prefrontal cortex in two main areas: the dorsolateral prefrontal cortex, which is part of the architectonic trend originating in the hippocampus; and the ventral prefrontal cortex, which is part of the paleocortical trend emerging from the caudal orbitofrontal (olfactory) cortex. Dorsolateral prefrontal cortex comprises Brodmann areas 9 and 46. Ventral prefrontal cortex comprises the inferior prefrontal convexity (Brodmann areas 45, 12, and 47) and the orbital surfaces (Brodmann areas 11, 13, and 14). Three functional areas have been described in the ventral prefrontal cortex:

(1) inferior (ventral) medial frontal regions;
(2) ventrolateral regions; and
(3) polar regions (Barbas, 1995, 2000; Bechara, Damasio, Tranel, & Anderson, 1998; Elliott, 2003).

The prefrontal cortex is a main cortical area receiving highly processed sensory information from all modalities. The diverse anatomical elements of the prefrontal cortex have distinct connections to other cortical and subcortical regions. With the exception of the connection between the prefrontal cortex and the basal ganglia, which receives unreciprocated connection from the prefrontal cortex, all prefrontal connections are reciprocal: structures sending fibers to the prefrontal cortex are the recipient of fibers from it. Different subareas of the prefrontal cortex have different connections (Fuster, 1997). The study of these connections is very helpful to understand the contribution of these subareas to cognition and behavior. Ventral prefrontal cortex is extensively interconnected with the hypothalamus, amygdala, hippocampus, and also with other paralimbic cortices in the temporal pole, insula, parahippocampal gyrus, and cingulate gyrus. The major connections of dorsolateral prefrontal cortex are with the other heteromodal and unimodal cortices in the parietal and temporal lobe as well as with orbitofrontal and related paralimbic areas, especially the cingulate gyrus. On the basis of the main connections of prefrontal cortex, we can differentiate a cognitive (dorsolateral) and an affective (ventral) prefrontal cortex.

Prefrontal cortex is situated in a critical position to establish a neural bridge between a stimulus and a response. Compared with other heteromodal cortices in the lateral temporal and posterior parietal lobes, the prefrontal heteromodal cortex (dorsolateral prefrontal cortex) receives connections not only from other heteromodal cortices processing sensory information from the outside, but also has important connections with paralimbic areas (orbitofrontal cortex and cingulate gyrus), a feature that may underlie its distinctive role in integrating extensively preprocessed sensory information with limbic and visceral states. In this way, prefrontal cortex establishes a bridge between a stimulus and a motor response and a bridge between the internal and external worlds, so the needs of the internal milieu can be discharged according to the opportunities and restrictions that prevail in the external milieu.

In the next section, we will review the main clinical symptoms associated with damage of prefrontal cortex. Finally, we will integrate neuroanatomical data with neuropsychological data to review current understanding of how prefrontal cortex controls behavior. This approach will allow us to illustrate how neuropsychology constitutes a window to the study of the function of the human brain

Clinical Aspects

Since the publication of the case of Phineas Gage by Harlow, several case reports have documented the existence of a plethora of behavior and cognitive symptoms following lesions in the prefrontal cortex (Mac Millian, 1996). Beside the existence of severe behavioral disorders, other case reports have illustrated that the main sequellae of prefrontal lesions concern the ability to plan and program a set of actions to attain a goal (Penfield & Evans, 1935). Although the literature tends to employ the term "frontal lobe syndrome" as if it is referred to a unitary entity, the examination of patients with prefrontal lesions reveals the existence of mixed patterns of troubles. The specific clinical picture in an individual patient is likely to be influenced by the lesion location, its progression rate, the onset age, and even the past personality of the patient (Mesulam, 2002). The main symptoms associated with prefrontal lesions can be divided into cognitive and behavioral disturbances.

Cognitive Dysfunction Associated with Prefrontal Damage

The study of performances of patients with prefrontal lesions in standard and experimental neuropsychology tests has revealed several cognitive disturbances. Characteristically, patients have disorders in planning, assessed by the Tower of London test, which require movement of a set of disks to a goal position, following certain rules that require planning a series of steps. In a seminal study, Penfield and Evans (1935) described planning difficulties of patients who had a right frontal glioma removed: "She had planned to get a simple supper for one guest and four members of her family. She looked forward to it with pleasure and had the whole day for preparation. When the appointed hour arrived she was in the kitchen, the food was all there, one or two things were on the stove, but the salad was not ready, the meat had not been started and she was distressed and confused by her long continued effort alone". Damasio and Anderson (2003) described that the ability to plan short- and long-term future behaviors often seems to be devastated. In these patients, modal response to questions regarding plans for the future is a recitation of the activities of the past several days. Consistent with this, their daily behavior often becomes a highly repetitive routine. Some patients may describe, generally in vague terms, long-term goals that are typically unrealistic or illogical. These verbalizations provide scant evidence of planning, and seem to have little or no impact on the guidance of behavior (Damasio & Anderson, 2003). In concordance with this observation, Slachevsky et al. (2003) have described that in an experimental paradigm generating a conflict between the action planned and the sensory-motor feedback, even though some patients with prefrontal lesions reported the correct strategy to solve the conflict, they had poor performance in motor adaptation. This observation suggested a knowing/doing dissociation, or a dissociation between strategy awareness and the failure to apply it. This result suggests that the knowing/doing dissociation (i.e., the failure of verbalization to guide behavior), could reflect a poorly established explicit knowledge of action (Slachevsky et al., 2003).

Other cognitive disorders include deficits in response initiation (assessed by figural and verbal fluency tests), response suppression (assessed by go/no-go, Stroop, and Wisconsin Card Sorting Test, WCST), conceptualization (assessed by the WCST), rule deduction (assessed by the Brixton test), generation of abstract principles (assessed by the California Card Sorting Test), using strategies (assessed by the six elements or the hotel tests), knowledge and feedback to regulate behavior (assessed by the California Card Sorting Test), and switching cognitive sets (assessed by the Trail Making Test, TMT, Part B)(Delis, Squire, Bihrle, & Massman, 1992; Godefroy, 2003; Milner, 1963; Slachevsky et al., 2001).

The TMT has been described as a visuomotor task with different processing demands for each part. Part A requires the subject to draw a line as rapidly as possible joining consecutive numbers pseudorandomly arranged on a page. In Part B, the participant draws a line alternating between consecutive numbers and letters, which are also pseudorandomly arranged on a page (Lezak, 1995). In a sample of patients with frontal lobe and posterior lesions, Stuss et al. (2001) have shown that damage of specific frontal lobe regions yields differential results on the TMT, depending on which measure of this multicomponent task is used. Although patients with frontal lobe damage tended to be slower than the control and nonfrontal groups, patients with left frontal lobe damage were the slowest. Patients who made more errors in TMT Part B had primary damage in dorsolateral areas (Stuss, Bisschop et al., 2001). Evaluation of patients with prefrontal lobe lesions with the WCST[1] have shown that (1) patients with dorsolateral prefrontal lesions were impaired in extradimensional shifting (shifting across perceptual dimensions, such as from color to form, on the basis of feedback) and (2) patients with ventral prefrontal cortex lesions were not impaired in set-shifting, but were prone to loss of set (Stuss et al., 2000)

Patients with prefrontal lobe lesions also have deficits in language. Excluding motor deficits, i.e., articulation problems, and Broca's aphasia, the language deficits related to the frontal lobes can be classified as activation and formulation (paralinguistic) deficits:

- Activation deficits are characterized by low fluency language, i.e., truncated spontaneous language and/or low performance in letter-based or semantic fluency tests (Alexander, Benson, & Stuss, 1989).
- Formulation problems, or discourse disorders, are generative and narrative in nature. Left-sided lesions result in simplification and repetition (perseveration) of sentence forms, and omission of elements. The narrative coherence discourse of patients with right-sided lesions is loss due to amplification of details, wandering from the topic and insertion of irrelevant elements, and disprosody (Alexander et al., 1989; Joanette, Goulet, & Hannequin, 1990).

Frontal lobe damage does not result in classical amnesia; however, patients with frontal lesions may have a variety of memory impairments, such as loss of source memory, disturbed memory for temporal order, and high rates of false recognition and confabulation (Alexander, Stuss, & Benson, 1979; Alexander, Stuss, & Fansabedian, 2003; Parkin, Bindschaedler, Harsent, & Metzler, 1996). Evaluation of patients with prefrontal damage with most standardized neuropsychological memory tests reveals little or no impairment. Nevertheless, the study of patients with an appropriate memory test such as the California Verbal Learning Test, which includes measures of serial position learning, semantic organization, interference effects, cued recall, recognition, and response bias, has revealed the effect of frontal brain damage in memory tasks. For example, Kramer et al. (2005), in a group of patients with neurodegenerative dementia, showed that smaller frontal volumes were associated with less semantic clustering and response bias, which is the tendency to favor "yes" or

---

[1] In the WCST, subjects must determine the established sorting criterion (color, form, or number) through a process of trial and error, then shift to a new criterion according to a change in examiner feedback.

"no" responses, particularly when there is uncertainty about the correct responses. Patients with prefrontal lesions had inefficient learning in the California Verbal Learning Test owing to poor implementation of a strategy of subjective organization, i.e., the consistent recall of words paired together across sequential trials. However, only patients with lesions centered either on the left posterior dorsolateral frontal region or on the posterior medial frontal region had overall impaired learning and recall (Alexander et al., 2003). Prefrontal damage is also associated with trouble in recognition memory, when tests had an organizational component such as categorized lists (Wheeler, Stuss, & Tulving, 1995). In agreement with these results, relatives and carers of patients with frontal lobe damage often describe them as forgetful in the execution of daily activities, despite relatively normal scores on standard tests of anterograde memory. One factor contributing to this discrepancy is impairment in various aspects of attention (Damasio & Anderson, 2003).

Frontal lobe damage also results in diminished attention to novel events and increased susceptibility to distraction. Patients with prefrontal damage show increased vulnerability to distracting stimuli. In agreement with this deficit in selective attention, Holmes (1938) suggested that an important role of prefrontal cortex is suppression of reflexive ocular behavior. Indeed, damage to the inferior dorsolateral prefrontal cortex has been associated with impairments on the "antisaccade" paradigm, which requires inhibition of reflexive glances to peripheral stimuli (Guitton, Buchtel, & Douglas, 1985). Impairment in continuous attention has also been described following frontal lobe damage. Rueckert and Grafman (1996) have shown that patients with right prefrontal damage had longer reaction time, missed more targets in tests of sustained attention, and got worse with time in the Continuous Performance Test. These results suggest a special role for the right frontal lobe in sustaining attention over time (Rueckert & Grafman, 1996).

Finally, the abovementioned cognitive disturbances, i.e., impairments in working memory, attention, and inhibitory control, may result in major dissociations between well-preserved memory capacities (as demonstrated by normal in standardized memory tests) and severely impaired utilization of those abilities in real-life situations (Damasio & Anderson, 2003). Indeed, in a pivotal study, Shallice and Burguess (1991) described two patients without impairment in standard frontal tests who had severe difficulties in two tests which required carrying out a number of fairly simple but open-ended tasks over a 15–30-min period (the multiple errand test and the six elements test). They spent too long on individual tasks, did not complete all the tasks, and violated rules. These authors suggested that difficulty in performing the multiple subgoal tasks arose from an inability to reactivate, after a delay, previously generated intentions when they are not directly signaled by the stimulus situation (Shallice & Burgess, 1991). The result of this study is very important because it illustrates the existence of dissociation in performances in patients with prefrontal damage according to task settings: performances are preserved when facing tasks with a single explicit problem to tackle at any one time, short trials, task initiation prompted by another person and what constitutes successful trial completion is clearly characterized. In contrast, performances are impaired in real life because task demands imposed by everyday life are quite different from task demands of assessment in

Table 12.1   Difference between clinical settings and real life (adapted from Ackert, 1990)

Clinical settings	Everyday life
Structured by examiner	Unstructured
Assisted in task focus by examiner	Little task focus provided
Nonpunitive setting	Negative feedback on errors
Planning aided by the examiner	Planning by individual
Motivation aided by the examiner	Self-motivation necessary
Persistence encouraged	Persistence up to individual
Failure not emphasized	Fear of failure
Protected environment	Minimally protective milieu
Inadequacies not exposed	Inadequacies visible to others
Competition absent	Competition present

clinical settings (Ackert, 1990) (see Table 12.1). As we will discuss in "From Clinic to Prefrontal Function," the existence of dissociation in performance according to task setting is very important to understand the role of prefrontal cortex in behavior.

In summary, prefrontal damage is associated with memory impairment, which is due to defects of strategic processes of memory, i.e., organizational aspects of memory at encoding and retrieval (Kramer et al., 2005). Prefrontal lesions are associated with troubles in working memory tasks that involve bridging of temporally separate elements and the comparison or manipulation of several pieces of information (Damasio & Anderson, 2003). Indeed, a meta-analysis found no evidence for an effect of frontal lobe lesions on forward digit or spatial span, probably because these tasks are important for determining working memory storage capacity, but do not provide information relating to rehearsal or executive control, i.e., control or manipulation of information held on-line. In contrast, prefrontal cortex lesions disrupted reversal of the sequences (e.g., digits backwards), self-ordered pointing, conditional associative learning tasks, and delayed-response performance when the delay was filled with a distractor. All these tasks measured manipulation of information held on-line (D'Esposito & Postle, 1999; Owen, Downes, Sahakian, Polkey, & Robbins, 1990; Petrides, 1989).

Behavioral Disturbances Associated with Prefrontal Damage

Prefrontal lesions are associated with alterations of emotions and personality, including changes in behavioral response patterns, moods, and attitudes (Damasio & Anderson, 2003). Descriptions of affective and emotional changes in frontal lobe patients include pseudodepressive state and euphoric state. The euphoria of the frontal patient is neither constant in time, nor is it always characterized by a pure feeling of elation. Rather, it usually occurs in sporadic or recurrent fashion (Fuster, 1997). Patients with mesial frontal lobe lesions often appeared to have blunted emotional response, as if their affects had been neutralized (Damasio & Van Hoesen, 1983). Paradiso et al. (1999) have described that patients with dorsolateral prefrontal lesions frequently experienced depressive symptoms and social unease 3 months after damage from stroke or trauma (Paradiso, Chemerinski, Yazici, Tartaro, & Robinson, 1999). Characteristically, many patients with prefrontal lesions have instability of humor, i.e., a patient who appeared euphoric would look indifferent some later time. External circumstances may "set" the patient's emotional tone, but frequently the reaction will be found inappropriate to them (Damasio & Anderson, 2003).

Besides disturbances in emotion and affect, patients with prefrontal lesions have behavioral symptoms. Abulia, apathy, aspontaneity, mutism, lack of drive, poor motivation, distractibility, impulsivity, disinhibition, irritability, restlessness, and indifference have been reported following prefrontal damage (Godefroy, 2003). Prefrontal lobe lesions could lead to excessive adherence to environmental stimuli: patients imitate the examiner's gestures, although not instructed to do so (imitation behavior) and presentation of objects implies the order to grasp and use them (utilization behavior) (Lhermitte, 1986; Lhermitte, Pillon, & Serdaru, 1986). Some patients have pathologic patterns of collecting: they have an irrepressible need either to seize surrounding objects and store them (hoarding behavior) or to collect specific items. For example, Volle et al. (2002) reported a patient who collected household electrical appliances following bilateral damage of orbito- and polar-prefrontal cortex. Likewise, Cohen et al. (1999) reported a patient who borrowed cars following prefrontal damage secondary to the rupture of an aneurysm.

Patients also tend to violate social rules. Prefrontal damage acquired early in childhood has been associated with psychopathic state (Anderson, Bechara, Damasio, Tranel, & Damasio, 1999). Patients with prefrontal damage acquired in adulthood may present an "acquired sociopathy" (Blair & Cipolotti, 2000): for example, Burns and Swerdlow (2003) reported a patient who displayed impulsive sexual behavior with pedophilia associated with a right orbitofrontal tumor. Interestingly, the behavioral symptoms resolved following tumor resection (Burns & Swerdlow, 2003). Notably, patients with "acquired sociopathy" have preserved access and processing of social knowledge (Saver & Damasio, 1991). More recently, Koenigs et al. (2007) showed that patients with focal bilateral damage to the ventromedial prefrontal cortex produce an abnormally "utilitarian" pattern of judgments on moral dilemmas that pit compelling considerations of aggregate welfare against highly emotionally aversive behaviors (for example, having to sacrifice one person's life to save a number of other lives). According to Churchland (2003), lesions in prefrontal cortex are "followed by significant changes in self-control, and particularly in the capacity to inhibit unwise impulses, despite normal functioning of many other self-representational capacities. Personality changes commonly occur with prefrontal damage. Hitherto quiet and self-controlled, a person with lesions in the ventromedial region of frontal cortex is apt to be more reckless in decision making, impaired in impulse control, and socially insensitive."

Evaluation of patients with bilateral prefrontal damage and behavioral disturbances revealed that they had abnormal autonomic responses to socially meaningful stimuli, despite normal autonomic responses to elementary and uncondicionated stimuli

(Damasio, Tranel, & Damasio, 1990). The perception or comprehension of emotional information also appears to be altered in some patients. Indeed, families with frontal lobe injuries often complain that the patient has impaired empathy (Grattan, Bloomer, Archambault, & Eslinger, 1994). Patients with behavioral disturbances frequently are impaired in a card task that mimics real-life decision making in the way that the task factors uncertainty, reward, and punishment components of the decision in real life (the Iowa Gambling Task). Subjects are given a loan of money, four decks of card (face down), and are asked to draw cards in a manner to lose the least amount of money and win the most. Turning each card results in an immediate reward ($100 for decks A and B, and $50 for decks C and D). At an unpredictable point, the turning of some cards results in a financial penalty (larger for decks A and B, and smaller for decks C and D). Playing mostly from decks with the larger initial payoff (A and B) is disadvantageous in the long run because of the larger penalties. Playing mostly from decks C and D, with the smaller initial reward, turns out to be advantageous, i.e., leading to financial gain. There is no way for a subject to predict when a penalty will arise or to calculate with precision the net gain or loss from each deck. Normal subjects tend to initially sample from all decks, but choose significantly more from decks C and D while avoiding decks A and B over the course of selection of 100 cards. In contrast, patients with prefrontal damage select more the disadvantageous decks (A and B) than the advantageous ones (C and D) (Bechara, Damasio, Damasio, & Anderson, 1994). The impairment in decision making has been tied with a failure to respond autonomically to anticipated future outcomes, as has been suggested by the absence in patients with prefrontal damage of anticipatory skin conductance responses prior to the selection of risky cards in the Iowa Gambling Task (Bechara, Tranel, Damasio, & Damasio, 1996). Nevertheless, other authors suggested that behavioral disturbances of patients with prefrontal damage are tied to an impairment in self-awareness (see later) (Stuss & Levine, 2002). In the next section, we will review some of the troubles in awareness reported in patients with prefrontal damage.

Troubles in Consciousness and Prefrontal Cortex

Prefrontal damage is associated with impairment of awareness. Indeed, unawareness of acquired cognitive impairment (anosognosia), alterations in behavior, emotions, and thought processes are common consequences of frontal lobe dysfunction, especially in right frontal lobe lesions. Indeed, self-awareness has been described as the highest cognitive attribute of the frontal lobes (Stuss & Benson, 1986). Janowsky et al. (1989) showed that subjects with prefrontal lobe lesions were impaired in a task in which they had to judge the probability that they recognize the correct answer to a multiple-choice question (Janowsky, Shimamura, & Squire, 1989). Prefrontal damage has been associated with an impairment in the conscious monitoring of actions (Slachevsky et al., 2001). Levine et al. (1998) described a patient who had lost all personal past memories after a right ventro-prefrontal lesion that affected frontotemporal connectivity (uncinate fasciculus). This subject had virtually no episodic memory for preinjury events, whereas semantic facts about his life were retained as if relearned. He could learn new memories about his life after the injury but these were recalled without any affective balance; that is, they were not episodic in nature (Levine et al., 1998). A notable clinical feature observed was a self-regulatory disorder, a deficit in behavior regulation according to internal goals and constraints affecting his functioning in a variety of real-life unstructured situations (Levine, Freedman, Dawson, Black, & Stuss, 1999). Specifically, he became error-prone in making decisions involving his children and his work. He was unable to resume his former position and failed in a work trial in a less demanding position. According to the authors, the episodic impairment in memory (i.e., inability to recollect past episodes from a specific time and place prior to the injury) and the self-regulatory disorder were not coincidental, but rather causally related. They are tied to impaired autonoetic (self-knowing) consciousness, a capacity that facilitates awareness of the self as a continuous entity across time (autonoetic awareness). Impaired autonoetic consciousness should affect behavior with respect to both past (memory performance) and future (self-regulation) (Levine et al., 1999).

The above review illustrates that patients with prefrontal damage presented a variety of clinical symptoms, either disturbances in the cognitive domain and/or behavior troubles. The identifications of these disturbances could be quite difficult because "sizeable frontal lobe lesions can remain clinically silent for many years. Even after massive bifrontal lesions in monkeys, chimpanzees, and humans change can often be detected only in comparison with the previous personality of that individual rather than in reference to any set of absolute behavioral standards. In fact, many of the alterations associated with prefrontal lesions appear to overlap with the range of normal human behavior" (Mesulam, 2002).

To simplify the identification of clinical disturbances associated with prefrontal damage, a French cooperative group has proposed a list of the main behavioral and cognitive changes where highly suggestive and supportive disorders are separately depicted (Tables 12.2, 12.3) (Godefroy, 2003).

Table 12.2 Main behavioral disorders suggestive of prefrontal cortex damage (adapted from Godefroy, 2003)

Highly suggestive:
Global hypoactivity with abulia, apathy, aspontaneity
Global hyperactivity with distractibility, impulsivity, disinhibition
Perseveration and stereotyped behavior
Syndrome of environmental dependency (imitation and utilization behavior)
Other supportive features:
Confabulation and reduplicative paramnesia
Anosognosia and anosodysphoria
Disturbances of emotion and social behavior
Disorders of sexual behavior and control of micturition

Table 12.3 Main cognitive disorders suggestive of prefrontal cortex damage (adapted from Godefroy, 2003)

Highly suggestive:
Response initiation; response suppression and focused attention
Rule deduction, maintenance, and set-shifting
Problem solving and planning
Information generation
Other supportive features:
Task coordination and divided attention; sustained attention
Strategic mnemonic processes
Theory of mind

In the next section, we will discuss the contribution of the clinical manifestations associated with prefrontal damage to unravel the role of prefrontal cortex in behavior control.

From Clinic to Prefrontal Function

First of all, it is important to note that unlike functional neuroimaging and all physiological measures of an intact system, which only support the engagement of a brain region by a cognitive process and not its necessity for this process, lesion studies, i.e., the study of deficit associated with damage to a specific brain region, provide convincing evidence about the necessity of this brain region in a cognitive process and/or the control of a specific behavior (Sarter, Berntson, & Cacioppo, 1996). Moreover, the study of patients with appropriate tasks, built on current cognitive psychological thinking and the newest anatomical findings in the discovery and dissociation of cognitive processes, is very useful in establishing fractionation of specific processes and systems in the brain (Stuss & Levine, 2002). Nevertheless, it is important to be very cautious in interpreting lesions studies since several variables can interfere with the results. For example, studies of patients with lesions restricted to the prefrontal cortex typically included several causes, such as strokes, focal cerebral trauma, tumor resection, or epileptic patients following frontal lobe excisions. Aside from difficulties in interpreting disparity in cause, extent of dysfunction, and time course of neural changes, the impact that compensatory plasticity has on these studies is unclear. But when significant and reproducible deficits are documented in patients with focal prefrontal cortex lesions the results are compelling (Gazzaley & D'Esposito, 2007).

The above revision of clinical manifestations associated with prefrontal damage leads to some important conclusions:

There is not a unitary frontal lobe syndrome. Indeed, there is an important diversity in the clinical presentation of patients with prefrontal damage:

1. Some patients exhibit only behavioral disturbances, whereas their performance in cognitive tasks is not so impaired (Eslinger & Damasio, 1985).
2. In the cognitive domain, performances in various tests assessing executive functions may be dissociated. For example, Burguess and Shallice (1996) studied the performance of patients with prefrontal lobe lesions in the Hayling test. Subjects were given a sentence with its final highly significant constraint word removed: in part A, they had to complete the sentence as quickly as possible (e.g., "He mailed a letter without a...stamp"), and in part B, they had to complete the sentence with any word that make no sense (e.g., "Most cats see very well at... talk"). Frontal damage resulted in slower responses in part A, suggesting a deficit of response initiation, and higher error rate in part B, suggesting a deficit of response suppression. However, double dissociation emerged, suggesting that some patients had only deficit in response initiation and others had only deficit in response suppression (Burgess & Shallice, 1996).
3. Damage in specific regions of the frontal lobe yields differential results in different neuropsychological tests such as the TMT, the WCST, the Stroop test, the California Verbal Learning Test, and verbal fluency (Stuss et al., 2000; Stuss, Floden, Alexander, Levine, & Katz, 2001; Stuss et al., 1998; Stuss, Bisschop et al., 2001; Alexander et al., 2003). The corollary of this differential impairment in neuropsychological tests according to damage localization is that the frontal lobes are functionally heterogeneous (Stuss, Bisschop et al., 2001).
4. Some studies have found a relationship between behavioral disorders and cognitive deficit. For example, Godefroy et al. (1996) found that distractibility and impulsivity were related to attentional deficits as measured in neuropsychological tests (Godefroy, Lhullier, & Rousseaux, 1996). Burguess et al. (1998) also found multiple correlation between neuropsychological tests (modified card sorting test, TMT, verbal fluency, and six elements test) and behavioral abnormalities assessed by a questionnaire covering common symptoms of frontal lobe syndrome (Burgess, Alderman, Evans, Emslie, & Wilson, 1998). However, others studies have reported patients with important behavioral disturbances without cognitive troubles (Blair & Cipolotti, 2000; Eslinger & Damasio, 1985; Saver & Damasio, 1991). Moreover, in a study of 13 patients with prefrontal lobe lesions, Sarazin et al. (1998) showed that executive-function test performance was significantly correlated with regional cerebral glucose metabolism, measured with PET in the dorsolateral prefrontal cortex (Brodmann areas 8, 9, 45, 46, and 47) and the anterior cingulate cortex (Brodmann area 32). In contrast, behavioral scores were significantly correlated with regional cerebral glucose metabolism in the frontopolar (Brodmann area 10) and orbitofrontal cortex (Brodmann areas 11, 12, 13, and 14) (Sarazin et al., 1998).
5. Prefrontal damage symptoms vary greatly according to setting. Indeed, patients with prefrontal damage can perform quite well in an important number of tasks and they can even cope with most activities of daily life. Symptoms manifested themselves when patients face a conflict, when the environment contains distractors,

when prepotent response must be inhibited, or when subjects must elaborate a strategy to solve a conflict or an unusual situation (Mesulam, 2002; Slachevsky et al., 2003). In other words, prefrontal cortex damage leads to distinctive sets of impairments that are context-dependent (contingent), i.e., manifestation of prefrontal damage varies according to the situation (Mesulam, 2002).

The clinical manifestations associated with prefrontal damage are quite different from impairments observed with damage to other regions of the cerebral cortex. Indeed, many cortical areas in the human brain are devoted to the modality-specific representation of events, faces, words, and locations. Others mediate the transmodal binding of these representations so that faces can lead to recognition, words to comprehension, intentions to actions, and events to memory (Mesulam, 1998). Focal brain lesions that interfere with these processing streams lead to disconnection syndromes such as apraxia, prosopagnosia, and amnesia (Mesulam, 2000). As Mesulam (2002) wrote, "prefrontal cortex is not essential for encoding any of these representations and is not implicated in the pathogenesis of any traditional disconnection syndrome, i.e., prefrontal damage do not lead to troubles such as agnosia or amnesia." Thus, the manifestations of these syndromes generally do not vary according to the situation. That is, for example, a patient with anaphasia presented signs of his language disturbances independently of the context. In contrast, damage to prefrontal cortex leads to impairments that are context-dependent (see above) (Mesulam, 2002).

On the other hand, manifestations of prefrontal damage are not limited to patients with lesions in the prefrontal lobe. The clinical manifestations of prefrontal lobe dysfunction may be produced by lesions in the white matter (i.e., disrupting ascending and descending to and from the prefrontal cortex), and lesions in subcortical structures that constitutes frontal-subcortical circuits (Duffy & Campbell, 1994). Indeed, Duffy and Campbell (1994) have suggested that the "validity of the term frontal lobe syndrome warrants inspection. The term appears to denote a constellation of clinical signs and symptoms that are referable to a specific neuroanatomical focus, the frontal lobe. The persistence of this anatomically based descriptive term is an anachronism that draws support from a strict localizationist approach to brain-behavior relationships."

The main characteristics of clinical manifestations associated with prefrontal cortex damage (i.e., heterogeneity of symptoms and the variability of these manifestations according to context and task demand) illustrate that damage to this cerebral region does not affect overall intelligence in humans, but has a detrimental effect on the control and regulation of behavior. On the other hand, the existence of prefrontal symptoms due to either lesions in the prefrontal cortex or disconnection of prefrontal cortex from other brain regions suggested that processes and representations being controlled by prefrontal cortex are primary localized to other brain regions (Gazzaley & D'Esposito, 2007; Hogan, Vargha-Khadem, Saunders, Kirkham, & Baldeweg, 2006). The neuropsychological evidence led to the conceptualization of frontal lobe function as executive functions, a subset of cognitive functions concerned with the management of human behavior across time. These functions enable human beings to develop and carry out plans, form analogies, obey social rules, solve problems, adapt to unexpected circumstances, do many tasks simultaneously, and place episodes in time and place, which ensures that memories can be retrieved (Grafman, 1995). Executive function encompasses a diverse collection of processes, including divided attention and sustained attention, working memory, set-shifting, flexibility, planning, and the regulation of goal-directed behavior, and can be defined as a brain function underlying the human faculty to act or think not only in reaction to external events but also in relation to internal goals and states (Gazzaley & D'Esposito, 2007). As we have reviewed, prefrontal dysfunction can affect principally cognition and behavior. Grafman and Livan (1999) have proposed dividing executive function into two main aspects: "cold" executive functions, subtended by the dorsolateral prefrontal cortex and mediating abilities such as verbal reasoning, mechanistic planning, or problem solving and "hot" executive functions, subtended by the ventromedial prefrontal cortex and mediating functions, such as obeying the rules of interpersonal social behavior, the experience of reward and punishment, and the interpretation of complex emotions (Grafman & Litvan, 1999). In the same way, Stuss and Levine (2002) have proposed that behavioral and cognitive disorders reflect two different facets of frontal lobe functions. Cognitive troubles corresponded to disturbances of what is referred to as executive functioning strictly speaking, subtended by dorso-lateral prefrontal cortex. Behavioral disturbances reflected a trouble from what these authors referred to as behavioral self-regulation, subtended by ventral medial/ orbital prefrontal cortex (see above) (Stuss & Levine, 2002).

Finally, one of the main points have discussed about prefrontal lobe dysfunction is the preservation of primary deficits of motility, sensation, or major cognitive domain (for example, prefrontal lobe lesions do not lead to agnosia, aphasia, or a true amnesic syndrome). Therefore, executive control means control of mental processes whose primary operative sites are localized elsewhere in the brain. Indeed, prefrontal cortex controls sensory and motor representations, which are predominantly the domain of primary sensory motor and unimodal association. As we have reviewed, some authors proposed fragmenting control processes of prefrontal cortex into several dimensions. More recently, Gazzaley and D'Esposito (2007) suggested that the parcelling of prefrontal cortex in several dimensions does not help us to understand the role, and proposed a unifying theory of prefrontal cortex function, i.e., a common function that underlies prefrontal cortex in diverse mental processes, control of sensory input, motor output, cognition, and emotion (Gazzaley & D'Esposito, 2007). This function may be the control and dynamic integration of the external and the internal environment. In fact, there is no consensus about how prefrontal cortex controls cognition and regulates behavior. Moreover, there are still important questions to be resolved to understand how prefrontal lobe function leads to such a variety of symptoms in patients with prefrontal lobe dysfunction.

In our opinion, some of the main questions are:

1. What are the best ways to evaluate patients with prefrontal lobe dysfunction? As we discussed, the frontal lobe syndrome is not a uniform entity, but is a constellation of behavioral alterations that occur and evolve in different patterns in

patients with prefrontal lesions. Moreover, manifestations of prefrontal cortex dysfunction vary greatly according to the setting. Patients can perform several tasks quite well, but troubles arise when novel tasks are involved or existing habits must be overridden. Hence, it is fundamental to evaluate patients with adequate neuropsychological tasks to reveal daily-life difficulties of patients in the laboratory. Improvement of evaluation will certainly help to clarify the different factors intervening in heterogeneity of clinical manifestations of prefrontal cortex dysfunction.

2. Which processes mediate between troubles in control and emergence of clinical manifestations of patients with prefrontal cortex dysfunction? Undoubtedly if we are able to understand the processes mediating the emergence of clinical manifestations of frontal lobe syndrome, we will be able to develop better diagnostic methods and therapeutic strategy.

Acknowledgements This work was funded by Fondecyt 1050155. We are grateful to Beatriz Luna and Francisco Aboitiz for feedback on an early draft.

References

Ackert, M. B. (1990). A review of the ecological validity of neuropsychological tets. In D. E. Tupper & K. E. Cicerone (Eds.), *The neuropsychology of everyday life: Assessment and basic competencies* (pp. 19–56). Boston: Kluwer.

Alexander, M. P., Benson, D. F., & Stuss, D. T. (1989). Frontal lobes and language. *Brain and Language, 37*, 656–691.

Alexander, M. P., Stuss, D. T., & Benson, D. F. (1979). Capgras syndrome: A reduplicative phenomenon. *Neurology, 29*, 334–339.

Alexander, M. P., Stuss, D. T., & Fansabedian, N. (2003). California Verbal Learning Test: Performance by patients with focal frontal and non-frontal lesions. *Brain, 126*, 1493–1503.

Anderson, S. W., Bechara, A., Damasio, H., Tranel, D., & Damasio, A. R. (1999). Impairment of social and moral behavior related to early damage in human prefrontal cortex. *Nature Neurosciences, 2*, 1032–1037.

Barbas, H. (1995). Anatomic basis of cognitive-emotional interactions in the primate prefrontal cortex. *Neurosciences and Biobehavioral Review, 19*, 499–510.

Barbas, H. (2000). Connections underlying the synthesis of cognition, memory, and emotion in primate prefrontal cortices. *Brain Research Bulletin, 52*, 319–330.

Bechara, A., Damasio, A. R., Damasio, H., & Anderson, S. W. (1994). Insensitivity to future consequences following damage to human prefrontal cortex. *Cognition, 50*, 7–15.

Bechara, A., Damasio, H., Tranel, D., & Anderson, S. W. (1998). Dissociation Of working memory from decision making within the human prefrontal cortex. *Journal of Neurosciences, 18*, 428–437.

Bechara, A., Tranel, D., Damasio, H., & Damasio, A. R. (1996). Failure to respond autonomically to anticipated future outcomes following damage to prefrontal cortex. *Cerebral Cortex, 6*, 215–225.

Blair, R. J., & Cipolotti, L. (2000). Impaired social response reversal. A case of 'acquired sociopathy'. *Brain, 123*, 1122–1141.

Burgess, P. W., Alderman, N., Evans, J., Emslie, H., & Wilson, B. A. (1998). The ecological validity of tests of executive function. *Journal of the International Neuropsychological Society, 4*, 547–558.

Burgess, P. W., & Shallice, T. (1996). Response suppression, initiation and strategy use following frontal lobe lesions. *Neuropsychologia, 34*, 263–272.

Burns, J. M., & Swerdlow, R. H. (2003). Right orbitofrontal tumor with pedophilia symptom and constructional apraxia sign. *Archives of Neurology, 60*, 437–440.

Cohen, L., Angladette, L., Benoit, N., & Pierrot-Deseilligny, C. (1999). A man who borrowed cars. *Lancet, 353*, 34.

Churchland, P. S. (2003). Self-representation in nervous systems. *Annals of the New York Academy of Sciences, 1001*, 31–38.

D'Esposito, M., & Postle, B. R. (1999). The dependence of span and delayed-response performance on prefrontal cortex. *Neuropsychologia, 37*, 1303–1315.

Damasio, A., & Anderson, S. (2003). The frontal lobes. In K. M. Heilman & E. Valenstein (Eds.), *Clinical neuropsychology* (4 ed., pp. 404–446). Oxford: Oxford University Press.

Damasio, A. R., Tranel, D., & Damasio, H. (1990). Individuals with sociopathic behavior caused by frontal damage fail to respond autonomically to social stimuli. *Behavioral Brain Research, 41*, 81–94.

Damasio, A. R., & Van Hoesen, G. W. (1983). Emotional disturbances associated with focal lesions of the frontal lobe. In K. M. Heilman & P. Satz (Eds.), *The neurophysiology of human emotion: Recent advances* (pp. 85–108). New York: Guilford Press.

Damasio, H. (1996). Human neuroanatomy relevant to decision-making. In A. R. Damasio, H. Damasio, & Y. Christen (Eds.), *Neurobiology of decision making* (pp. 1–12). Berlin: Springer.

Delis, D. C., Squire, L. R., Bihrle, A., & Massman, P. (1992). Componential analysis of problem-solving ability: Performance of patients with frontal lobe damage and amnesic patients on a new sorting test. *Neuropsychologia, 30*, 683–697.

Duffy, J. D., & Campbell, J. J., 3rd. (1994). The regional prefrontal syndromes: A theoretical and clinical overview. *Journal of Neuropsychiatry and Clinical Neurosciences, 6*, 379–387.

Elliott, R. (2003). Executive functions and their disorders. *British Medical Bulletin, 65*, 49–59.

Eslinger, P. J., & Damasio, A. R. (1985). Severe disturbance of higher cognition after bilateral frontal lobe ablation: Patient EVR. *Neurology, 35*, 1731–1741.

Fuster, J. (1997). *The prefrontal cortex*. Philadelphia: Lippincott-Raven.

Gazzaley, A., & D'Esposito, M. (2007). Unifying Prefrontal Cortex Function. Executive control, neural networks, and top-down modulation. In B. Miller & J. Cummings (Eds.), *The human frontal lobes. Functions and disorders* (2 ed., pp. 187–204). New York: The Guilford Press.

Godefroy, O. (2003). Frontal syndrome and disorders of executive functions. *Journal of Neurology, 250*, 1–6.

Godefroy, O., Lhullier, C., & Rousseaux, M. (1996). Non-spatial attention disorders in patients with frontal or posterior brain damage. *Brain, 119*, 191–202.

Grafman, J. (1995). Similarities and distinctions among current models of prefrontal cortical functions. *Annals of the New York Academy of Sciences, 769*, 337–368.

Grafman, J., & Litvan, I. (1999). Importance of deficits in executive functions. *Lancet, 354*, 1921–1923.

Grattan, L. M., Bloomer, R. H., Archambault, F. X., & Eslinger, P. J. (1994). Cognitive flexibility and empathy after frontal lobe lesion. *Neuropsychiatry Neuropsychology and Behavioral Neurology, 7*, 251–259.

Guitton, D., Buchtel, H. A., & Douglas, R. M. (1985). Frontal lobe lesions in man cause difficulties in suppressing reflexive glances and in generating goal-directed saccades. *Experimental Brain Research, 58*, 455–472.

Hogan, A. M., Vargha-Khadem, F., Saunders, D. E., Kirkham, F. J., & Baldeweg, T. (2006). Impact of frontal white matter lesions on performance monitoring: ERP evidence for cortical disconnection. *Brain, 129*, 2177–2188.

Holmes, G. (1938). The cerebral integration of ocular mouvements. *British Medical Journal, 2*, 107–112.

Janowsky, J. S., Shimamura, A. P., & Squire, L. R. (1989). Memory and metamemory: Comparisons between patients with frontal lobe lesions and amnesic patients. *Psychobiology, 17,* 3–11.

Joanette, Y., Goulet, P., & Hannequin, D. (1990). *Right hemisphere and verbal communication.* New York: Springer.

Koenigs, M., Young, L., Adolphs, R., Tranel, D., Cushman, F., Hauser, M., et al. (2007). Damage to the prefrontal cortex increases utilitarian moral judgements. *Nature, 446,* 908–911.

Kramer, J. H., Rosen, H. J., Du, A. T., Schuff, N., Hollnagel, C., Weiner, M. W., et al. (2005). Dissociations in hippocampal and frontal contributions to episodic memory performance. *Neuropsychology, 19,* 799–805.

Levine, B., Black, S. E., Cabeza, R., Sinden, M., McIntosh, A. R., Toth, J. P., et al. (1998). Episodic memory and the self in a case of isolated retrograde amnesia. *Brain, 121,* 1951–1973.

Levine, B., Freedman, M., Dawson, D., Black, S., & Stuss, D. T. (1999). Ventral frontal contribution to self-regulation: Convergence of episodic memory and inhibition. *Neurocase, 5,* 263–275.

Lezak, M. D. (1995). *Neuropsychological assessment* (3 ed.). New York: Oxford University Press.

Lhermitte, F. (1986). Human autonomy and the frontal lobes. Part II: Patient behavior in complex and social situations: The "environmental dependency syndrome". *Annals of Neurology, 19,* 335–343.

Lhermitte, F., Pillon, B., & Serdaru, M. (1986). Human autonomy and the frontal lobes. Part I: Imitation and utilization behavior: A neuropsychological study of 75 patients. *Annals of Neurology, 19,* 326–334.

Mac Millian, M. (1996). Phineas gage: A case for all reason. In C. Code, C. Wallesch, Y. Joannette, & A. RochLecours (Eds.), *Classic cases in neuropsychology* (pp. 243–262). Hove: Psychology Press.

Mesulam, M. M. (1998). From sensation to cognition. *Brain, 121,* 1013–1052.

Mesulam, M. M. (2000). Behavioral neuroanatomy: Large cortical networks, association cortex, frontal syndromes, the limbic system and hemspheric specialization. In M. M. Mcsulam (Ed.), *Principles of behavioral and cognitive Neurology* (2 ed., pp. 1–120). New York: Oxford University Press.

Mesulam, M. M. (2002). The human frontal lobes: Transcending the default mode through contingent encoding. In D. Stuss & R. Knight (Eds.), *Principles of frontal lobe function* (pp. 8–30). Oxford: Oxford University Press.

Milner, B. (1963). Effects on different brain regions on card sorting. *Archives of Neurology, 9,* 100–110.

Owen, A. M., Downes, J. J., Sahakian, B. J., Polkey, C. E., & Robbins, T. W. (1990). Planning and spatial working memory following frontal lobe lesions in man. *Neuropsychologia, 28,* 1021–1034.

Pandya, D. N., & Yeterian, E. H. (1996). Morphological correlates of human and monkey frontal lobe. In A. R. Damasio, H. Damasio, & Y. Christen (Eds.), *Neurobiology of decision making* (pp. 13–46). Berlin: Springer.

Paradiso, S., Chemerinski, E., Yazici, K. M., Tartaro, A., & Robinson, R. G. (1999). Frontal lobe syndrome reassessed: Comparison of patients with lateral or medial frontal brain damage. *Journal of Neurology, Neurosurgery and Psychiatry, 67,* 664–667.

Parkin, A. J., Bindschaedler, C., Harsent, L., & Metzler, C. (1996). Pathological false alarm rates following damage to the left frontal cortex. *Brain and Cognition, 32,* 14–27.

Penfield, W., & Evans, J. (1935). The frontal lobe in man: A clinical study of maximum removals. *Brain, 58,* 115–133.

Petrides, M. (1989). Frontal lobes and memory. In F. Boller & J. Grafman (Eds.), *Handbook of neuropsychology* (pp. 75–90). Amsterdam: Elsevier.

Rolls, E. (2002). The functions of the orbitofrontal cortex. In D. Stuss & R. T. Knight (Eds.), *Principles of frontal lobe functions* (pp. 354–375). Oxford: Oxford University Press.

Rueckert, L., & Grafman, J. (1996). Sustained attention deficits in patients with right frontal lesions. *Neuropsychologia, 34,* 953–963.

Sarazin, M., Pillon, B., Giannakopoulos, P., Rancurel, G., Samson, Y., & Dubois, B. (1998). Clinicometabolic dissociation of cognitive functions and social behavior in frontal lobe lesions. *Neurology, 51,* 142–148.

Sarter, M., Berntson, G. G., & Cacioppo, J. T. (1996). Brain imaging and cognitive neuroscience. Toward strong inference in attributing function to structure. *American Psychologist, 51,* 13–21.

Saver, J. L., & Damasio, A. R. (1991). Preserved access and processing of social knowledge in a patient with acquired sociopathy due to ventromedial frontal damage. *Neuropsychologia, 29,* 1241–1249.

Shallice, T., & Burgess, P. W. (1991). Deficits in strategy application following frontal lobe damage in man. *Brain, 114,* 727–741.

Slachevsky, A., Pillon, B., Fourneret, P., Pradat-Diehl, P., Jeannerod, M., & Dubois, B. (2001). Preserved adjustment but impaired awareness in a sensory-motor conflict following prefrontal lesions. *Journal of Cognitive Neurosciences, 13,* 332–340.

Slachevsky, A., Pillon, B., Fourneret, P., Renie, L., Levy, R., Jeannerod, M., et al. (2003). The prefrontal cortex and conscious monitoring of action. An experimental study. *Neuropsychologia, 41,* 655–665.

Stuss, D. T., Alexander, M. P., Hamer, L., Palumbo, C., Dempster, R., Binns, M., et al. (1998). The effects of focal anterior and posterior brain lesions on verbal fluency. *Journal of the International Neuropsychological Society, 4,* 265–278.

Stuss, D. T., & Benson, D. F. (1986). *The frontal lobes.* New York: Raven Press.

Stuss, D. T., Bisschop, S. M., Alexander, M. P., Levine, B., Katz, D., & Izukawa, D. (2001). The Trail Making Test: A study in focal lesion patients. *Psychological Assessment, 13,* 230–239.

Stuss, D. T., Floden, D., Alexander, M. P., Levine, B., & Katz, D. (2001). Stroop performance in focal lesion patients: Dissociation of processes and frontal lobe lesion location. *Neuropsychologia, 39,* 771–786.

Stuss, D. T., & Levine, B. (2002). Adult clinical neuropsychology: Lessons from studies of the frontal lobes. *Annual Review of Psychology, 53,* 401–433.

Stuss, D. T., Levine, B., Alexander, M. P., Hong, J., Palumbo, C., Hamer, L., et al. (2000). Wisconsin Card Sorting Test performance in patients with focal frontal and posterior brain damage: Effects of lesion location and test structure on separable cognitive processes. *Neuropsychologia, 38,* 388–402.

Volle, E., Beato, R., Levy, R., & Dubois, B. (2002). Forced collectionism after orbitofrontal damage. *Neurology, 58,* 488–490.

Wheeler, M. A., Stuss, D. T., & Tulving, E. (1995). Frontal lobe damage produces episodic memory impairment. *Journal of the International Neuropsychological Society, 1,* 525–536.

Chapter 13
The Maturation of Cognitive Control and the Adolescent Brain

B. Luna

Abstract Cognitive control, which allows us to guide behavior in a planned and voluntary fashion, continues to improve through adolescence in parallel with refinements in brain processes including synaptic pruning and myelination. The adolescent period is of special significance because the shift to mature adult-level cognitive processing begins to occur and because this period is vulnerable to errors in cognitive control evident in the emergence of major psychopathologic dysfunction and in risk-taking behavior. In this chapter, we review the literature characterizing the nature of developmental change in cognitive control of behavior and its relation to the brain maturational processes that occur at this time that affect brain function. Evidence from studies characterizing developmental improvements in speed of information processing, voluntary response inhibition, and working memory indicate that the ability to have cognitive control is present early in development and what continues to improve through adolescence is the ability to use this tool in a controlled and flexible manner. Functional magnetic resonance imaging and diffusion tensor imaging studies provide evidence that concurs with behavior results, indicating that the basic circuitry that supports cognitive control, including prefrontal systems, are on-line early in development and that a shift to increased functional integration throughout the brain underlies mature adult-level executive control. Viewed in the light of the current literature, the adolescent period is beginning to be understood as a necessary period of transition when there is a shift to integrated brain function that supports efficient and flexible control of behavior.

B. Luna
Departments of Psychiatry and Psychology, Laboratory of Neurocognitive Development,
Western Psychiatric Institute and Clinic, University of Pittsburgh Medical Center, Pittsburgh,
PA 15213, USA, e-mail: lunab@upmc.edu

F. Aboitiz and D. Cosmelli (eds.) *From Attention to Goal-Directed Behavior.*
© Springer-Verlag Berlin Heidelberg 2009

Abbreviations

ACC   Anterior cingulate cortex
DA   Dopamine
DLPFC   Dorsolateral prefrontal cortex
EEG   Electroencephalography
fMRI   Functional magnetic resonance imaging
PET   Positron emission tomography
VLPFC   Ventrolateral prefrontal cortex

Introduction

This chapter reviews the literature on the interaction of brain and behavior underlying development through adolescence. First, we will define the nature of cognitive control to understand the precise mechanisms that are at the core of maturity. Second, we will review behavioral studies that depict the changes in cognitive control that occur at this stage. Third, we will review what is known about changes to brain structure during this period of development. Functional magnetic resonance imaging (fMRI) has provided a novel, noninvasive method for investigating the interface between brain and behavior which is appropriate for pediatric populations and affords an unprecedented spatial resolution to localize function at the whole-brain level. We will review the available fMRI data regarding changes in brain function that contribute to the integration of brain and behavior.

Adolescence typically refers to ages 12–17, taking into consideration variability in factors such as puberty and gender (Spear, 2000). Cognitive and brain structural changes, as well as pubertal/hormonal events, take place during this period and have a significant effect on behavior. We have a limited understanding of what processes support the developmental transition into peak cognitive performance and stabilization in young adulthood. The changes known to occur in adolescence suggest that plasticity and biologically determined mechanisms have a significant influence on development. Importantly, understanding normative development allows us to discern impaired or abnormal development. Major psychiatric disorders, such as affective disorders, anxiety, and schizophrenia, emerge during adolescence (Barlow, 1988; Angold, Costello, & Worthman, 1998; Waddington, Torrey, Crow, & Hirsch, 1991; Luna & Sweeney, 2004a), suggesting that vulnerabilities specific to this stage of development may be central to these disorders. Identifying the processes underlying normative maturation would allow us to investigate atypical processes in psychopathologic dysfunction, such as those described in Chap. 15 by Gaspar et al. Adolescence is also a time when sensation- and novelty-seeking behaviors peak across species and cultures; these behaviors may be necessary to develop the social skills needed to gain independence in adulthood (Steinberg, 2004; Arnett, 1992; Spear, 2000; Chambers, Taylor, & Petenza, 2003). Unfortunately, it is also during adolescence and young adulthood that experimental use and abuse of substances often begins (Chambers et al., 2003; Grant, 1997; Warner, Kessler, Hughes, Anthony, & Nelson, 1995). Investigation of the normative processes underlying the transition to mature behavior, including incentive processes, may inform our understanding of "risky" behavior in adolescence.

Cognition

Behavior can be elicited by endogenous or exogenous cues. *Exogenous* behavior is elicited by extraneous stimuli such as perceptual or emotional states. While not a reflex, which implies only spinal chord involvement, exogenously driven behavior is reflexive in nature, that is, it does not require a plan or goal. It is a fast, reactive response to an external cue, for example, looking at a suddenly appearing light. *Endogenously* driven behavior is behavior that is under *voluntary* control. Executive function and cognitive control refer to this type of behavior, which is goal-oriented and requires a plan to be executed. It is this type of behavior that the justice system uses to determine culpability. For example, you are cognitively controlling your behavior when you choose to reach for an object in the presence of alternative choices or choose not to look at a visual stimulus that would otherwise be compelling. Two main processes of cognitive control are response inhibition and working memory. At any given time we are confronted with a range of possible responses. Voluntary response inhibition allows a goal-directed response to override a more compelling, yet goal-inappropriate, response. Working memory is the ability to temporarily store and manipulate information for the purpose of a goal-directed response. These core cognitive processes support endogenously driven behavior and are crucial to cognitive development (Kail, 1993; Fry & Hale, 1996; Case, 1996; Dempster, 1993). Cognitive control is known to be supported by a widely distributed brain circuitry, in which prefrontal cortex plays a primary role (Fuster, 1997; Goldman-Rakic, Chafee, & Friedman, 1993). Whereas exogenously driven behavior is mature early in life, endogenously driven behavior continues to improve through adolescence.

Cognitive Maturation

Core executive abilities that are present early in life (Diamond & Goldman-Rakic, 1989; Levin, Culhane, Hartmann, Evankovich, & Mattson, 1991) demonstrate continued improvement through adolescence (Demetriou, Christou, Spanoudis, & Platsidou, 2002; Anderson, Anderson, Northam, Jacobs, & Catroppa, 2001; Davies & Rose, 1999; Luna, Garver, Urban, Lazar, & Sweeney, 2004; Asato, Sweeney, & Luna, 2006). Whereas much is known about the cognitive milestones achieved during infancy and childhood, relatively less information is available regarding the maturation of cognition into adolescence. Recent studies, however, have begun to elucidate the nature of this prolonged development in cognitive processes such as response inhibition and working memory, as well as in the complex cognitive abilities evident in typical neuropsychological assessments.

Speed of Processing

The speed at which information processing takes place is an essential component of cognitive control. The system must perceive the task demands, plan a behavior, and enact a response. Processing speed is important in both exogenously and endogenously driven responses, and it is believed to reflect the integrity of the brain processes underlying connectivity and functional integration (Kail, 1993; Kail, 1991; Luna et al., 2004). Processing speed has been found to decrease exponentially during childhood and adolescence across cognitive domains (Hale, 1990; Luna et al., 2004; Kail, 1993). The influence of processing speed, measured by reaction time, is evident in simple tasks with minimal cognitive demands, such as visually guided saccades (Luna et al., 2004; Fischer, Biscaldi, & Gezeck, 1997; Fukushima, Hatta, & Fukushima, 2000), manual reaction time (Elliott, 1970), and visual matching tasks (Klein & Foerster, 2001; Luna et al., 2004; Fry et al., 1996), as well as tasks that require cognitive planning (Luna et al., 2004). For example, reaction times for making an eye movement to a suddenly appearing visual stimulus reach maturity by 14–15 years, the same age at which subjects demonstrate maturity in the reaction time to initiate a cognitively driven eye movement (Luna et al., 2004) (Fig. 13.1). The age at which processing speed matures is highly consistent across tasks and

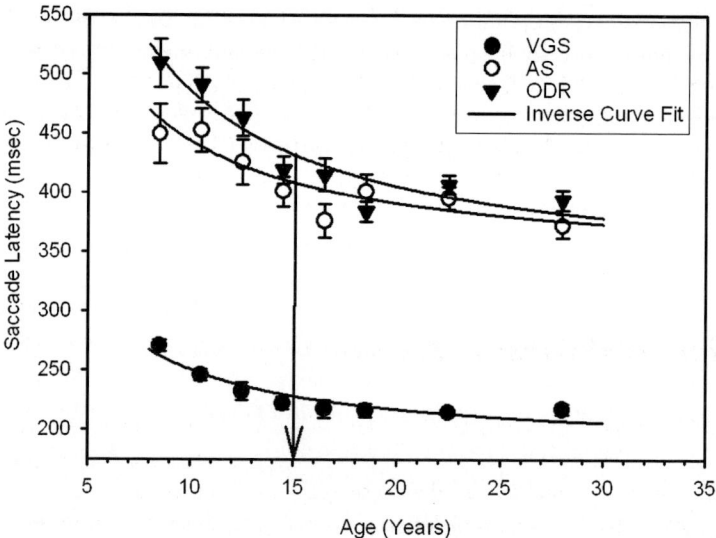

Fig. 13.1 Mean ± one standard error of the mean of the latency to initiate a saccade in the visually guided saccade task (*solid circles*), antisaccade (*open circles*), and the oculomotor delayed response (*triangles*). *Lines* indicate the inverse curve fit on the mean latency to initiate an eye movement response in milliseconds by age in years. *Arrows* depict the age at which change-point analyses indicate adult levels of performance were reached. (Reprinted with permission from Luna et al. 2004)

cognitive loads, suggesting that it is independent of cognitive level even though increasing difficulty results in longer reaction times. Processing speed contributes to the ability to engage cognitive processes, and its maturation may reflect increased efficiency from brain developmental processes, such as synaptic pruning and myelination (see later), that support improvements in cognitive processes, such as response inhibition and working memory.

Development of Voluntarily Guided Behavior from Late Childhood to Adolescence

Neuropsychological tests demonstrate that higher-order cognitive processes, including response inhibition and working memory, continue to develop through late childhood. This process is believed to be supported by prefrontal circuitry (Levin et al., 1991; Demetriou et al., 2002; Fry et al., 1996; Luna et al., 2004; Asato et al., 2006). Specifically, voluntary response suppression and working memory are executive processes (Fuster, 1997) that are essential to the emergence of adult-level cognitive control of behavior (Kail, 1993; Dempster, 1993). The ability to voluntarily suppress a response (i.e., response inhibition) underlies goal-directed executive behavior by allowing an individual to exert self-regulation to control task-irrelevant behavior by stopping prepotent responses and filtering distractors (Bjorklund & Harnishfeger, 1995; Dempster, 1992; Dempster, 1992). Working memory, the maintenance of information on-line (Baddeley, 1986), is crucial for goal-directed executive behavior by supporting response planning and preparation (Case, 1995). These executive processes are seen to break down during psychiatric illness (Sweeney, Takarae, Macmillan, Luna, & Minshew, 2004). Since cognitive maturation occurs concurrently with brain maturation, determining the neurobiological basis of cognitive development is crucial to understanding the biological influences on healthy and impaired development.

Development of Voluntary Response Suppression

The ability to voluntarily inhibit a reactive, task-irrelevant response improves into adolescence. Inhibitory tasks require that a reactive response be suppressed in favor of a planned response. In the go/no-go task, subjects are asked to press a button in response to a series of letters except in response to a target letter (e.g., *x*). The Stroop task requires subjects to suppress established reading responses and instead say the color of the ink of a word written in an incompatible color ("green" would be the correct response for the word *red* written in green). The Simon task requires subjects to respond to a color cue while suppressing the interfering location of the cue. In flanker tasks, subjects must suppress a response to the arrows surrounding a central arrow and instead report the direction of the central stimuli. In the stop-signal task, subjects are asked to voluntarily stop a reactive response after it has started. Fixation tasks require subjects to refrain from making reflexive eye movements away from fixation towards suddenly appearing peripheral visual targets. Finally, the antisaccade task is an inhibition task that has been used extensively in developmental, clinical, and monkey studies. In this task, subjects must suppress an eye movement to a suddenly appearing peripheral visual cue and instead make an eye movement to the mirror location. Errors are usually followed by an eye movement to the correct location, indicating that subjects understand the instructions but are unable to inhibit the reflexive response to look at the light. There is consistent improvement throughout childhood and adolescence across these tasks (Wise, Sutton, & Gibbons, 1975; Ridderinkhof, Band, & Logan, 1999; Bedard et al., 2002; Williams, Ponesse, Schachar, Logan, & Tannock, 1999; Luna et al., 2004).

Although younger subjects have a higher rate of inhibitory failures than older subjects, they are able to perform some trials correctly. Developmental improvement is seen in an increased number of correct responses. Since response inhibition cannot be performed by chance, the fact that children do perform some trials correctly indicates that younger subjects *can* suppress responses, but are unable to do so in a *consistent* manner. These results suggest that improvements in both the ability to compute inhibitory responses and the ability to sustain an inhibitory set support adult-level response inhibition. They also indicate that the ability to inhibit a response is present early in life and that development involves increased flexibility and control in using this ability. Late integration of frontal regions is believed to play an important role in the development of response inhibition, as evidenced by the fact that adults with frontal lesions are impaired on the antisaccade task (Guitton, Buchtel, & Douglas, 1985). However, it is the ability to integrate a large circuitry including frontal regions that is crucial to voluntary response inhibition. Inhibitory control is dependent on efficient top-down frontal modulation of behavior in which executive regions modulate the subcortical regions supporting reactive behavior. The circuitry underlying performance of the antisaccade task has been extensively delineated in single-cell monkey studies, making this task a well-informed, brain-based model of response inhibition (Everling & Munoz, 2000; Funahashi, Chafee, & Goldman-Rakic, 1993). Antisaccade studies have shown continued improvements in response suppression through adolescence, with maturation occurring around age 15 (Fischer et al., 1997; Munoz, Broughton, Goldring, & Armstrong, 1998; Fukushima et al., 2000; Klein et al., 2001; Luna et al., 2004). Single-cell studies in nonhuman primates indicate that preparatory attenuation of saccade-related neuronal activity in the superior colliculus (Everling, Dorris, Klein, & Munoz, 1999) and frontal eye fields (Everling et al., 2000) is crucial to a successful inhibitory response. These results suggest that response *planning* and preparation are essential for efficient top-down modulation of reflexive acts and that these abilities are still immature in adolescence. Taken together, these results indicate that voluntary response inhibition continues to develop through adolescence.

Development of Working Memory

Working memory is the ability to maintain and manipulate on-line information about stimuli that are no longer present in the external environment. Its primary role is to support goal-directed responses (Baddeley, 1986), and it has been found to underlie adult-level higher-order cognition (Bjorklund & Harnishfeger, 1990; Nelson et al., 2000; Dempster, 1993; Case, 1992). Similar to voluntary response suppression, working memory has a prolonged development (Swanson, 1999; Luciana & Nelson, 1998; Demetriou et al., 2002; Luna et al., 2004; Gathercole, Pickering, Ambridge, & Wearing, 2004; Zald & Iacono, 1998). Spatial working memory tasks require that a response be guided by the on-line representation of a goal; the accuracy of the response is used as an index of working memory integrity. In many of these tasks, the location of a cue must be kept in working memory during a delay period, which may or may not include an interference stimulus or manipulation requirement (Swanson, 1999; Kwon, Reiss, & Menon, 2002). In the study reported by Zald el al. (1998), subjects touched the location of a previously presented cue after varying delays during which centrally presented words had to be verbalized. The results showed that across task delays and interference conditions, 14-year-olds showed immature performance. The memory-guided saccade task, also known as the oculo-motor delayed response task, requires an eye movement response to be guided by the representation in working memory of the location of a previously presented visual cue. This task does not require manipulation of the representation in working memory and, as such, is a direct measure of maintenance. The memory-guided saccade task has been used in nonhuman primates to identify delay-dependent cells that sustain working memory throughout the brain, including in prefrontal cortex (Funahashi, Inoue, & Kubota, 1997). This task also has a protracted developmental trajectory (Luna et al., 2004). Luna et al. (2004) showed that the accuracy of the *initial* working memory response matures around 15 years of age; however, *subsequent* corrective eye movement responses, which are more precise, continue to mature into the second decade of life. A widely distributed brain circuitry underlies spatial working memory in adults and includes dorsolateral prefrontal cortex (DLPFC), the cortical eye fields, anterior cingulate cortex (ACC), insula, basal ganglia, thalamus, and lateral cerebellum (Hikosaka & Wurtz, 1983; Sweeney et al., 1996).

While infants demonstrate working memory abilities (Diamond et al., 1989), a range of working memory tasks concur in indicating a protracted development in working memory through adolescence. Variability is present in the timing within late childhood and adolescence when adult-level working memory performance is reached. This variability is due to differences in stimuli, domain, opportunity for using strategies, and amount of interference control. There is evidence for different developmental trajectories in spatial versus object working memory tasks that sometimes reflect earlier development in the spatial domain (van Leijenhorst, Crone, & van der Molen, 2007), while at other times indicate earlier development in the object domain (Conklin, Luciana, Hooper, & Yarger, 2007; Luciana, Conklin, Hooper, & Yarger, 2005). The ability to use different types of strategies can often account for discrepancies regarding the time of development. Tasks where verbal rehearsal strategies can be used often show protracted development that may be independent of working memory processes (Cowan, Saults, & Morey, 2006; van Leijenhorst et al., 2007; Conklin et al., 2007). Planning based on working memory information also appears to enhance working memory performance for a longer developmental trajectory that shows continued benefit into adulthood. For example, while young adolescents can perform at adult levels when required to identify the location of a previously presented target, improvements continue into adulthood when organized search of several targets is required (Luciana et al., 2005). Inhibitory control is integral to working memory performance especially in tasks where distractors are used during periods of maintenance (Roncadin, Pascual-Leone, Rich, & Dennis, 2007). Therefore, improved response inhibition can enhance working memory performance (Bjorklund et al., 1990; Dempster, 1981). Although working memory and inhibitory processes are often considered part of the same executive process, their relative contribution can be manipulated to discern their unique influences to development (Luna et al., 2004; Asato et al., 2006; Luciana et al., 2005). Another important factor that can affect working memory performance is the ability to detect errors and use these to inform future responses. There is growing evidence that there is a protracted development of processes used to monitor performance and errors that are still immature in adolescence (van Leijenhorst et al., 2007; Velanova, Wheeler, & Luna, 2008; Crone, Jennings, & van der Molen, 2004). These results indicate that while basic aspects of working memory are present early in development, there is continued specialization of working memory processes that continues into adulthood. Throughout the life span, working memory directed behavior follows an inverted-U shape: there is improvement through adolescence, stabilization in adulthood, and a decrement in elderly years (Cowan, Naveh-Benjamin, Kilb, & Saults, 2006).

Given the delayed development of core cognitive processes, it is not surprising that there is protracted development in the performance of complex cognitive tasks (for reviews see Diamond, 2002; Welsh, 2002). Complex cognitive tasks are those that involve multiple cognitive processes, including attention, response inhibition, and working memory as well as language processing and mathematical abilities. Typical tasks include those that require planning and rule-guided behavior, for example, the Wisconsin Card Sorting Test, which requires subjects to deduce the randomly changing rule that determines correct responses; the Tower of London, in which subjects must arrange stimuli in a certain predetermined order in a minimal number of steps; and the Contingency Naming Test, in which subjects are given rules to produce a Stroop-like response. Failures in set maintenance and in planning are often indicated to underlie the protracted development of performance in complex cognitive tasks (Chelune & Thompson, 1987; Levin et al., 1991; Welsh, Pennington, & Groisser, 1991). However, developmental improvements in core cognitive abilities such as processing speed, working memory, and inhibition have been found to modulate development of these tasks (Asato et al., 2006).

In sum, there are aspects of working memory that appear early in development. However, there are improvements throughout adolescence in the ability to perform complex tasks, be more precise, use strategies, and control distraction, resulting in an efficient and adaptable working memory system that is able to support higher-order cognitive processes, such as abstract thought, and enhances decision making. Although an immature system may be able to process simple working memory demands, inefficient processing of more complex demands may result in poor decision making.

Maturation of Brain Structure

At the same time that improvements in cognitive control are emerging, structural brain changes are occurring that are believed to support mature behavior. The gross morphology of the brain, including the degree of cortical folding (Armstrong, Schleicher, Omran, Curtis, & Zilles, 1995), overall size, weight, and regional functional specialization, is adult-like by early childhood (Caviness, Kennedy, Bates, & Makris, 1996; Giedd, Snell, et al., 1996a; Reiss, Abrams, Singer, Ross, & Denckla, 1996). The skull thickens throughout childhood, accounting for increases in head size. Once the basic morphology is in place, brain processes occur that are believed to support the molding of brain structure to fit the biological and external environments of the organism. Some well-characterized mechanisms that occur throughout adolescence are synaptic pruning and increased myelination (Huttenlocher, 1990; Yakovlev & Lecours, 1967; Jernigan, Trauner, Hesselink, & Tallal, 1991). These processes lead to more efficient neuronal processing, which supports mature cognitive control of behavior. We are born with an excess of synaptic connections, which decrease in number throughout development owing to activity-dependent stabilization, that is, the connections that are used remain and those that are not used are eliminated (Rauschecker & Marler, 1987). Synaptic pruning promotes efficient integration of regional circuitry, enhancing the capacity and speed of information processing (Huttenlocher, 1990; Huttenlocher & Dabholkar, 1997). Myelination, which speeds neuronal transmission, also increases the efficiency of information processing and supports the integration of the widely distributed circuitry needed for complex behavior (Goldman-Rakic et al., 1993). These structural changes are believed to underlie the functional integration of frontal regions with the rest of the brain (Thatcher, Walker, & Giudice, 1987; Chugani, 1998; Luna & Sweeney, 2004b). Immature myelination in adolescence may limit top-down modulation of behavior, affecting cognitive control (Casey, Tottenham, Liston, & Durston, 2005; Luna et al., 2004; Thatcher, 1991; Chugani, Phelps, & Mazziotta, 1987). Recent structural neuroimaging studies have indicated reductions in gray matter throughout cortical association areas, notably the frontal and temporal regions (Gogtay et al., 2004; Toga, Thompson, & Sowell, 2006; Giedd et al., 1999; Sowell, Thompson, Tessner, & Toga, 2001; Paus et al., 1999), as well as the basal ganglia (Sowell, Thompson, Holmes, Jernigan, & Toga, 1999). This pattern continues past adolescence and may reflect both reduced synaptic connections and increased myelination. Diffusion tensor imaging studies indicate a continued increase in frontal white matter integrity throughout childhood, providing evidence for continued myelination with age (Klingberg, Vaidya, Gabrieli, Moseley, & Hedehus, 1999). These findings provide a new outlook on brain development, which had originally been thought to progress from posterior to anterior regions, with prefrontal cortex developing last (Hudspeth & Pribram 1990; Stuss 1992; Diamond & Taylor 1996; Luciana & Nelson, 1998). Instead results suggest that the functional integration of a widely distributed circuitry characterizes late development into adulthood (Luna et al., 2004).

Late development is also evident in the basal ganglia (Giedd, Vaituzis, et al., 1996b; Sowell et al., 1999; Toga et al., 2006) and orbitofrontal cortex (Gogtay et al., 2004), which may limit adolescents' ability to properly assess valence (reward and punishment). Evidence indicates that during adolescence, there is greater activity in excitatory dopamine (DA) systems than in inhibitory 5-hydroxytryptamine systems, resulting in an imbalance between reward and suppression mechanisms (Lambe, Krimer, & Goldman-Rakic, 2000; Chambers et al., 2003). Nigrostriatal DA neurons and components of the basal ganglia show their highest activity in childhood, then decrease exponentially during the first three decades of life (Segawa, 2000). D1 receptors in ventral striatum develop earlier than D2 receptors, which continue to decrease throughout adolescence (Meng, Ozawa, Itoh, & Takashima, 1999). D2 receptors help guide voluntary saccades (Kato et al., 1995). The number of DA transporters, which remove DA from the synapse and are primarily found in the striatum, peaks during adolescence, as does myelination in this region (Meng et al., 1999). Projections from the basal ganglia that ascend to the thalamus via striatal pathways, which provide loops with frontal systems, continue to mature in adolescence (Segawa, 2000). Animal studies indicate that DA inputs to prefrontal cortex peak in adolescence, which increase inhibitory inputs to prefrontal cortex, resulting in decreased excitatory outputs (Lewis, 1997; Spear, 2000). The effects of brain immaturities have lead to two schools of thought regarding reward processing in adolescence. One proposal is that adolescence is a time of low motivation which leads to the search for more salient rewards, perhaps resulting in substance abuse (Spear, 2000; Castellanos & Tannock, 2002). Others believe that adolescence is a time of exaggerated reward response which leads to sensation seeking and risk-taking behavior (Chambers et al., 2003). Both arguments are supported by the effects of the immaturity of the system. An alternative proposal is that adolescents demonstrate low motivation when there is a low cognitive load requiring little effort, such as in reaction time tasks (Bjork et al., 2004), but overactivity of the reward system when cognitive effort is required (Ernst et al., 2005; Galvan et al., 2006) (see later).

Taken together, these studies indicate that the brain systems that are crucial for exerting cognitive control over behavior and processing rewards are still immature during adolescence. These immaturities result in a system that is able to exert cognitive control, but in an inconsistent manner with limited flexibility and motivational control. In other words, the basic elements are established, but refinements are needed to support the necessary efficiency in circuit processing to establish reliable executive control.

Pubertal Maturation and Cognitive Development

Concurrent with cognitive development and brain maturation in late childhood and adolescence is pubertal development. During puberty, neurotransmitters and glial cells regulate neuronal control of gonadotropin-releasing hormone, which stimulates the secretion of hormones in the pituitary gland, resulting in ovarian development in females and spermatogenesis in males (Ojeda, Ma, & Rage, 1995). The timing of this process is determined by age as well as metabolic and neuronal factors. Changes in gonad hormone levels have direct effects on molecular mechanisms throughout the brain (McEwen, 2001). There is evidence that gonad hormones may influence cortical development, which may help account for sex differences in cognitive development (Clark & Goldman-Rakic, 1989). A direct way to determine pubertal timing is by assessing bone age using X-rays, which is counterindicated for normative pediatric studies. Most methods for determining pubertal timing, such as breast and testicular growth in pediatric examination, age of menarche onset, age of mutation (voice change for men), increased testosterone in saliva, peak height growth velocity, Tanner staging, Tanner physical examination, and Tanner self-report (Duke, Litt, & Gross, 1980; Tanner, 1962), measure the emergence of secondary characteristics that result from puberty and occur later than the onset of hormonal changes. Tanner staging, though limited owing to its subjective nature which can lead to variability, is the least invasive measure of pubertal timing. Hormonal changes are well recognized to affect mood and social processes (Alsaker, 1996). Hormonal changes affect sex steroid receptors in the hippocampus which support novelty seeking and regulate DA in the nucleus accumbens (Chambers et al., 2003). Pubertal changes may thus have a direct link to reward sensitivity. On the other hand, it has been harder to establish direct links between pubertal changes and cognitive processes. Some studies have found a link between spatial abilities and pubertal timing (Petersen, 1976; Waber, Mann, Merola, & Moylan, 1985), whereas others have not (Strauss & Kinsbourne, 1981; Orr, Brack, & Ingersoll, 1988). To characterize changes during adolescence, it is important to delineate the parallel systems that undoubtedly contribute to how cognitive control will manifest itself in behavior and, more importantly, to how it may play a role in risk-taking behavior and the vulnerability of this stage to psychopathologic dysfunction.

Development of Brain Function

The investigation of how the brain functions during the exertion of cognitive control is central to understanding the interface between cognitive control and immaturities in brain structure. In this manner, we can begin to understand the consequences of brain immaturities and the basis of behavior in adolescence. Moreover, characterizing a normative template of brain function can provide a window into the impaired development evident in major psychiatric disorders.

The main approaches that have been used to understand this relationship are electroencephalography (EEG) and neuroimaging techniques, such as positron emission tomography (PET) and, more prominently, fMRI. Whereas EEG is used extensively to study the brain basis of cognition in infancy, it has rarely been used to assess changes throughout the life span. A landmark study investigated changes in EEG coherence throughout childhood and into adulthood (Thatcher et al., 1987). The results showed an increase in coherence of EEG activity across neocortical regions throughout adolescence, primarily between frontal and other cortical areas, with differentiation in the left hemisphere and integration in the right hemisphere. These findings have been supported by PET results demonstrating immaturities in the local cerebral resting metabolic rates of frontal, parietal, and temporal regions that only begin to mature in adolescence (Chugani, 1998). Taken together, the results indicate the late integration of the frontal lobe into the widely distributed brain circuitry supporting cognitive control of behavior. However, these procedures have limited spatial resolution and do not allow the identification of circuit-based changes in the brain. PET is particularly counterindicated in pediatric studies because of the invasive nature of radioactive isotopes that are injected to localize brain function. In contrast, fMRI is a noninvasive neuroimaging procedure that allows in vivo investigation of brain activity underlying cognitive function in pediatric populations. Furthermore, fMRI provides high spatial resolution, localizing brain function on the order of millimeters. It measures regional changes in blood oxygenation levels produced by increases in metabolic demands resulting from neuronal activity that supports the cognitive processes of interest. It is an indirect measure of neuronal activity, but one that has increasingly been found to be tightly coupled with neuronal activity (Logothetis & Pfueffer, 2004).

A burgeoning literature has begun to characterize adolescent development using fMRI. As fMRI requires subjects to remain still and follow instructions, it is usually performed on children older than 6–8 years. This caveat has opened the door for the emergence of developmental studies that investigate later stages of development. Cognitive control has been investigated by probing the developmental differences supporting inhibitory control, working memory, and rule-guided behavior. The interpretation of fMRI results is an area of debate as age differences may result in either increases or decreases in activity. One theory of age-related decreases suggests that maturation is characterized by a focalization of activity due to increased sophistication of neuronal processes (Durston et al., 2006). However, many studies also show age-related increases in activity and the recruitment of regions that do not participate at younger ages. One approach is to predefine function in the adult system as the mature goal state and function that deviates from this standard as immature. Age-related decreases could indicate focalization of activity as well as a decreased need for recruitment of particular regions. Age-related increases could reflect immaturity in the structure of the circuits supporting mature performance.

Development of Brain Function Underlying Voluntary Response Inhibition

Studies investigating developmental differences in response inhibition have consistently found that activity in prefrontal cortex increases from childhood to adulthood. A failure to recruit ventrolateral prefrontal cortex (VLPFC), a key region supporting response inhibition (Blasi et al., 2006), was found in subjects of 8–12 years of age compared with adults while performing a mixed go/no-go (inhibition of an established response) and flanker (interference control) task (Bunge, Dudukovic, Thomason, Vaidya, & Gabrieli, 2002 a,b). These results are supported by another study that found increased activity in prefrontal regions in adults (22–40 years old) compared with adolescents (12–19 years old) in a task requiring the suppression of an established response (Rubia et al., 2000). Age-related increases in prefrontal function have also been found during a Stroop interference task (Adleman et al., 2002), as adolescents demonstrate reduced prefrontal recruitment in comparison with adults, whereas their parietal activity is mature. On the other hand, age-related decreases in the inferior frontal gyrus and increases in the middle frontal gyrus were found from 8 to 20 years of age while subjects were performing a go/no-go task (Tamm, Menon, & Reiss, 2002). These results were interpreted as indicating that decreased activity in inferior frontal gyrus are due to developmental improvements in the ease of performing the task, whereas increases in middle frontal gyrus reflect the specialization of this region for inhibitory control. Studies that test subjects across the life span indicate overall age-related increases in prefrontal regions that are associated with improved inhibitory performance. In a study looking at 7–57-year-olds, overall age-related increases in activity of inferolateral prefrontal cortex and frontostriatal systems were correlated with improved performance during a Stroop task (Marsh et al., 2006). Similarly, in a study looking at 10–47-year-olds performing a stop-signal task, overall age-related increases in right inferior prefrontal cortex were found during successful inhibition (Rubia, Smith, Taylor, & Brammer, 2007). Additionally, correlational analyses indicated age-related integration of frontostriatothalamic and frontocerebellar neural pathways for inhibitory control. Another study comparing 10–17-year-olds with 20–43-year-old adults found similar increases in frontostriatal regions that correlated with improvements in a battery of inhibitory tasks (go/no-go, Simon, and attentional set-shifting (Rubia et al., 2006). These results confirm earlier findings indicating that frontal regions supporting improvements in performance show age-related increases (Rubia et al., 2000).

Previous studies have assumed that the developmental progression from childhood to adolescence to adulthood is linear; however, there is evidence that development is stagelike (Thatcher et al., 1987) and that adolescence may be qualitatively different from either childhood or adulthood. Taking a systems-level approach at defining developmental differences in the circuit-based processes supporting response inhibition,

Luna et al. (2001) reported both increases and decreases that differed depending on when in development they occurred. Subjects 8–30 years of age were studied using the antisaccade task and analyses compared children (8–12 years old), adolescents (13–17 years old), and adults (18–30 years old). Activity in DLPFC, which was defined as the middle frontal gyrus, showed an increase from childhood to adolescence but a decrease from adolescence to adulthood. Adults showed increased activity in premotor areas and parietal eye fields, as well as cerebellum. Age-related increases in parietal and cerebellar regions have also been found in other developmental studies of inhibition (Rubia et al., 2006; Bunge et al., 2002 a, b). These results suggest that there may be stagelike changes in development: from childhood to adolescence the DLPFC, which is crucial for inhibitory control (Fuster, 1997; Fassbender et al., 2004), shows increased participation matching increased performance; however, from adolescence to adulthood, when performance does not change dramatically, there is a decrease in prefrontal activity, indicating a specialization of this region for cognitive control. In other words, although performance is stable from adolescence to adulthood, adolescents must exert more executive effort to do the task. In adult studies, prefrontal activity increases with higher cognitive loads. In this manner, adolescent brain function during inhibition tasks is similar to adult brain function during more difficult tasks. Difficult tasks have a higher probability of error; thus, although adolescents may behave like adults, they require more effort, which makes them more vulnerable to error (Luna et al., 2004).

An important aspect of inhibitory performance is the ability to detect errors and use this information to influence subsequent responses. Error detection and performance monitoring are integral to cognitive control and the ACC has been identified as the primary structure where error is processed (Braver, Barch, Gray, Molfese, & Snyder, 2001). Developmental studies have begun to find that there are immaturities in error processing during response inhibition through adolescence (Rubia et al., 2007; Velanova et al., 2008). Using a stop task where performance was equivalent across ages, Rubia et al. (2007) found increases for adults relative to children in rostral ACC and right inferior prefrontal cortex during error commission. Our own studies have indicated not only increased activity in ACC during errors of inhibition, but we also found that this was specific to different stages of error processing. We found that ventral ACC had equivalent activation in initial stages of error processing across childhood and adolescence; however, a later second stage of error processing showed that only adults recruited the dorsal ACC, reflecting immaturities even in adolescence (Velanova et al., 2008) (Fig. 13.2). These results imply that while the detection of errors may be on-line early in development, later stages that may influence subsequent behavior may have a protracted development.

Taken together, developmental fMRI studies indicate that prefrontal cortex is involved in response inhibition early in childhood and that its relative participation becomes more robust and focal with maturity. The results support the early contribution of prefrontal cortex to response inhibition. However, the ability to consistently inhibit responses characterizes the transition to mature performance, and this process necessitates a widely distributed circuitry and integrated brain system (Goldman-Rakic, 1988). The fact that functional integration across neocortex is

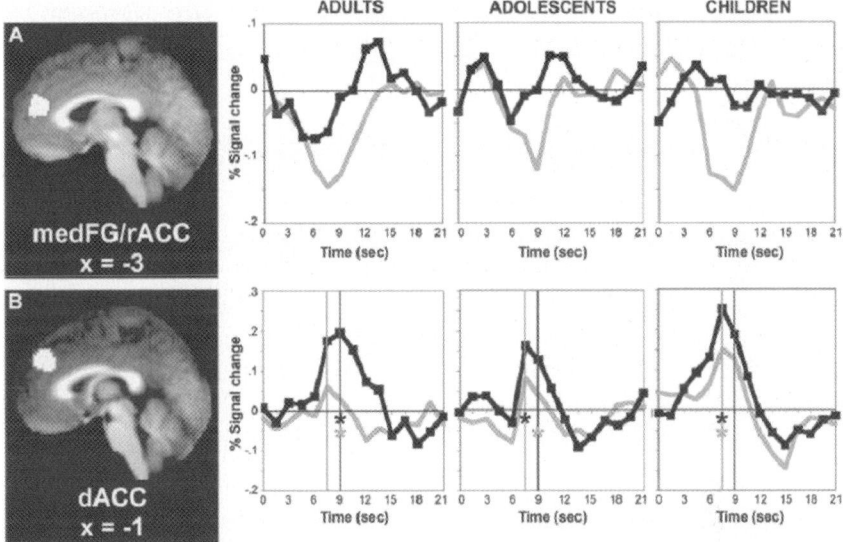

Fig. 13.2 (**a**) No age differences in the rostral anterior cingulate cortex region during correct (*black lines*) and incorrect (*gray lines*) inhibitory responses. (**b**) Greater activity in adults in the dorsal anterior cingulate cortex region for incorrect inhibitory responses. *rACC* rostral anterior cingulate cortex, *dACC* dorsal anterior cingulate cortex. (Reprinted with permission Velanova et al. 2008)

present throughout adolescence (Thatcher et al., 1987; Chugani, 1998) indicates that in conjunction with developmental improvements in the intrinsic computational capacity of prefrontal cortex there is increased integration of this and other brain regions during adolescence. Several studies have indicated that additional areas become active in adulthood, including parietal, striatal, and cerebellar regions (Luna et al., 2001; Tamm et al., 2002; Rubia et al., 2007; Bunge, Dudukovic, Thomason, Vaidya & Gabrieli, 2002 a, b), which may establish an efficient circuitry supporting mature response inhibition through connections with prefrontal cortex.

While there is variability across studies in the specific structures showing developmental differences in brain function, there is agreement that some functional immaturities are present throughout adolescence. Immaturities in brain function in the presence of mature performance could indicate a fragile system that may be particularly vulnerable to impairment, as evident in psychopathologic dysfunction and risk-taking behavior. Characterizing a normative system allows the investigation of the impaired development usually associated with failures of cognitive control, as seen, for example, in conditions such as attention deficit hyperactivity disorder, Tourette's syndrome, and autism (Luna & Sweeney, 1999; Luna et al., 2002). It also helps us begin to understand the mechanisms underlying risk-taking behavior in adolescence.

Development of Brain Function Underlying Working Memory

Many studies have characterized the changes in brain function that underlie the development of working memory. Working memory tasks involve different stages of information processing: encoding task demands, maintaining information on-line, and, following Baddeley's influential model (Baddeley, 1986), manipulating information during maintenance. Most studies have focused on characterizing maintenance systems, but some studies have looked at developmental changes during manipulation. Collectively, developmental studies indicate that the integration of prefrontal and parietal systems is crucial to the maturation of working memory. Early studies, focusing on children younger than 12 years, identified areas that were recruited by different age groups. These studies used spatial working memory tasks in which subjects had to remember the location of a visual cue (Thomas et al., 1999; Nelson et al., 2000) or simple n-back tasks in which subjects had to press a button in response to a letter that was repeated with one intervening letter (e.g., A-S-A; Casey et al., 1995). The results showed that children, like adults, are able to recruit prefrontal and parietal regions involved in working memory; however, the degree of activation was not compared. Using a categorical working memory task that requires multiple cognitive components (object processing and language), one study found that children recruit premotor/striatal/cerebellar networks rather than the ventral prefrontal and inferior temporal regions recruited by adults (Ciesielski, Lesnik, Savoy, Grant, & Ahlfors, 2006). These results suggest that core working memory systems are available early in childhood but may fail to be recruited in more complex working memory tasks. In older samples, increased participation of DLPFC and parietal cortex has been found from childhood to adolescence (Crone, Wendelken, Donohue, van Leijenhorst, & Bunge, 2006; Scherf, Sweeney, & Luna, 2006; Klingberg, Forssberg, & Westerberg, 2002; Olesen, Macoveanu, Tegner, & Klingberg, 2006a). Klingberg et al. (2002) found age-related increases in activity of superior frontal and intraparietal cortex when 9–18-year-old subjects had to remember the location of a visual target. Crone et al. (2006) found that although 8–12-year-olds recruited VLPFC for maintenance, they failed to recruit DLPFC during the manipulation of working memory when they had to reorder the presentation of three pictures of objects. By adolescence, activation in DLPFC was adult-like. In a study that required subjects to remember locations of circles, age-related increases in prefrontal activity were found in adults compared with 13-year-old children; however, when a distractor was present, children showed more prefrontal activity, reflecting their difficulty inhibiting competing stimuli (Olesen, Macoveanu, Tegner, & Klingberg, 2006b). Scherf et al. (2006) used the oculomotor delayed response task to focus on the maintenance aspect of working memory. Activation in DLPFC was found only minimally in children, and its highest level of participation was in adolescents. Children relied more on basal ganglia and insula, whereas adults recruited additional regions, including temporal regions. This study considered brain function at the systems level and found that with maturity a larger and more distributed circuitry supported brain function (Fig. 13.3). Increases in activation have also been associated with increases

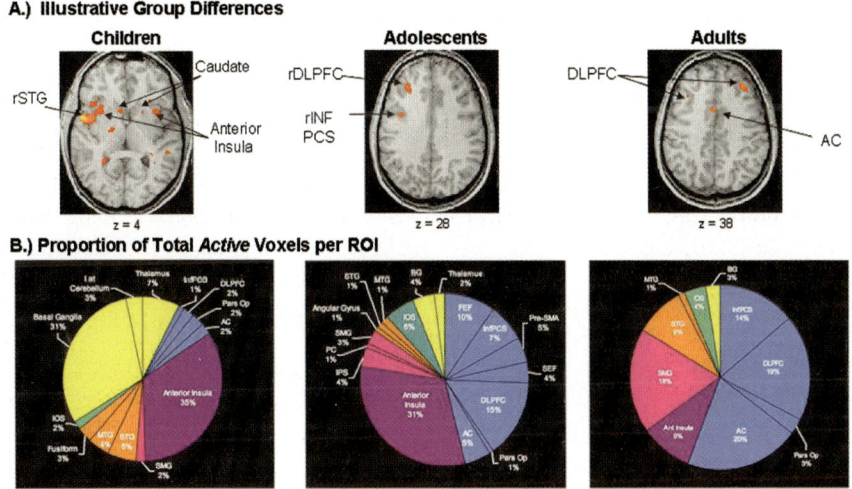

Fig. 13.3 (**a**) Axial slices illustrating regions of age group differences. (**b**) Distribution of regions with working memory related activity for each age group. While children showed predominance in the recruitment of basal ganglia and insula, adolescents showed a bias towards insula and right dorsolateral prefrontal cortex. By adulthood, activity is more distributed with several regions sharing control in working memory. *DLPFC* dorsolateral prefrontal cortex, *rDLPFC* right dorsolateral prefrontal cortex, *ROI* region of interest, *AC, PCS, rINF*. (Reprinted with permission from Scherf et al. 2006)

in the integrity of white matter in frontoparietal regions (Olesen, Nagy, Westerberg, & Klingberg, 2003); however, the interaction of parietal and prefrontal regions appears to be even more strongly associated with performance (Edin, Macoveanu, Olesen, Tegner, & Klingberg, 2007).

Development of Brain Function Underlying Reward Processing

A crucial aspect of cognitive control is the manner in which motivation influences optimal performance. Striatal and DA systems known to support motivation continue to mature during adolescence. fMRI studies have begun to characterize the developmental trajectory of reward/motivation processing through adolescence (van Leijenhorst, Crone, & Bunge, 2006; Galvan et al., 2006; Eshel, Nelson, Blair, Pine, & Ernst, 2007; Galvan, Hare, Voss, Glover, & Casey, 2007). Several studies have found that rewards differentially affect activity in reward-related circuitry during adolescence. Using a gambling task where subjects had to guess the winning response, Ernst et al. (2005) found that adolescents have *elevated* activity in the nucleus accumbens during reward anticipation, whereas adults have *decreased* activity in the amygdala to reward omission. During the selection of risky choices, adolescents showed decreased activity of orbital frontal cortex, VLPFC, and ACC, indicating limited executive control (Eshel et al., 2007). Similarly, Galvan ct al. (2006) used a reward-association task and found that adolescents showed elevated activity in the nucleus accumbens compared with adults, who showed elevated activity in orbital frontal cortex. Activity in the nucleus accumbens was highest in individuals who rated themselves as engaging in risky behavior (Galvan et al., 2007). In contrast to the findings in previous reports, Bjork et al., (2004), using a reaction time task with relatively low cognitive load, found that adolescents showed *decreased* activity compared with adults in the ventral striatum when they were anticipating responding for a reward, but were not different from adults in activity supporting gain outcomes. Discrepant results may be due to differences in the cognitive load of the response: one study had very low demands in a reaction time task (Bjork et. al., 2004) that may lead to underactivity in the reward system and the others had higher cognitive demands when subjects had to anticipate and choose a response and showed overactivity of the reward system (Ernst et al., 2005; Galvan et al.,2006). Work in the rat model system suggests that the nucleus accumbens is predominantly involved in goal-directed behavior associated with rewards (Carelli, 2004). Alternatively, adolescents may have delayed striatal processing that is underactive initially during reward assessment but is overactive during reward feedback. This possibility still needs to be explored.

Cognitive and Brain Maturation

Traditional theory indicates that higher-order cognitive development is due to a late maturation of frontal circuitry (Hudspeth & Pribram, 1990; Stuss, 1992; Diamond & Taylor, 1996; Luciana et al., 1998). This viewpoint is based on early studies (Huttenlocher, 1990; Yakovlev et al., 1967) showing a late stabilization of synaptic pruning and myelination in frontal, relative to visual, cortex in humans and is not consistent with nonhuman primate studies that show concurrent development throughout the neocortex. This view suggests that executive abilities do not emerge until the brain is fully mature and that abnormalities in development are localized in the frontal cortex. In contrast, a careful review of these studies, along with the integration of recent magnetic resonance (Giedd et al., 1999; Sowell et al., 2001; Paus et al., 1999) and histological (Benes, 1998) data, demonstrates a dynamic late maturation *throughout* association cortex (Rakic, Bourgeois, Eckenhoff, Zecevic, & Goldman-Rakic, 1986). The last regions to show thinning of gray matter are the association areas of each lobe, with temporal regions showing the most protracted development, presumably due to the continued specialization of language (Gogtay et al., 2004). In this manner, the expansion of functional integration coupled with enhanced regional computational capacity allows for the more efficient use of the widely distributed circuitry underlying adult-level higher-order cognition (Goldman-Rakic et al., 1993). Adolescence may mark the beginning of a qualitatively different stage in the brain–behavior relationship. Newly established distributed circuitry, rather than regional circuitry, may govern behavior, reflecting the transition from exogenous to endogenous control. This approach suggests that although development in childhood is characterized by gaining abilities, maturation in adolescence consists of a qualitative difference in the use of existing capabilities. In this new framework, the emergence of psychiatric illness can be viewed not as a loss of cognitive abilities, but as an inability to shift to this new mode of operation.

Summary

Behavioral studies indicate that there are important improvements in cognitive control throughout adolescence. Although the basic tools that enable cognitive control are present early in development, the ability to use these tools in a consistent, flexible manner continues to improve through adolescence. Despite the often adult-like cognitive control of behavior exhibited by adolescents, these processes are still not supported by mature brain systems. There is evidence of immaturity, namely, in the ability of prefrontal systems to integrate with other brain regions to efficiently coordinate a planned action. Voluntary control of behavior underlies decision making and is supported by top-down modulation of behavior, which enables prefrontal systems to implement goal-driven behavior by efficiently coordinating with regions involved in processing behavioral demands and enacting responses. The transition to mature behavior is characterized by a shift to distributed function supporting cognitive control. Synaptic pruning enhances the ability to support complex computations necessary for higher-order behavior, such as the ability to enact plans that allow for response inhibition and retaining representations of goals in working memory. Myelination supports the ability to tap into a widely distributed circuitry that allows executive regions to control other regions that determine responses. This transition in adolescence may be a necessary stage of development that supports the effective sculpting of the brain to optimally match the unique environment of the individual. There are still many issues that need to be understood, such as the precise mechanisms that limit mature cognitive control, the immaturities in the brain that hamper efficient functional integration, and the ways in which these factors go awry in clinical populations.

Acknowledgements Research presented in this review was supported by the National Institutes of Mental Health (NIMH RO1 MH067924). We thank everyone in the Laboratory of Neurocognitive Development and Krista Garver and Chuck Geier for editorial comments.

References

Adleman, N. E., Menon, V., Blasey, C. M., White, C. D., Warsofsky, I. S., Glover, G. H. et al. (2002). A developmental fMRI study of the Stroop Color-Word task. *NeuroImage, 16*, 61–75.

Alsaker, F. D. (1996). Annotation: The impact of puberty. *Journal of Child Psychology and Psychiatry, 37*, 249–258.

Anderson, V. A., Anderson, P., Northam, E., Jacobs, R., & Catroppa, C. (2001). Development of executive functions through late childhood and adolescence in an Australian sample. *Developmental Neuropsychology, 20*, 385–406.

Angold, A., Costello, E. J., & Worthman, C. M. (1998). Puberty and depression: The roles of age, pubertal status and pubertal timing. *Psychological Medicine, 28*, 51–61.

Armstrong, E., Schleicher, A., Omran, H., Curtis, M., & Zilles, K. (1995). The ontogeny of human gyrification. *Cerebral Cortex, 5*, 56–63.

Arnett, J. (1992). Reckless behavior in adolescence: A developmental perspective. *Developmental Reveiw, 12*, 339–373.

Asato, M. R., Sweeney, J. A., & Luna, B. (2006). Cognitive processes in the development of TOL performance. *Neuropsychologia, 44*, 2259–2269.

Baddeley, A. (1986). *Working memory*. New York: Oxford University Press.

Barlow, D. H. (1988). *Anxiety and its disorders: The nature and treatment of anxiety and panic*. New York: Guilford Press.

Bedard, A. C., Nichols, S., Barbosa, J. A., Schachar, R., Logan, G. D., & Tannock, R. (2002). The development of selective inhibitory control across the life span. *Developental Neuropsychology, 21*, 93–111.

Benes, F. M. (1998). Brain development, VII. Human brain growth spans decades. *American Journal of Psychiatry, 155*, 1489.

Bjork, J. M., Knutson, B., Fong, G. W., Caggiano, D. M., Bennett, S. M., & Hommer, D. W. (2004). Incentive-elicited brain activation in adolescents: Similarities and differences from young adults. *The Journal of Neuroscience: The Official Journal of the Society for Neuroscience, 24*, 1793–1802.

Bjorklund, D. F. & Harnishfeger, K. K. (1990). The resources construct in cognitive development: Diverse sources of evidence and a theory of inefficient inhibition. *Developmental Review, 10*, 48–71.

Bjorklund, D. F. & Harnishfeger, K. K. (1995). The evolution of inhibition mechanisms and their role in human cognition and behavior. In F. N. Dempster & C. J. Brainerd (Eds.), *Interference & inhibition in cognition* (pp. 141–173). San Diego, CA: Academic.

Blasi, G., Goldberg, T. E., Weickert, T., Das, S., Kohn, P., Zoltick, B. et al. (2006). Brain regions underlying response inhibition and interference monitoring and suppression. *European Journal of Neuroscience, 23*, 1658–1664.

Braver, T. S., Barch, D. M., Gray, J. R., Molfese, D. L., & Snyder, A. (2001). Anterior cingulate cortex and response conflict: Effects of frequency, inhibition and errors. *Cerebral Cortex, 11*, 825–836.

Bunge, S. A., Dudukovic, N. M., Thomason, M. E., Vaidya, C. J., & Gabrieli, J. D. E. (2002a). Development of frontal lobe contributions to cognitive control in children: Evidence from fMRI. *Neuron, 33*, 301–311.

Bunge, S. A., Dudukovic, N. M., Thomason, M. E., Vaidya, C. J., & Gabrieli, J. D. (2002b). Immature frontal lobe contributions to cognitive control in children: Evidence from fMRI. *Neuron, 33*, 301–311.

Carelli, R. M. (2004). Nucleus accumbens cell firing and rapid dopamine signaling during goal-directed behaviors in rats. *Neuropharmacology, 47*, 180–189.

Case, R. (1992). The role of the frontal lobes in the regulation of cognitive development. *Brain and Cognition, 20*, 51–73.

Case, R. (1995). Capacity-based explanations of working memory growth: A brief history and reevaluation. In F. E. Weinert & W. Schneider (Eds.), *Memory performance and competencies: Issues in growth and development.* (pp. 23–44).

Case, R. (1996). Modeling the process of conceptual change in a continuously evolving hierarchical system. *Monogr Soc Res Child Dev, 61,* 283–295.

Casey, B. J., Cohen, J. D., Jezzard, P., Turner, R., Noll, D. C., Trainor, R. J. et al. (1995). Activation of prefrontal cortex in children during a nonspatial working memory task with functional MRI. *NeuroImage, 2,* 221–229.

Casey, B. J., Tottenham, N., Liston, C., & Durston, S. (2005). Imaging the developing brain: What have we learned about cognitive development? *Trends Cogn Sci, 9,* 104–110.

Castellanos, F. X. & Tannock, R. (2002). Neuroscience of attention-deficit/hyperactivity disorder: The search for endophenotypes. *Nature Reviews.Neuroscience, 3,* 617–628.

Caviness, V. S., Kennedy, D. N., Bates, J. F., & Makris, N. (1996). The Developing Human Brain: A Morphometric Profile. In R. W. Thatcher, G. Reid Lyon, J. Rumsey, & N. A. Krasnegor (Eds.), *Developmental Neuroimaging: Mapping The Development of Brain and Behavior* (pp. 3–14). New York: Academic.

Chambers, R. A., Taylor, J. R., & Petenza, M. N. (2003). Developmental neurocircuitry of motivation in adolescence: A critical period of addiction vulnerability. *American Journal of Psychiatry, 160,* 1041–1052.

Chelune, G. J. & Thompson, L. L. (1987). Evaluation of the general sensitivity of the Wisconsin Card Sorting Test among younger and older children. *Developental Neuropsychology, 3,* 81–89.

Chugani, H. T. (1998). A critical period of brain development: Studies of cerebral glucose utilization with PET. *Preventive Medicine, 27,* 184–188.

Chugani, H. T., Phelps, M. E., & Mazziotta, J. C. (1987). Positron emission tomography study of human brain functional development. *Annals of Neurology, 22,* 487–497.

Ciesielski, K. T., Lesnik, P. G., Savoy, R. L., Grant, E. P., & Ahlfors, S. P. (2006). Developmental neural networks in children performing a Categorical N-Back Task. *NeuroImage, 33,* 980–990.

Clark, A. S. & Goldman-Rakic, P. S. (1989). Gonadal hormones influence the emergence of cortical function in nonhuman primates. *Behavioral Neuroscience, 103,* 1287–1295.

Conklin, H. M., Luciana, M., Hooper, C. J., & Yarger, R. S. (2007). Working memory performance in typically developing children and adolescents: Behavioral evidence of protracted frontal lobe development. *Developmental Neuropsychology, 31,* 103–128.

Cowan, N., Naveh-Benjamin, M., Kilb, A., & Saults, J. S. (2006). Life-span development of visual working memory: When is feature binding difficult? *Developmental Psychol, 42,* 1089–1102.

Cowan, N., Saults, J. S., & Morey, C. C. (2006). Development of Working Memory for Verbal-Spatial Associations. *J Mem Lang, 55,* 274–289.

Crone, E. A., Jennings, J. R., & van der Molen, M. W. (2004). Developmental change in feedback processing as reflected by phasic heart rate changes. *Developmental Psychol, 40,* 1228–1238.

Crone, E. A., Wendelken, C., Donohue, S., van Leijenhorst, L., & Bunge, S. A. (2006). Neurocognitive development of the ability to manipulate information in working memory. *Proceedings of the National Academy of Sciences of the United States of America, 103,* 9315–9320.

Davies, P. L. & Rose, J. D. (1999). Assessment of cognitive development in adolescents by means of neuropsychological tasks. *Developmental Neuropsychology, 15,* 227–248.

Demetriou, A., Christou, C., Spanoudis, G., & Platsidou, M. (2002). The development of mental processing: Efficiency, working memory, and thinking. *Monographs of the Society for Rersearch in Child Development, 67,* 1–155; discussion 156.

Dempster, F. N. (1981). Memory span: Sources of individual and developmental differences. *Psychological Bulletin, 89,* 63–100.

Dempster, F. N. (1992). The rise and fall of the inhibitory mechanism: Toward a unified theory of cognitive development and aging. *Developmental Review, 12,* 45–75.

Dempster, F. N. (1993). Resistance to Interference: Developmental changes in a basic processing mechanism. In M. L. Howe & R. Pasnak (Eds.), *Emerging themes in cognitive development, volume I: foundations* (pp. 3–27). New York: Springer.

Diamond, A. (2002). Normal development of prefrontal cortex from birth to young adulthood: Cognitive functions, anatomy, and biochemistry. In D. T. Stuss & R. T. Knight (Eds.), *Principles of frontal lobe function* (pp. 466–503). New York: Oxford University Press.

Diamond, A. & Goldman-Rakic, P. S. (1989). Comparison of human infants and rhesus monkeys on Piaget's AB task: Evidence for dependence on dorsolateral prefrontal cortex. *Experimental Brain Research, 74*, 24–40.

Diamond, A. & Taylor, C. (1996). Development of an aspect of executive control: Development of the abilities to remember what I said and to "do as I say, not as I do". *Developmental Psychobiology, 29*, 315–334.

Duke, P. M., Litt, I. F., & Gross, R. T. (1980). Adolescents' self-assessment of sexual maturation. *Pediatrics, 66*, 918–920.

Durston, S., Davidson, M. C., Tottenham, N., Galvan, A., Spicer, J., Fossella, J. A. et al. (2006). A shift from diffuse to focal cortical activity with development. *Developmental Science, 9*, 1–8.

Edin, F., Macoveanu, J., Olesen, P. J., Tegner, J., & Klingberg, T. (2007). Stronger synaptic connectivity as a mechanism behind development of working memory-related brain activity during childhood. *Journal of Cognitive Neuroscience, 19*, 750–760.

Elliott, R. (1970). Simple reaction time: Effectsassociated with age, preparatory interval, incentive-shift and mode of presentation. *Journal of Experimental Child Psychology, 9*, 86–107.

Ernst, M., Nelson, E. E., Jazbec, S., McClure, E. B., Monk, C. S., Leibenluft, E. et al. (2005). Amygdala and nucleus accumbens in responses to receipt and omission of gains in adults and adolescents. *NeuroImage, 25*, 1279–1291.

Eshel, N., Nelson, E. E., Blair, R. J., Pine, D. S., & Ernst, M. (2007). Neural substrates of choice selection in adults and adolescents: Development of the ventrolateral prefrontal and anterior cingulate cortices. *Neuropsychologia, 45*, 1270–1279.

Everling, S., Dorris, M. C., Klein, R. M., & Munoz, D. P. (1999). Role of primate superior colliculus in preparation and execution of anti-saccades and pro-saccades. *Journal of Neuroscience, 19*, 2740–2754.

Everling, S. & Munoz, D. P. (2000). Neuronal correlates for preparatory set associated with pro-saccades and anti-saccades in the primate frontal eye field. *Journal of Neuroscience, 20*, 387–400.

Fassbender, C., Murphy, K., Foxe, J. J., Wylie, G. R., Javitt, D. C., Robertson, I. H. et al. (2004). A topography of executive functions and thier interactions revealed by functional magnetic resonance imaging. *Brain research.Cognitive Brain Research, 20*, 132–143.

Fischer, B., Biscaldi, M., & Gezeck, S. (1997). On the development of voluntary and reflexive components in human saccade generation. *Brain Research, 754*, 285–297.

Fry, A. F. & Hale, S. (1996). Processing speed, working memory, and fluid intelligence: Evidence for a developmental cascade. *Psychological Science, 7*, 237–241.

Fukushima, J., Hatta, T., & Fukushima, K. (2000). Development of voluntary control of saccadic eye movements. I. Age-related changes in normal children. *Brain & Development, 22*, 173–180.

Funahashi, S., Chafee, M. V., & Goldman-Rakic, P. S. (1993). Prefrontal neuronal activity in rhesus monkeys performing a delayed anti-saccade task. *Nature, 365*, 753–756.

Funahashi, S., Inoue, M., & Kubota, K. (1997). Delay-period activity in the primate prefrontal cortex encoding multiple spatial positions and their order of presentation. *Behavioural Brain Research, 84*, 203–223.

Fuster, J. M. (1997). *The prefrontal cortex* (3 ed.). New York: Raven Press.

Galvan, A., Hare, T. A., Parra, C. E., Penn, J., Voss, H., Glover, G. et al. (2006). Earlier development of the accumbens relative to orbitofrontal cortex might underlie risk-taking behavior in adolescents. *Journal of Neuroscience, 26*, 6885–2692.

Galvan, A., Hare, T., Voss, H., Glover, G., & Casey, B. J. (2007). Risk-taking and the adolescent brain: Who is at risk? *Developmental Science, 10*, F8–F14.

Gathercole, S. E., Pickering, S. J., Ambridge, B., & Wearing, H. (2004). The structure of working memory from 4 to 15 years of age. *Developmental Psychology, 40*, 177–190.

Giedd, J. N., Blumenthal, J., Jeffries, N. O., Castellanos, F. X., Liu, H., Zijdenbos, A. et al. (1999). Brain development during childhood and adolescence: A longitudinal MRI study. *Nature Neuroscience, 2*, 861–863.

Giedd, J. N., Snell, J. W., Lange, N., Rajapakse, J. C., Casey, B. J., Kozuch, P. L. et al. (1996). Quantitative magnetic resonance imaging of human brain development: Ages 4–18. *Cerebral Cortex, 6*, 551–560.

Giedd, J. N., Vaituzis, A. C., Hamburger, S. D., Lange, N., Rajapakse, J. C., Kaysen, D. et al. (1996). Quantitative MRI of the temporal lobe, amygdala, and hippocampus in normal human development: Ages 4–18 years. *The Journal of Comparative Neurology, 366*, 223–230.

Gogtay, N., Giedd, J. N., Lusk, L., Hayashi, K. M., Greenstein, D., Vaituzis, A. C. et al. (2004). Dynamic mapping of human cortical development during childhood through early adulthood. *Proceedings of the National Academy of Sciences of the United States of America, 101*, 8174–8179.

Goldman-Rakic, P. S. (1988). Topography of cognition: Parallel distributed networks in primate association cortex. *Annual Review of Neuroscience, 11*, 137–156.

Goldman-Rakic, P. S., Chafee, M., & Friedman, H. (1993). Allocation of function in distributed circuits. In T. Ono, L. R. Squire, M. E. Raichle, D. I. Perrett, & M. Fukuda (Eds.), *Brain Mechanisms of Perception and Memory: From Neuron to Behavior* (pp. 445–456). New York: Oxford University Press.

Grant, B. F. (1997). Prevalence and correlates of alcohol use and DSM-IV alcohol dependence in the United States: Results of the National Longitudinal Alcohol Epidemiologic Survey. *Journal of Studies on Alcohol, 58*, 464–473.

Guitton, D., Buchtel, H. A., & Douglas, R. M. (1985). Frontal lobe lesions in man cause difficulties in suppressing reflexive glances and in generating goal-directed saccades. *Experimental Brain Research, 58*, 455–472.

Hale, S. (1990). A global developmental trend in cognitive processing speed. *Child Development, 61*, 653–663.

Hikosaka, O. & Wurtz, R. H. (1983). Visual and oculomotor function in monkey substantia nigra pars reticulata. I. Relation of visual and auditory responses to saccades. *Journal of Neurophysiology, 49*, 1230–1253.

Hudspeth, W. J. & Pribram, K. H. (1990). Stages of brain and cognitive maturation. *Journal of Educational Psychology, 82*, 881–884.

Huttenlocher, P. R. (1990). Morphometric study of human cerebral cortex development. *Neuropsychologia, 28*, 517–527.

Huttenlocher, P. R. & Dabholkar, A. S. (1997). Regional differences in synaptogenesis in human cerebral cortex. *The Journal of Comparative Neurology, 387*, 167–178.

Jernigan, T. L., Trauner, D. A., Hesselink, J. R., & Tallal, P. A. (1991). Maturation of human cerebrum observed in vivo during adolescence. *Brain, 114*, 2037–2049.

Kail, R. (1991). Development of processing speed in childhood and adolescence. In H. W. Reese (Ed.), *Advances in child development and behavior* (pp. 151–185). San Diego, CA: Academic.

Kail, R. (1993). Processing time decreases globally at an exponential rate during childhood and adolescence. *Journal of Experimental Child Psychology, 56*, 254–265.

Kato, M., Miyashita, N., Hikosaka, O., Matsumura, M., Usui, S., & Kori, A. (1995). Eye movements in monkeys with local dopamine depletion in the caudate nucleus. I. Deficits in spontaneous saccades. *Journal of Neuroscience, 15*, 912–927.

Klein, C. & Foerster, F. (2001). Development of prosaccade and antisaccade task performance in participants aged 6 to 26 years. *Psychophysiology, 38*, 179–189.

Klingberg, T., Forssberg, H., & Westerberg, H. (2002). Increased brain activity in frontal and parietal cortex underlies the development of visuospatial working memory capacity during childhood. *Journal of Cognitive Neuroscience, 14*, 1–10.

Klingberg, T., Vaidya, C. J., Gabrieli, J. D. E., Moseley, M. E., & Hedehus, M. (1999). Myelination and organization of the frontal white matter in children: A diffusion tensor MRI study. *Neuroreport, 10*, 2817–2821.

Kwon, H., Reiss, R. L., & Menon, V. (2002). Neural basis of protracted developmental changes in visuo-spatial working memory. *Proceedings of the National Academy of Sciences of the United States of America, 99*, 13336–13341.

Lambe, E. K., Krimer, L. S., & Goldman-Rakic, P. S. (2000). Differential postnatal development of catecholamine and serotonin inputs to identified neurons in prefrontal cortex of rhesus monkey. *Journal of Neuroscience, 20*, 8780–8787.

Levin, H. S., Culhane, K. A., Hartmann, J., Evankovich, K., & Mattson, A. J. (1991). Developmental changes in performance on tests of purported frontal lobe functioning. *Developmental Neuropsychology, 7*, 377–395.

Lewis, D. A. (1997). Development of the prefrontal cortex during adolescence: Insights into vulnerable neural circuits in schizophrenia. *Neuropsychopharmacology, 16*, 385–398.

Logothetis, N. K. & Pfueffer, J. (2004). On the nature of the BOLD fMRI contrast mechanism. *Magnetic Resonance Imaging, 22*, 1517–1531.

Luciana, M., Conklin, H. M., Hooper, C. J., & Yarger, R. S. (2005). The development of nonverbal working memory and executive control processes in adolescents. *Child Development, 76*, 697–712.

Luciana, M. & Nelson, C. (1998). The functional emergence of prefrontally-guided working memory systems in four- to eight-year-old children. *Neuropsychologia, 36*, 273–293.

Luna, B., Garver, K. E., Urban, T. A., Lazar, N. A., & Sweeney, J. A. (2004). Maturation of cognitive processes from late childhood to adulthood. *Child Development, 75*, 1357–1372.

Luna, B., Minshew, N. J., Garver, K. E., Lazar, N. A., Thulborn, K. R., Eddy, W. F. et al. (2002). Neocortical system abnormalities in autism: An fMRI study of spatial working memory. *Neurology, 59*, 834–840.

Luna, B. & Sweeney, J. A. (1999). Cognitive functional magnetic resonance imaging at very-high-field: Eye movement control. *Topics in Magnetic Resonance Imaging, 10*, 3–15.

Luna, B. & Sweeney, J. A. (2004a). Cognitive development: fMRI studies. In M. S. Keshavan, J. L. Kennedy, & R. M. Murray (Eds.), *Neurodevelopment and Schizophrenia* (pp. 45–68). London: Cambridge University Press.

Luna, B. & Sweeney, J. A. (2004b). The emergence of collaborative brain function: fMRI studies of the development of response inhibition. *Annals of the New York Academy of Sciences, 1021*, 296–309.

Luna, B., Thulborn, K. R., Munoz, D. P., Merriam, E. P., Garver, K. E., Minshew, N. J. et al. (2001). Maturation of widely distributed brain function subserves cognitive development. *NeuroImage, 13*, 786–793.

Marsh, R., Zhu, H., Schultz, R. T., Quackenbush, G., Royal, J., Skudlarski, P. et al. (2006). A developmental fMRI study of self-regulatory control. *Human Brain Mapping, 27*, 848–863.

McEwen, B. S. (2001). Invited review: Estrogens effects on the brain: Multiple sties and molecular mechanisms. *Journal of Applied Psychology, 91*, 2785–2801.

Meng, S. Z., Ozawa, Y., Itoh, M., & Takashima, S. (1999). Developmental and age-related changes of dopamine transporter, and dopamine D1 and D2 receptors in human basal ganglia. *Brain Research, 843*, 136–144.

Munoz, D. P., Broughton, J. R., Goldring, J. E., & Armstrong, I. T. (1998). Age-related performance of human subjects on saccadic eye movement tasks. *Experimental Brain Research, 121*, 391–400.

Nelson, C. A., Monk, C. S., Lin, J., Carver, L. J., Thomas, K. M., & Truwitt, C. L. (2000). Functional neuroanatomy of spatial working memory in children. *Developmental Psychology, 36*, 109–116.

Ojeda, S. R., Ma, Y. M., & Rage, F. (1995). A role for TGF in the neuroendocrine control of female puberty. In T. M. Plant & P. A. Lee (Eds.), *The Neurobiology of Puberty* (pp. 103–117). Bristol: Journal of Endocrinology Limited.

Olesen, P. J., Macoveanu, J., Tegner, J., & Klingberg, T. (2006b). Brain Activity Related to Working Memory and Distraction in Children and Adults. *Cerebral Cortex, [Epub ahead of print]*.

Olesen, P. J., Macoveanu, J., Tegner, J., & Klingberg, T. (2006a). Brain Activity Related to Working Memory and Distraction in Children and Adults. *Cerebral Cortex, [Epub ahead of print]*.

Olesen, P. J., Nagy, Z., Westerberg, H., & Klingberg, T. (2003). Combined analysis of DTI and fMRI data reveals a joint maturation of white and grey matter in a fronto-parietal network. *Cognitive Brain Research, 18,* 48–57.

Orr, D. P., Brack, C. J., & Ingersoll, G. (1988). Pubertal maturation and cognitive maturity in adolescents. *Journal of Adolescent Health Care, 9,* 273–279.

Paus, T., Zijdenbos, A., Worsley, K., Collins, D. L., Blumenthal, J., Giedd, J. N. et al. (1999). Structural maturation of neural pathways in children and adolescents: In vivo study. *Science, 283,* 1908–1911.

Petersen, A. C. (1976). Physical androgyny and cognitive functioning in adolescence. *Developmental Psychology, 12,* 524–533.

Rakic, P., Bourgeois, J. P., Eckenhoff, M. F., Zecevic, N., & Goldman-Rakic, P. S. (1986). Concurrent overproduction of synapses in diverse regions of the primate cerebral cortex. *Science, 232,* 232–235.

Rauschecker, J. P. & Marler, P. (1987). What signals are responsible for synaptic changes in visual cortical plasticity? In J. P. Rauschecker & P. Marler (Eds.), *Imprinting and Cortical Plasticity* (pp. 193–200). New York: Wiley.

Reiss, A. L., Abrams, M. T., Singer, H. S., Ross, J. L., & Denckla, M. B. (1996). Brain development, gender and IQ in children. A volumetric imaging study. *Brain, 119,* 1763–1774.

Ridderinkhof, K. R., Band, G. P. H., & Logan, G. D. (1999). A study of adaptive behavior: Effects of age and irrelevant information on the ability to inhibit one's actions. *Acta Psychologica, 101,* 315–337.

Roncadin, C., Pascual-Leone, J., Rich, J. B., & Dennis, M. (2007). Developmental relations between working memory and inhibitory control. *Journal of the International Neuropsychological Society, 13,* 59–67.

Rubia, K., Overmeyer, S., Taylor, E., Brammer, M., Williams, S. C., Simmons, A. et al. (2000). Functional frontalisation with age: Mapping neurodevelopmental trajectories with fMRI. *Neuroscience and Biobehavioral Reviews, 24,* 13–19.

Rubia, K., Smith, A. B., Taylor, E., & Brammer, M. (2007). Linear age-correlated functional development of right inferior fronto-striato-cerebellar networks during response inhibition and anterior cingulate during error-related processes. *Human Brain Mapping*.

Rubia, K., Smith, A. B., Woolley, J., Nosarti, C., Heyman, I., Taylor, E. et al. (2006). Progressive increase of frontostriatal brain activation from childhood to adulthood during event-related tasks of cognitive control. *Human Brain Mapping, 27,* 973–993.

Scherf, K. S., Sweeney, J. A., & Luna, B. (2006). Brain basis of developmental change in visuospatial working memory. *Journal of Cognitive Neuroscience, 18,* 1045–1058.

Segawa, M. (2000). Development of the nigrostriatal dopamine neuron and the pathways in the basal ganglia. *Brain & Development, 22,* S1–4.

Sowell, E. R., Thompson, P. M., Holmes, C. J., Jernigan, T. L., & Toga, A. W. (1999). In *vivo* evidence for post-adolescent brain maturation in frontal and striatal regions. *Nat Neurosci, 2,* 859–861.

Sowell, E. R., Thompson, P. M., Tessner, K. D., & Toga, A. W. (2001). Mapping continued brain growth and gray matter density reduction in dorsal frontal cortex: Inverse relationships during postadolescent brain maturation. *Journal of Neuroscience, 21,* 8819–8829.

Spear, L. P. (2000). The adolescent brain and age-related behavioral manifestations. *Neuroscience and Biobehavioral Reviews, 24,* 417–463.

Steinberg, L. (2004). Risk taking in adolescence: What changes, and why? *Annals of the New York Academy of Sciences, 1021,* 51–58.

Strauss, E. & Kinsbourne, M. (1981). Does age of menarche affect the ultimate level of verbal and spatial skills? *Cortex, 17,* 323–326.

Stuss, D. T. (1992). Biological and psychological development of executive functions. *Brain and Cognition, 20,* 8–23.

Swanson, H. L. (1999). What develops in working memory? A life span perspective. *Developmental Psychology, 35,* 986–1000.

Sweeney, J. A., Mintun, M. A., Kwee, S., Wiseman, M. B., Brown, D. L., Rosenberg, D. R. et al. (1996). Positron emission tomography study of voluntary saccadic eye movements and spatial working memory. *Journal of Neurophysiology, 75,* 454–468.

Sweeney, J. A., Takarae, Y., Macmillan, C., Luna, B., & Minshew, N. J. (2004). Eye movements in neurodevelopmental disorders. *Current Opinion in Neurology, 17,* 37–42.

Tamm, L., Menon, V., & Reiss, A. L. (2002). Maturation of brain function associated with response inhibition. *Journal of the American Academy of Child and Adolescent Psychiatry, 41,* 1231–1238.

Tanner, J. M. (1962). *Growth of adolescence* (2nd ed.). Oxford, England: Blackwell Scientific Publications.

Thatcher, R. W. (1991). Maturation of the human frontal lobes: Physiological evidence for staging. *Developmental Neuropsychology, 7,* 397–419.

Thatcher, R. W., Walker, R. A., & Giudice, S. (1987). Human cerebral hemispheres develop at different rates and ages. *Science, 236,* 1110–1113.

Thomas, K. M., King, S. W., Franzen, P. L., Welsh, T. F., Berkowitz, A. L., Noll, D. C. et al. (1999). A developmental functional MRI study of spatial working memory. *NeuroImage, 10,* 327–338.

Toga, A. W., Thompson, P. M., & Sowell, E. R. (2006). Mapping brain maturation. *Trends in Neurosciences, 29,* 148–159.

van Leijenhorst, L., Crone, E. A., & Bunge, S. A. (2006). Neural correlates of developmental differences in risk estimation and feedback processing. *Neuropsychologia, 44,* 2158–2170.

van Leijenhorst, L., Crone, E. A., & van der Molen, M. W. (2007). Developmental trends for object and spatial working memory: A psychophysiological analysis. *Child Development, 78,* 987–1000.

Velanova, K., Wheeler, M. E., & Luna, B. Maturational Changes in Anterior Cingulate and Frontoparietal Recruitment Support the Development of Error Processing and Inhibitory Control. *Cerebral Cortex,* (in press).

Waber, D. P., Mann, M. B., Merola, J., & Moylan, P. M. (1985). Physical maturation rate and cognitive performance in early adolescence: A longitudinal examiniation. *Developmental Psychology, 21,* 666–681.

Waddington, J. L., Torrey, E. F., Crow, T. J., & Hirsch, S. R. (1991). Schizophrenia, neurodevelopment and disease. *Archives of General Psychiatry, 48,* 271–273.

Warner, L. A., Kessler, R. C., Hughes, M., Anthony, J. C., & Nelson, C. B. (1995). Prevalence and correlates of drug use and dependence in the United States. Results from the National Comorbidity Survey. *Archives in General Psychiatry, 52,* 219–229.

Welsh, M. C. (2002). Developmental and clinical variations in executive functions. In D. L. Molfese & V. J. Molfese (Eds.), *Developmental variations in learning: Applications to social, executive function, language, and reading skills* (pp. 139–185). Mahwah: Lawrence Erlbaum.

Welsh, M. C., Pennington, B. F., & Groisser, D. B. (1991). A normative-developmental study of executive function: A window on prefrontal function in children. *Developmental Neuropsychology, 7,* 131–149.

Williams, B. R., Ponesse, J. S., Schachar, R. J., Logan, G. D., & Tannock, R. (1999). Development of inhibitory control across the life span. *Developmental Psychology, 35,* 205–213.

Wise, L. A., Sutton, J. A., & Gibbons, P. D. (1975). Decrement in Stroop interference time with age. *Perceptual and Motor Skills, 41,* 149–150.

Yakovlev, P. I. & Lecours, A. R. (1967). The myelogenetic cycles of regional maturation of the brain. In A. Minkowski (Ed.), *Regional Development of the Brain in Early Life* (pp. 3–70). Oxford: Blackwell Scientific.

Zald, D. H. & Iacono, W. G. (1998). The development of spatial working memory abilities. *Developmental Neuropsychology, 14,* 563–578.

Chapter 14
Electrophysiological and Genetic Markers of Attention Deficit–Hyperactivity Disorder: Boundary Conditions for Normal Attentional Processing and Behavioral Control

X. Carrasco, M. Henríquez, F. Zamorano, P. Rothhammer, F. Daiber, and F. Aboitiz

Abstract Electrophysiological markers of attention deficit–hyperactivity disorder (ADHD) imply a deficit in frontoparietal function, evidenced by alterations in the P300 wave associated with oddball-type tasks. There are also differences between ADHD and control subjects in potentials related to monitoring and detection of errors, such as the frontomedial negativity, which has been attributed to activation of the ventromedial prefrontal cortex and anterior cingulate cortex. Finally, quantitative electroencephalography (measuring absolute spectral power and frequency distribution) has reappeared in the last few years with consistent findings that would permit the discrimination of clinical subtypes. Genetic studies of ADHD have been oriented to linkage analysis in affected families and to the study of candidate genes. Most of these genes codify proteins related to neurotransmitter transmission, specifically dopaminergic, serotoninergic, and noradrenergic. In accordance with the neurobiological basis for ADHD, consisting of a relative prefrontal dysfunction, those genes showing a more consistent association with this condition relate to dopaminergic transmission, a selective regulator of frontal lobe dynamics. An example of this is the dopamine receptor DRD4, but there are other genes which will be reviewed in this chapter. The combination of genetic and electrophysiological analyses promises to be an excellent approach to the neurobiological bases of diverse neuropsychiatric disorders, among them ADHD.

X. Carrasco

Servicio de Neurología y Psiquaitría, Hospital de Niños Luis Calvo Mackenna, Antonio Varas 360, Providencia, Santiago, Chile, e-mail: xcarrasc@med.uchile.cl

F. Aboitiz and D. Cosmelli (eds.) *From Attention to Goal-Directed Behavior.*
© Springer-Verlag Berlin Heidelberg 2009

Introduction

Attention deficit–hyperactivity disorder (ADHD) is among the most common child neuropsychiatric conditions, affecting some 8–12% of children of school age, and affecting boys more than girls in a ratio of 3:1 (American Psychiatric Association, 1994; Faraone, Sergeant, Gillberg, & Biederman, 2003). ADHD is usually subdivided into three categories, with predominance in inattention, in hyperactivity and impulsivity, or the combined type that includes inattention and hyperactivity/impulsivity. In about 50% of the cases, this condition is maintained through adult age (Barkley, 2002). This clinical condition impacts strongly in several areas of social development, such as education, family life, and job stability (Fisher & Beckley, 1999).

ADHD is considered to be a multifactorial trait in which genetic factors account for about 80% of the cases, while environmental factors such as premature delivery, perinatal hypoxia, maternal smoking, alcohol consumption, and others may explain the remaining 20% (Swanson et al., 2007). Although great efforts have been devoted to identify specific genes associated with this condition, these have had only limited success, some of the most relevant associations being found between genes related to catecholaminergic or monoaminergic neurotransmission (see later). This evidence, and the therapeutic effect that stimulants have in these patients, has led to the dopaminergic hypothesis of ADHD, where a dysfunction in cathecolaminergic transmission, or perhaps more specifically in dopaminergic transmission, is seen as a core deficit in this condition (Swanson et al., 2007; see also Chap. 11 by Aboitiz), notwithstanding some evidence that links other neurotransmitter and receptor systems to ADHD (see later). The dopaminergic proposal implies that a deficit in dopaminergic transmission results in a malfunction of frontostriatal and frontocerebellar circuits involved in predicting events and behavioral outcomes, plus a deficit in the frontoamygdalar loop that assigns emotional significance to the predicted and detected events (reviewed in Swanson et al., 2007). Furthermore, a recent proposal points to a distinction between "hot" and "cool" executive functions, depending on whether these involve orbital and medial prefrontal cortex networks connected with the amygdala and related to emotional content ("hot") or involve mainly networks associated with dorsolateral prefrontal regions involved in analytical processing ("cool"). In this context, "cool" dysfunction, implying purely cognitive deficits, would be characteristic to the inattention symptoms of ADHD, while hyperactivity/impulsivity symptoms would reflect "hot" deficits in executive functions (Castellanos, Sonuga-Barke, Milham, & Tannock, 2006).

Despite intense research in the last few years, the identification of ADHD-related cognitive signs has not been without difficulty. Some authors point to a deficit in inhibitory mechanisms (Barkley, 1997; but see Castellanos et al., 2006), in the mechanisms of behavioral conditioning (Sagvolden, Johansen, Aase, & Russell, 2005), or in time perception (Castellanos & Tannock, 2002). Nigg (2005) surveyed the performance of ADHD subjects in several neuropsychological tasks, pointing out that no specific deficit is sufficient to account for ADHD, but there are key cognitive domains in which deficits are manifested: vigilance attention, cognitive control (in particular, working memory and response suppression), and motivation (specifically, approach to reinforcement objective).

Moreover, despite the term "attention deficit," recent evidence indicates that despite a strong impairment in sustained attention mechanisms (perhaps a key inattention signature), ADHD children have no deficit in spatial orienting (Huang-Pollock and Nigg, 2003), and perform better than controls in divided-attention tasks (Koschack, Kunert, Derichs, Weniger, & Irle, 2003). This suggests that ADHD is characterized by a different pattern of allocation of attentional resources, having an attentional framework more expanded in space but with difficulties in sustained attention in time. This condition fits other ADHD-related traits such as lack of inhibition, delay avoidance, or distorted time perception, and may reflect a disbalance in dopaminergic signaling mechanisms (Aboitiz, López, López-Calderón, & Carrasco, 2006; see also Chap. 11 by Aboitiz).

Thus, there are no absolute diagnostic markers for ADHD, making this condition a controversial trait in which there is active research on its neurobiological and genetic substrates, as well as on its clinical characterization. In this sense, this chapter will present a review of the state of the art in these lines of research, focusing on the postulated electrophysiological and genetic markers of ADHD, and including studies that use a combination of these markers.

Electrophysiological Markers of ADHD

Electrophysiological studies of ADHD go back to the 1930s and have accompanied the multiple conceptualizations that have been associated with this condition, for example, the early terms "minimal brain dysfunction" or "childhood hyperkinetic reaction" (Strauss & Lehtinen, 1947; American Psychiatric Association, 1968). The term "ADHD" was coined in 1980 (American Psychiatric Association, 1980), and the diagnostic criteria for this condition have undergone only minor changes since then. We will review the studies performed since that date, although some earlier studies can also provide important information. Although EEG techniques have permitted us to open a window on the neurobiological processes underlying the cognitive deficits of ADHD, it may be still premature to use these techniques as useful diagnostic or monitoring tools. However, quantitative EEG (qEEG) is rapidly becoming an important complementary diagnostic element.

Event-Related Potentials in ADHD

There is a host of publications on event-related potentials (ERPs) in ADHD since the 1970s (reviewed in Barry, Johnstone, & Clarke, 2003). Some consistent differences from normal subjects have been observed in tasks involving sustained attention to auditory and visual stimuli, although several researchers have underlined the absence of correlation between electrophysiological abnormalities and behavioral performance in ADHD subjects who score as well and controls (Karayanidis et al., 2000).

According to several reports, the positive deflection in the ERP around 300 ms after stimulus (P300) during visual or auditory sustained attention tasks shows decreased amplitude in comparison with that for normal subjects (Strandburg et al., 1996; Jonkman et al., 1997). The same results are seen when the P300 response is generated in visual or auditory oddball paradigms, where the subject must respond to an infrequent target stimulus among multiple presentations of standard stimuli; these differences are claimed to be age-dependent (Satterfield, Schell, Nicholas, Satterfield, & Freese, 1990; Satterfield, Schell, & Nicholas, 1994; Johnstone, Barry, & Anderson, 2001). The cognitive meaning of the P300 wave is not yet clear. According to one interpretation, this potential reflects the sum of neural activities involved in working memory devoted to solving the task, including the stimulus and its context. Another possibility, in our view incompatible with the first one, is that the P300 implies the phenomenon of task closure (Verleger, 1988), which consists of the activation of cognitive networks associated with the event of problem solving.

A recent study showed that ADHD patients show indemnity in the early attentional filters (evidenced in the early potentials P100/N100) triggered by unattended peripheral stimuli, but display an increased cortical excitability in late ERP waves (i.e., the P300 potential) in response to these stimuli (López et al., 2006). Thus, despite being apparently successful in initially filtering the unattended stimulus, the latter is nevertheless able to enter the working memory buffer and trigger an activation usually related to cognitive processing. An additional study (López et al., 2008) took advantage of the attentional blink phenomenon, where perception of a stimulus is impaired if it is presented in close temporal proximity to a previous, attended stimulus. In normal subjects, we confirmed the elicitation of a P300 wave every time the second stimulus was consciously detected and the absence of P300 when there was no conscious report of this stimulus (Raymond, Shapiro, & Amell, 1992). However, ADHD subjects elicited P300 (although of lower amplitudes) both when they consciously detected the stimulus and when they were unable to detect it. Again, this indicates that there is a decreased threshold to trigger the activation of cortical ensembles, even if they may not reach conscious levels. This implies a different pattern of cortical dynamics in ADHD subjects, perhaps marked by increased low-scale excitability but decreased large-scale excitability, the latter associated with engagement of large-scale networks associated with attentional processes and working memory during goal-directed behavior (see Chap. 11 by Aboitiz).

Additional waves of interest are the error-related negativities, the error positivity, and the frontomedial negativity or its equivalent feedback-related negativity, whose source is considered to be related to the anterior cingulate cortex in the medial prefrontal cortex. These potentials belong to a family of waves generated at different times after the detection of an error, in paradigms where the subject is repeatedly exposed to events of gain or loss such as gambling simulations or any task where there is feedback in relation to performance. The neurobiological role of the anterior cingulate cortex has been the focus of considerable attention; a common interpretation is that its activity reflects cognitive and behavioral regulation on the basis of contextual cues, either by conflict or outcome monitoring. This cortical region shows increased activation in situations that require more cognitive effort (for a review see Botvinick, Cohen, & Carter, 2004).

A recent study reported a lower error positivity amplitude in ADHD children, which is abolished with methylphenidate administration (Jonkman et al., 2007). Additionally, the feedback-related negativity showed larger amplitude in ADHD subjects than in controls, even though this difference was not confirmed in behavioral performance (Van Meel, Oosterlaan, Heslenfeld, & Sergeant, 2005). Likewise, the error-related negativity has been reported to be higher in the combined ADHD subtype than in controls, despite similar behavioral performances (Burgio-Murphy et al., 2007). This evidence suggests that a core deficit in ADHD is a dysfunction of error monitoring systems (Van Meel, Heslenfeld, Oosterlaan, & Sergeant, 2007).

Polysomnography

Sleep disorders in ADHD have been recognized for several decades, mainly on the basis of parental reports to the clinician, the most commonly reported symptom being restlessness during sleep. The DSM-III included this observation among the diagnostic criteria for ADHD; this criterion was subsequently retracted, thus being eliminated in newer versions of the DSM (American Psychiatric Association, 1980). However, in a recent meta-analysis of the polysomnography literature published since 1980, which selected 12 articles with a total of 333 ADHD subjects and 231 controls, the main conclusion was that the only consistent alteration in ADHD was the presence of periodic limb movements (Sadeh, Pergamin, & Bar-Haim, 2006). A significant variable in this study was the use of the previous night for adaptation to the study, to mitigate the effects of the first night. Additional associations that may be important are a higher incidence of enuresis and nonspecific alterations in sleep architecture in ADHD children (Van der Heijden, Smits, & Gunning, 2005). Another report describes polysomnography in children and adolescents between 8 and 16 years old, subdivided into four groups: Control, combined ADHD, tic disorder (TD) and ADHD + TD (Kirov, Kinkelbur, Banaschewski, & Rothenberger, 2007). The ADHD group displayed an increase in total sleep time, a shorter latency in sleep onset, a larger number of sleep cycles, and an increase of total duration of REM sleep, while in the TD group there was an increase in nocturnal awakenings. The ADHD + TD group evidenced the combination of the alterations of each group. The increase in REM sleep in ADHD has been reported by several authors and might be related to a dopaminergic dysfunction associated with ADHD.

qEEG in ADHD

Generalities of the qEEG Technique

EEG is considered to reflect the ongoing extracellular voltage changes that take place in cortical tissue and are possible to detect in the scalp surface. These are thus field potentials with a major contribution of synaptic potentials on the superficial cortical layers. Perhaps the main contributors to these potentials are pyramidal cells of layer III that generate corticocortical association fibers, project their dendrites to layers II and I and, in the prefrontal cortex, are targets of dopaminergic axons. However, there is also a contribution from deeper brain regions that connect to the neocortex, such as hippocampal areas and thalamocortical projections, and there is activity modulation by different monoaminergic brainstem systems such as dopamine via the mesocortical and mesolimbic pathways, and noradrenaline from the locus coeruleus (Niedermeyer & Lopes da Silva, 1993).

qEEG consists of a frequency analysis into segments of the digital EEG to extract the relative contribution of different frequency bands to the signal. The main frequency bands used are delta (1.5–4 Hz), theta (4–8 Hz), alpha (8–13 Hz), beta (13–30 Hz), and gamma (above 30 Hz). This analysis is performed in specific behavioral conditions (sleep/wakefulness) and in different topographical regions in the scalp. Predominance of slow, high-amplitude waves (delta, theta, and alpha) has been classically considered to reflect a so-called synchronized activity, while higher frequencies (beta, gamma) correspond to a "desynchronization" of the EEG signal.

In general terms, a "synchronized" EEG signal is considered to be a consequence of the simultaneous activation of large numbers of neurons whose individual synaptic activities generate changes in the extracellular potential. These synchronous potentials add together and acquire enough amplitude to be recorded at the surface of the skull. For instance, multiple interconnected neuronal ensembles in the limbic system, including septal nuclei, hippocampus, and entorhinal cortex, are proposed to generate theta activity. In addition, repetitive pacemaker-like discharges in the GABAergic thalamic reticular nucleus might cause the spindle-like EEG activity during sleep, via multiple connections with other thalamic nuclei and the neocortex. On the other hand, EEG "desynchronization" is considered to be a consequence of events that interrupt this basal slow activity. For example, in occipital electrodes, alpha activity is substituted by beta oscillations when the subject opens his/her eyes (which activates the primary visual cortex); in the REM phase of sleep, rich in oniric activity, spindles and slow waves become replaced by beta activity; and voluntary motor activation is marked by the transformation of mu activity (8–13 Hz) into beta activity. Beta and especially low gamma activity have been proposed to reflect the activation of intrinsic cortical circuits involved in conscious cognitive events that imply sustained attention, although this is a subject in discussion (Steriade, Gloor, Llinás, Lopes da Silva, & Mesulam, 1990). More recent findings suggest that highly relevant cognitive processes such as working memory are related to theta activity

(Jensen & Tesche, 2002; Raghavachari et al., 2001; Raghavachari et al., 2006). In these studies, an increase in theta activity (defined here as between 4 and 12 Hz) was observed in parietotemporooccipital cortices, hippocampus, and parahippocampus during the retention phase of a Sternberg task; furthermore, this activity decayed with task closure. In a similar line, it has been postulated that attentional function consists of a complex sequence of activation of different but partly overlapping networks, comprising different anatomical regions and operating at different oscillatory frequencies (Fan, McCandliss, Sommer, Raz, & Posner, 2002; Fan et al., 2007; for review see Kahana, 2006).

Therefore, as it apparently reflects the global workings of the cerebral cortex in different behavioral situations, qEEG is among the most used strategies to analyze neurobiological processes associated with neuropsychiatric conditions. An alteration in cortical dynamics due to dopaminergic dysfunction (as has been postulated for ADHD) will most likely be reflected in disturbances in the frequency components of the qEEG during distinct behavioral tasks. In fact, there is a corpus of consistent findings in ADHD which points to discrimination among subgroups of this condition on the basis of qEEG analyses. The National Institute of Mental Health suggested the use of this technique as a complement to clinical diagnosis (Jensen et al., 1993), especially considering that this is a highly available, noninvasive technique, and at relatively low cost. Current imaging analyses based on hemodynamic and metabolic responses may be insufficient to reveal the highly dynamic processes of assembly and reassembly of cortical ensembles during a cognitive process. In this context, the techniques of qEEG and frequency-domain analyses may provide an important insight into these processes.

Some Results from qEEG Applied to ADHD Subjects

A pioneering study in this field compared EEG spectral power in 25 right-handed ADHD boys aged 9–12 years with that in 27 matched controls (Mann, Lubar, Zimmerman, Miller, & Muenchen, 1992) under three conditions: eyes open in fixation (baseline), silent reading, and drawing figures from the Bender-Gestalt Visual Motor Test. While in basal conditions there were only minor differences, during both tasks, reading and drawing, ADHD boys presented a significant increment in frontal theta activity and a decrease in temporal beta activity (from 13 to 21.75 Hz). This pattern differed from that found in dyslexic patients, who show an increase in left temporal theta activity (Duffy, Denckla, Bartels, & Sandini, 1980; Duffy, Denckla, Bartels, Sandini, & Kiessling, 1980).

In the same decade, Chabot and Serfontein (1996) analyzed a larger sample of 407 attentionally impaired children aged 6–17 years, comprising ADHD patients (combined and inattentive subtypes, according to DSM-III criteria) and children with attentional problems who did not meet the criteria for ADHD. In the three groups there was also a division among children who had learning problems and children who did not. There was a control group of 310 children. Recordings were made in resting conditions, with eyes closed for a period of 20–30 min. There were two main types of anomalies. First, 46% of the cases showed a generalized increase in theta and alpha spectral power, mainly in frontocentral regions, with a normal alpha frequency average. The second abnormal finding was observed in 30% of the cases, and consisted of an increase in absolute theta and alpha power, but with a decreased alpha frequency average. The authors concluded that in both groups there was an alteration in cortical arousal, which was abnormally high in the first group and lower than normal (hypoarousal) in the second group. With this method, discrimination between subjects and controls reached a sensitivity of 93.7% and a specificity of 88%.

A more recent study examined 397 ADHD subjects (combined and inattentive subtypes) and 85 control subjects of both sexes, from 6 to 30 years old and divided into four age groups (6–11, 12–15, 16–20, and 21–30 years). qEEG was analyzed from electrode Cz (the most discriminant according to previous studies), in four behavioral conditions (eyes open and fixation, silent reading, listening to a text, and making a copy of geometrical figures). There were significant differences in the theta (4–8 Hz)/beta (13–21 Hz) ratio according to age, diagnosis, and condition, but there were no sex differences. In all age groups and in the four conditions, the theta/beta ratio was significantly higher in ADHD subjects than in controls. The sensitivity of this test was 86%, the specificity was 98%, and the predictive power reached 99% (Monastra et al., 1999). In a subsequent study (Monastra, Lubar, & Linden, 2001), the same group increased their sample size (a total of 469 subjects, both sexes, 6–20 years, subdivided into three groups: control, ADHD inattentive, and ADHD combined type) and confirmed the abovementioned findings but with decreased sensitivity (90%) and specificity (94%). It is noteworthy that these researchers have oriented their work in qEEG to therapeutic strategies, more specifically into the techniques of EEG-based biofeedback (Monastra et al., 2005; see also Chap. 7 by Hill et al.).

In agreement with the abovementioned findings, Clarke, Barry, McCarthy and Selikowitz (2001) reported that ADHD subjects of the combined type could be subdivided into different groups according to the qEEG activity patterns. These different patterns might reflect distinct mechanisms underlying similar clinical manifestations. In this study, 42.3% of the subjects (group 1) displayed an increase of high-amplitude theta waves, especially in frontal regions, with normal alpha and deficiency in beta and delta frequencies. A second group (37.5% of the subjects, group 2) was characterized by an increase in slow waves and a decrease in beta activity, while a third group (20.2% of the subjects, group 3) showed an excess of beta activity. According to the authors, the first group was characterized by frontal hypoactivation (hypoarousal), considering that beta activity normally increases with mental and physical activity. Group 2 was similar to group 1, but displayed an EEG signal that can be characterized as globally immature on the basis of the topographic distribution of activities; nevertheless, the theta activity observed in these patients is not fully consistent with this assumption. Finally, group 3 showed a pattern opposite to that of group 1 in terms of the frontal predominance of theta activity and was tentatively interpreted as displaying frontal hyperarousal. Groups 1 and 2 included subjects of both the combined and the inattentive subtypes, while group 3 consisted only of male subjects of the combined type, suggesting that it is more associated with impulsivity and hyperactivity symptoms. The EEG alterations described above are not modified in subjects with oppositional behavior comorbidity (Clarke, Barry, McCarthy, & Selikowitz, 2002a). However, comorbidity with learning disorders may increase the qEEG differences compared with controls, i.e., there is an even higher theta power with increased theta/alpha ratio in this comorbid population (Clarke, Barry, McCarthy, & Selikowitz, 2002b).

Likewise, in ADHD patients with good response to stimulants, the administration of the latter normalizes the qEEG signal when the alterations consist of hypoactivation or immaturity, owing in both cases to an increase in cortical arousal (Chabot, Orgill, Crawford, Harris, & Serfontein, 1999; Clarke, Barry, Bond, McCarthy, & Selikowitz, 2002; Clarke et al., 2003). In group 3, displaying an excess beta activity, stimulant use correlated with only a tendency to normalize qEEG activity, despite a good clinical response. This finding casts some doubt on the clinical relevance of the increased beta activity in this group. In ADHD patients of the inattentive type, good responders to psychostimulants are those who show a more pronounced reduction of relative beta power and a higher theta/beta ratio with respect to controls; similar findings were obtained with patients of the combined subtype (Clarke, Barry, McCarthy, Selikowitz, & Croft, 2002; Clarke, Barry, McCarthy, & Selikowitz, 2002c). Summarizing, the most consistent findings using qEEG in ADHD imply an increase in relative theta power and a decrease in alpha and beta power; this alteration continues in symptomatic adolescents and adults (Clarke, Barry, McCarthy, & Selikowitz, 2001; Lazzaro, et al., 1998; Bresnahan, Anderson, & Barry, 1999; Bresnahan & Barry, 2002; for a review see Barry, Clarke, & Johnstone, 2003).

The neurobiological significance of the observed abnormalities is under discussion. Most authors concur with the interpretation that these alterations represent a frontostriatal dysfunction (Swanson et al., 2007), which has been postulated to be a core deficit in ADHD, and is also supported by recent imaging studies. Despite methodological pitfalls in many of these studies, the most consistent findings using PET or single photon emission scintigraphy indicate frontostriatal hypoactivation (Zametkin et al., 1990; Ernst et al., 2003; Schweitzer et al., 2003; Kim, Lee, Shin, Cho, & Lee, 2002). Functional MRI studies have reported similar findings, including decreased function in frontomedial regions, including the dorsal anterior cingulate gyrus, striatum, cerebellar vermis, and dorsolateral prefrontal cortex (for a review see Bush, Valera, & Seidman, 2005). Although this possibility is of high interest, frontostriatal circuits include other components, such as thalamic and subthalamic regions, and evidence indicates that they are composed of several parallel corticostriatothalamocortical loops, all directed to the frontal lobe, which provides a yet largely unexplored complexity to this issue (see Chap. 9 by Hurtado). Furthermore, the link between qEEG rhythms and different cognitive functions that are affected in ADHD (many of them involving cortical regions other than the frontal lobe) still remains to be well defined (Fig. 14.1).

For ADHD, as well as for other neuropsychiatric impairments, the search for genetic and electrophysiological associations makes an emerging research line. A recent report describes a relationship between the risk alleles *DRD4-7R* and

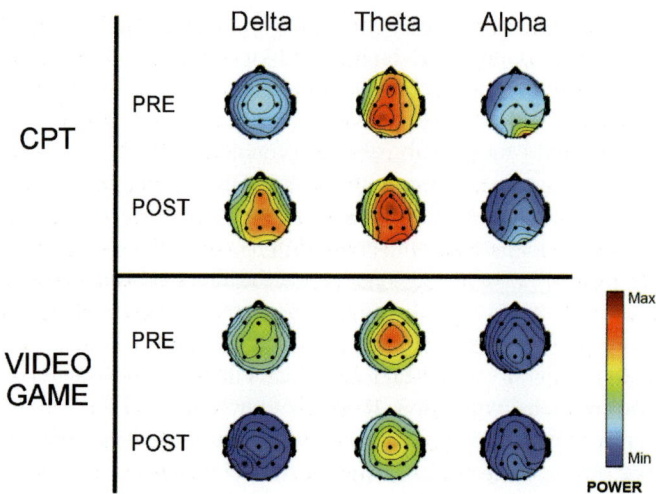

Fig. 14.1 Topographic representation of quantitative EEG during a continuous performance task (*top*) and while practicing a video game (*bottom*) in an attention deficit–hyperactivity disorder subject. Activities at delta (2.5–4 Hz), theta (4.5–7.5 Hz), and alpha (8–14 Hz) are shown in two conditions: no medication and under stimulant medication (dextroamphetamine)

SLC6A3-10R (see "Genetic Markers of ADHD") with gamma-band variations evoked with an auditory target detection task in 48 young adults (Demirlap et al., 2007). In this study, the authors reported an association between the presence of risk alleles and an increase in gamma band activity in response to both target and standard stimuli.

EEG Coherence

An elaboration of the qEEG information, especially when using high-density electrode recordings, is the analysis of coherence in the respective oscillatory activities in different electrodes. Coherence is defined as a correlation coefficient that estimates the covariance in signals recorded from two or more different electrodes, separated among them. This procedure can be performed for different frequency bands (see Chap. 1 by Bosman and Womelsdorf and Chap. 15 by Gaspar et al.). This measure, considered to be a measure of effective connectivity between different cortical regions, displays modifications during neurodevelopment which parallel changes in cognitive development, and has been claimed to present age-specific genetic and environmental determinants (Van Baal, Boomsma, & de Geus, 2001). Coherence can be observed within one hemisphere or across hemispheres, and depending on the distance between recording sites may permit one to analyze connections of different lengths, such as arcuate fibers, other longitudinal tracts of the subcortical white matter, or callosal connections (Shaw, 1981; Thatcher, Krause, & Hrybyk, 1986; Thatcher, Biver, & North, 2007; Aboitiz, López, & Montiel, 2003). The interpretation of differences in coherence across subjects is a matter of debate. In mentally retarded children, an increased coherence in some frequency bands has been reported (Barry, Clarke, McCarthy, & Selikowitz, 2002), suggesting that the differentiation and specialization of neuronal ensembles implies a reduction more than an increase of long-ranging coherence. Phase coherence analysis has been applied to ADHD populations, evidencing differences with control groups. Several studies performed in the 1990s explored phase coherence in ADHD patients. One of these showed, in 86% of the cases, a frontotemporal intrahemispheric hypercoherence within theta and alpha bands together with an interhemispheric incoherence, the latter more evident in low-IQ subjects (Chabot & Serfontein, 1996). In a sample of 40 right-handed subjects aged 8–10 years with normal IQ, including inattentive and combined ADHD subtypes and controls, an increase in phase coherences between short-distance electrodes in the theta band was found among nearby ipsilateral electrodes (such as FP1–F3) together with an increase in frontal interhemispheric coherence, especially in the theta band. Both differences implied significantly higher values in ADHD patients than in controls, and higher values in the combined versus the inattentive subtypes. The effect of methylphenidate on the alterations in EEG coherence was found only in conditions of task execution; no differences were observed in the resting state (Clarke et al., 2005). Some criticisms regarding the data on short-range coherence claim that part of this phenomenon could be due to volume conduction between closely located electrodes (Srinivasan, Nunez, & Silberstein, 1998; Srinivasan, Winter, Ding, & Nunez, 2007), and that the correct reference should be taken from the average array of electrodes in a high-density setup instead of using mastoid references (Nunez et al., 1999). With these considerations, Murias, Swanson and Srinivasan (2007) used 42 ADHD subjects and 21 controls, observing in ADHD subjects a higher than normal coherence around 8 Hz (low alpha), and a decreased coherence around 10–11 Hz (higher alpha). Furthermore, the latter alteration was found to be modified by the use of stimulant medication (Murias et al., 2007). These findings have been interpreted in the context of a lower brain size in ADHD subjects (Swanson et al., 2007), which could be related to a deficit in myelin content of encephalic tracts (Castellanos et al., 2002). However, such conclusions may be at this point somewhat premature.

Genetic Markers of ADHD

As well as with other approaches, a major challenge with qEEG studies in ADHD is the correlation of these findings with clinical diagnosis, which although following a strict interviewing schedule is not based on objective measures. Thus, a worthwhile strategy may be to correlate these analytical techniques with biological variables such as the presence of risk alleles for ADHD. Although this line of research may not fully comply with clinical requirements, it may provide important insights into the neurobiological mechanisms underlying this condition (Castellanos & Tannock, 2002). In this section, we will briefly review some aspects of the genetics of ADHD, pointing to genes that promise to offer interesting correlations with neurodynamical processes by virtue of their role in neuromodulatory processes of relevance to the cause of ADHD.

As mentioned, ADHD has a strong genetic basis. Family, twin, and adoption studies extensively support a genetic influence in the origin of ADHD. Family studies estimate that the overall risk of ADHD in first-degree relatives of an affected proband is between 4.0 and 5.4 times higher than in subjects without family history of ADHD (Faraone, Biederman, & Monuteaux, 2000). In addition, adoption studies are consistent in reporting that ADHD children are more similar on ADHD measures to their biological parents than to their adoptive parents, suggesting that the familiarity observed in ADHD has a clear genetic contribution (Joseph, 2002; Sprich, Biederman, Crawford, Mundy, & Faraone, 2000). Finally, twin studies have estimated this genetic contribution in about 76% of cases, ranging from 60 to 90% (Faraone et al., 2005; Waldman & Gizer, 2006).

Nevertheless, despite the consistent evidence in favor of a genetic basis in the origin of the disorder, molecular genetics studies have been only moderately successful in addressing which specific genes are related to ADHD inheritance and – in consequence – are eventually useful as risk markers for ADHD. It is now considered unlikely that genes conferring risk for ADHD strongly correlate with clinical diagnosis as established by the DSM-IV and ICD-10. Alternatively, it seems more probable that particular genes confer risk for specific subtypes, symptom dimensions, or cognitive deficits within ADHD. Up to now, more than 20 genes have been explored, with special focus on genes related to the catecholaminergic systems or involved in synaptic transmission. From these genes, four genes of the dopaminergic system (*DRD4, DRD5, SLC6A3, DBH*), one gene of the serotoninergic system (*5-HTT*), and one gene from the synaptic transmission system (*SNAP25*) have shown evidence of statistically significant association with ADHD in at least one meta-analysis study (Faraone & Khan, 2006; Faraone et al., 2005). Additionally, some new interesting findings suggest a strong association between ADHD and latrophilin 3 (*LPNH3*), a G-protein receptor expressed in the mesolimbic system (Muenke et al., 2007). These genes will be discussed in the next subsections.

Dopamine Receptor D4 Gene (DRD4)

The human dopamine receptor D4 gene (*DRD4*) is one of the most variable human genes known. Presumably, this high variability is a consequence of its subtelomeric location at chromosome 11p. The gene encodes for a seven-transmembrane domains, Gi-coupled receptor of 387–515 amino acids in length, expressed in human entorhinal and frontal cortex, cingulate cortex, substantia nigra, and cerebellum (Missale, Nash, Robinson, Jaber, & Caron, 1998; Paterson, Sunohara, & Kennedy, 1999). This pattern of expression is coincident with the neuronal pathways implicated in ADHD pathogenesis and includes structures involved in higher cognitive processes such as attention and inhibition (Durston, 2003).

The most widely studied polymorphic variant of *DRD4* is a variable number of tandem repeats (VNTR) of a 48-bp sequence located in exon 3, which encodes the third intracytoplasmic loop of the receptor, a region of interaction with G protein. In humans, one to ten copies of this tandem repeat motif have been reported. In 1996, La Hoste et al. (1996) reported for the first time the existence of an association between the 7-repeat variant of the 48-bp VNTR of the *DRD4* gene and ADHD. Since then there has been an intensive search for this variant in ADHD populations, which has concluded with a series of meta-analysis reviews in the last few years (Faraone, Doyle, Mick, & Biederman, 2001; Faraone & Khan, 2006; Faraone et al., 2005; Li, Sham, Owen, & He, 2006; Maher, Marazita, Ferrell, & Vanyukov, 2002; Purper-Ouakil et al., 2005). All of these meta-analyses are consistent in reporting a significant association between ADHD and the 7-repeat allele of the *DRD4* gene, with pooled odds ratios ranging from 1.1 to 1.4.

In a highly unusual finding, Swanson et al. (2000) reported that the 7-repeat "risk" allele apparently has protective effects in ADHD subjects instead of conferring risk to ADHD symptoms. When performance was probed in different cognitive tasks (Posner 2007; visual–spatial orienting, Stroop color–word conflict, and Logan go–stop), subjects with the 7-repeat allele ($n = 13$) performed the same as controls, while those without the allele ($n = 19$) showed the characteristic deficit in reaction times of ADHD. Furthermore, Durston et al. (2005) reported *DRD4* polymorphism effect in prefrontal gray matter volume, where the 7-repeat variant conferred protection, while other alleles were associated with decreased values (for a further review see Swanson et al., 2007). These findings may call for a reassessment of the role of the *DRD4* genotype in this disease, but might also be consistent with a role of this receptor in regulating tonic dopaminergic levels (see Chap. 11 by Aboitiz).

Dopamine Receptor D5 Gene (DRD5)

Several studies have reported an association between the 148-bp variant of the Gs-coupled *DRD5* and ADHD (Hawi et al., 2003; Lowe et al., 2004), which seems to be confirmed by meta-analysis studies (Faraone et al., 2005; Li, Sham, Owen, & He, 2006; Maher, Marazita, Ferrell, & Vanyukov, 2002; Purper-Ouakil et al., 2005).

Dopamine Transporter Gene (SLC6A3)

The dopamine transporter is responsible for the dopamine reuptake from the synaptic cleft to the presynaptic neuron and represents the principal mechanism for regulating the extracellular levels of dopamine in the striatum and nucleus accumbens (Ciliax et al., 1999). There are two main reasons for considering *SLC6A3* as a candidate gene for ADHD. First, the therapeutic effects of stimulant medication (such as methylphenidate) on ADHD symptoms seem to be mediated by the blockage of the dopamine transporter (Volkow et al., 1998). Second, the knockout and partial ablation models of the *SLC6A3* gene exhibit hyperactivity and inhibition deficits (Pogorelov, Rodriguiz, Insco, Caron, & Wetsel, 2005).

Cook et al. (1995) first reported an association between ADHD and the 10-repeat allele of the 40-bp 3′ untranslated region VNTR of *SLC63A*. Since then, there have been several reports either confirming or rejecting this association. This incongruence has not been solved by meta-analysis reviews. Two recent meta-analyses, including up to 800 informative meiosis, failed to detect association between ADHD and the 10-repeat allele of *SLC6A3* (Li, Sham, Owen, & He, 2006; Maher, Marazita, Ferrell, & Vanyukov, 2002), but a third one, including up to 1,300 informative meiosis from 18 linkage disequilibrium studies, reported preferential transmission of the 10-repeat allele to ADHD children (Yang et al., 2007).

There are reports of a higher than normal expression of the 10-repeat *SLC6A3* polymorphism in the striatum, temporal lobe, and cerebellum of ADHD patients (Dougherty et al., 1999; Mill, Asherson, Browes, D'Souza, & Craig, 2002a). Furthermore, HEK 293 cells transfected with the 10-repeat allele show higher expression of the receptor in comparison with cells transfected with the 9-repeat variant (VanNess, Owens, & Kilts, 2005). However, a recent study reported no differences in striatum, and even a lower dopamine transporter density in mesencephalon (Jucaite, Fernell, Halldin, Forssberg, & Farde, 2005). In this context, Volkow et al. (1998) suggested that dopamine transporter density might be highly changeable and dependent on homeostatic mechanisms that operate to maintain tonic levels of synaptic or extrasynaptic dopamine. Thus, dopamine transporter density may decrease in the presence of low dopamine levels and increase in the presence of high dopamine levels. Treatment with stimulant medication increases dopamine levels dramatically, and this may result in homeostatic increase in dopamine transporter density as observed in early studies (Swanson et al., 2007).

Dopamine β-Hydroxylase

Dopamine β-hydroxylase is the enzyme responsible for converting dopamine to norepinephrine, linking these two catecholaminergic systems. Two lines of evidence suggest that the dopamine β-hydroxylase gene (*DBH*) is an attractive candidate gene for ADHD: First, dopamine/norepinephrine dysregulation seems to be relevant in ADHD etiopathogenesis (Castellanos & Tannock, 2002) and, second, lower plasma β-hydroxylase activity has been reported in ADHD patients (Kopeckova, Paclt, & Goetz, 2006). In addition, the *DBH* gene strongly influences the plasma levels of the enzyme (Kopeckova, Paclt, & Goetz, 2006; Tang et al., 2006). A TaqI polymorphism in intron 5 of *DBH* has been investigated by several groups, with association or a trend of association between the Taq A2 allele and ADHD being detected in most of the studies. Two independent meta-analyses confirm this association, with an odds ratio of about 1.3 (Faraone & Khan, 2006; Wohl, Purper-Ouakil, Mouren, Ades, & Gorwood, 2005).

Serotonin Transporter (5-HTT)

Most of the genetics studies on ADHD have focused on the dopaminergic neuro-transmitter system. However, there is pharmacological and biochemical evidence suggesting that serotonine dysregulation is relevant in impulsive behavior and ADHD pathogenesis (Ribasés et al., 2007). Independent studies have reported an association between ADHD (or related subclinical characteristics) and a functional polymorphism of the transporter gene, consisting of a 44-bp insertion/deletion in its promoter region (long vs. short variants), where the long variant is associated with faster transport and with ADHD (Curran, Purcell, Craig, Asherson, & Sham, 2005; Faraone et al., 2005; Kent et al., 2002). Despite this promising evidence, it is worth noting that recent analyses in different ethnic populations have not found such association (Langley et al., 2003; Li et al., 2007; Ribases et al., 2007; Xu et al. 2005).

Synaptosomal Associated Protein 25

Synaptosomal associated protein 25 (SNAP25) is a presynaptic membrane protein which mediates calcium-dependent exocytosis of neurotransmitters from the synaptic vesicle. Interestingly, it is differentially expressed in the brain, especially in structures that have been implicated in the pathogenesis of ADHD (Durston, 2003; Mill et al., 2004). Furthermore, the Coloboma mouse strain, observed to be hyper-active and disordered in impulsivity and attention, presents a hemizygous deletion on chromosome 2q, a region comprising the *SNAP25* gene.

Two markers in the *SNAP25* gene, located at positions 1065 (T/C) and 1069 (T/G) respectively, have been associated with ADHD (Barr et al., 2000); however other reports found no association between these variables (Brophy, Hawi, Kirley, Fitzgerald, & Gill, 2002; Mill et al.2002a; Mill et al., 2004). A pooled analysis of these studies shows evidence of a significant association between the allele T1065C and ADHD (Faraone et al., 2005). Finally, there is some evidence that *SNAP25* may have a role in explaining the variance of continuous measures of ADHD symptoms, although a different marker was used in such analysis (Mill et al., 2005).

LPNH3

LPNH3 is a new promising candidate in the molecular genetics of ADHD. The gene is located in chromosome 4q and encodes a G-protein-coupled receptor expressed in the mesolimbic system. On the basis of a previous finding describing linkage of ADHD to the 4q.13 region in a genetic isolate from Colombia, Muenke and collaborators (Arcos-Burgos et al., 2004; Muenke et al., 2007) defined a minimal critical region of 327 kb within the *LPNH3* gene-coding region, in strong linkage disequilibrium with ADHD in sample of Colombian, German, and US families. These results are still to be replicated by independent studies and the role of latrophilin 3 receptor in the functional changes accompanying ADHD is largely unknown. However, the reported outstanding association with ADHD qualifies *LPHN3* as one of the most promising new candidate genes.

Gene Interactions

The interactions of different risk alleles may be more important in the determination of the ADHD phenotype than individual genes. In this context, we and others have observed a strong association between the combined presence of at least one 7-repeat *DRD4* allele and homozygosity for the 10-repeat *SLC6A3* alelle on one hand and ADHD on the other (Carrasco et al., 2006; Roman et al., 2001; Henríquez et al., 2007). Nevertheless, to date, genome scan analyses have evidenced only weak signals associated with ADHD, and furthermore these are not coincident across studies (Arcos-Burgos et al., 2004; Bakker et al., 2003; Brookes et al., 2006; Fisher et al., 2002; Ogdie et al., 2003). Some of these studies have confirmed association with *DRD4* and *DAT* genes.

Individual and Ethnic Diversity in Allelic Composition

ADHD is now considered to be a complex multifactorial trait, where the combination of several genes and environmental factors conflate to cause the development of this condition. Nevertheless, genetics seems to have a significant weight in this equation. Thus, ADHD appears as an excellent model to study the genetic bases of cognitive dysfunction in humans. In this context, a major challenge is to relate individual diversity in neurodynamic patterns with different genetic substrates in ADHD (Demirlap et al., 2007). The gene polymorphisms mentioned earlier are all likely to influence the activity of different neurotransmitter systems, particularly catecholamines and serotonin, which may profoundly influence the dynamical regimes and relative synaptic strengths of broad cortical regions, especially prefrontal, thus modulating behavioral planning, decision making, attention allocation, and working memory among other elements of goal-directed behavior. Nevertheless, this effect may not be restricted to cortical regions; there are important effects of dopamine and other neurotransmitters in other, subcortical components such as the ventral and dorsal striatum, and it is most likely that the interplay between all these systems determines short- and long-term behavioral outcomes. In this sense, the genetic variability in the systems modulating this complex architecture may provide the boundary conditions under which normal behavior takes place.

An additional dimension to this problem relates to ethnic differences in genetic composition, and the likelihood that these differences are reflected in different neurodynamical patterns among human populations. A clear example of this situation is posed by the *DRD4* gene. The allele frequencies of the ADHD-associated DRD4 receptor differ considerably among different populations, but only three variants account for more than 90% of the genetic pool. These variants are the 4-repeat, 7-repeat, and 2-repeat alleles, with global mean allele frequencies of 64.3, 20.6, and 8.2%, respectively. Interestingly, the 7-repeat allele appears quite frequently in the American population (48.3%) but only occasionally in East and South Asia (1.9%), while the 2-repeat allele presents an inverse pattern of distribution, being highly frequent in Asia (18.1%), but not in America (2.9%). The diversity of allele frequencies among different ethnic groups emphasizes the importance of population considerations in the design and interpretation of association studies pertaining to this polymorphism (Chang, Kidd, Livak, Pakstis, & Kidd, 1996). Interestingly, the association between ADHD and the 7-repeat allele of *DRD4* has failed to be replicated in Chinese, Korean, and Japanese populations (Brookes et al. 2005; Kim et al., 2005; Leung, Lee, Hung, Ho, Tang et al. 2005; Cheuk, Li, & Wong, 2006a, b). Instead, a higher frequency of the 2-repeat allele in ADHD children compared with controls has been reported in these and other populations (Comings, Gonzalez, Wu, Gade, Muhleman et al. 1999; Leung, Lee, Hung, Ho et al. 2005). It is worth noting that the proteins encoded by both alleles mediate a reduced response to dopamine in comparison with the 4-repeat allele variant (Asghari et al. 1995). It will be of the highest interest to determine if these population genetic differences reflect statistical differences in neurodynamic patterns.

Conclusions

Controversial as it is, ADHD requires us urgently to find objective markers to support diagnosis and an endophenotypic construct. Contributions from genetics and electrophysiology are promising but still insufficient. From the geneticist's viewpoint, ADHD appears more as a variant than as a specific morbid condition, possibly resulting from different allelic contributions of genes involved in circuits associated with behavioral control, especially but not exclusively frontostriatal circuits modulated by dopamine. On the other hand, qEEG evidence suggests an "immature" processing in some aspects. The notion of ADHD as a neurodevelopmental disorder is confirmed in the clinical practice in relation to similar-age peers and in consideration of imaging data indicating lower than normal brain volumes. There are also EEG-atypic subtypes displaying insufficient responses to first-rate stimulant medication. In this context, the study of the genetic substrates of variability in neurodynamical patterns is only in its infancy. The technique of evoked potentials also evidences differential processing mechanisms in ADHD, possibly indicating an antieconomic or inefficient use of cognitive resources, recruiting more neuronal ensembles than necessary for the tasks required. Behavioral performance is, however, similar to that of controls in many cases. This situation leads us to propose that ADHD requires a redefinition in the sense that the core problem may not be in the attentional system but rather in behavioral control. We hope that neuroscience and clinical neuropsychiatry will follow this line of enquiry, as the prevalence of ADHD and related disorders seems to increase each year and has strong implications for the understanding of normal and pathological human behavior.

Acknowledgments This work was funded by FONDECYT projects 1050721, 1070761, and 1080219, Comisión Nacional de Investigación Científica y Tecnológica, Gobierno de Chile.

References

Aboitiz, F., López, J., & Montiel, J. (2003). Long distance communication in the human brain: Timing constraints for interhemispheric synchrony and the origin of brain lateralization. *Biological Research, 36*, 89–99.

Aboitiz, F., López, V., López-Calderón, J., & Carrasco, X. (2006). Beyond endophenotypes: An interdisciplinary approach to attentional deficit-hyperactivity disorder. In M. Vanchevsky (Ed.), *Focus in cognitive psychology research* (pp. 183–205). New York: Nova Science.

American Psychiatric Association (1994). Diagnostic and statistical manual of mental disorders IV (DSM-IV).

American Psychiatric Association (1980). Diagnostic and statistical manual of mental disorders III (DSM-III).

American Psychiatric Association (1968). Diagnostic and statistical manual of mental disorders II (DSM-II).

Arcos-Burgos, M., Castellanos, F. X., Pineda, D., Lopera, F., Palacio, J. D., Palacio, L. G. et al. (2004). Attention-deficit/hyperactivity disorder in a population isolate: Linkage to loci at 4q13.2, 5q33.3, 11q22, and 17p11. *American Journal of Human Genetics, 75*, 998–1014.

Asghari, V., Sanyal, S., Buchwaldt, S., Paterson, A., Jovancovic, V. A., & Van Tol, H. H. (1995). Modulation of intracellular cyclic AMP levels by different human dopamine D4 receptor variants. *Journal of Neurochemistry, 65*, 1157–1165.

Bakker, S. C., Van der Meulen, E. M., Buitelaar, J. K., Sandkuijl, L. A., Pauls, D. L., Monsuur, A. J. et al. (2003). A whole-genome scan in 164 Dutch sib pairs with attention-deficit/hyperactivity disorder: Suggestive evidence for linkage on chromosomes 7p and 15q. *American Journal of Human Genetics, 72*(5), 1251–1260.

Barkley, R. A. (1997). Behavioral inhibition, sustained attention, and executive functions: Constructing a unifying theory of ADHD. *Psychological Bulletin, 121*, 65–94.

Barkley, R. A. (2002). Major life activity and health outcomes associated with attention-deficit/hyperactivity disorder. *Journal of Clinical Psychiatry, 63*(Suppl 12), 10–15.

Barr, C. L., Feng, Y., Wigg, K., Bloom, S., Roberts, W., Malone, M. et al. (2000). Identification of DNA variants in the SNAP-25 gene and linkage study of these polymorphisms and attention-deficit hyperactivity disorder. *Molecular Psychiatry, 5*(4), 405–409.

Barry, R. J., Clarke, A. R., McCarthy, R. A., & Selikowitz, M. (2002). EEG coherence in attention-deficit/hyperactivity disorder: A comparative study of two DSM-IV types. *Clinical Neurophysiology, 113*, 579–585.

Barry, R. J., Clarke, A. R., & Johnstone, S. J. (2003). A review of electrophysiology in attention-deficit/hyperactivity disorder: I. Qualitative and quantitative electroencephalography. *Clinical Neurophysiology, 114*, 171–183.

Barry, R. J., Johnstone, S. J., & Clarke, A. R. (2003). A review of electrophysiology in attention-deficit/hyperactivity disorder: II. Event related potentials. *Clinical Neurophysiology, 114*, 184–198.

Botvinick, M. M., Cohen, J. D., & Carter, C. S. (2004). Conflict monitoring and anterior cingulate cortex: An update. *Trends in Cognitive Sciences, 8*(12), 539–546.

Bresnahan, S. M., Anderson, J. W., & Barry, R. J. Age-related changes in quantitative EEG in attention-deficit/hyperactivity disorder (1999). *Biological Psychiatry, 46*(12), 1690–1697.

Bresnahan, S. M., & Barry, R. J. (2002). Specificity of quantitative EEG analysis in adults with attention deficit hyperactivity disorder. *Psychiatry Research, 112*(2), 133–144.

Brookes, K. J., Xu, Chen, X., C. K., Huang, Y. S., Wu, Y. Y., & Asherson, P. (2005). No evidence for the association of DRD4 with ADHD in a Taiwanese population within-family study. *BMC Medical Genetics, 6*, 31.

Brookes, K., Xu, X., Chen, W., Zhou, K., Neale, B., Lowe, N. A. et al. (2006). The analysis of 51 genes in DSM-IV combined type attention deficit hyperactivity disorder: Association signals in DRD4, DAT1 and 16 other genes. *Molecular Psychiatry, 11*(10), 934–953.

Brophy, K., Hawi, Z., Kirley, A., Fitzgerald, M., & Gill, M. (2002). Synaptosomal-associated protein 25 (SNAP-25) and attention deficit hyperactivity disorder (ADHD): Evidence of linkage and association in the Irish population. *Molecular Psychiatry, 7*(8), 913–917.

Burgio-Murphy, A., Klorman, R., Shaywitz, S. E., Fletcher, J. M., Marchione, K. E., Holahan, J. et al. (2007). Error-related event-related potentials in children with attention-deficit hyperactivity disorder, oppositional defiant disorder, reading disorder and math disorder. *Biological Psychology, 75*(1), 75–86.

Bush, G., Valera, E. M., & Seidman, L. (2005). Functional neuroimaging of attention-deficit/hyperactivity disorder: A review and suggested future directions. *Biological Psychiatry, 57*, 1273–1284.

Carrasco, X., Rothhammer, P., Moraga, M., Henriquez, H., Chakraborty, R., Aboitiz, F. (2006). Genotypic interaction between DRD4 and DAT1 loci is a high risk factor for attention-deficit/hyperactivity disorder in Chilean families. *American Journal of Medical Genetics B Neuropsychiatry Genetics, 141*(1), 51–54.

Castellanos, F. X., & Tannock, R. (2002). Neuroscience of attention-deficit/hyperactivity disorder: The search for endophenotypes. *Nature Review Neuroscience, 3*(8), 617–628.

Castellanos, F. X., Lee, P. P., Sharp, W., Jeffries, N. O., Greenstein, D. K., Clasen, L. S. et al. (2002). Developmental trajectories of brain volume abnormalities in children and adolescents with attention-deficit/hyperactivity disorder. *JAMA, 288*(14), 1740–1748.

Castellanos, F. X., Sonuga-Barke, E. J. S., Milham, M. P., & Tannock, R. (2006). Characterizing cognition in ADHD: Beyond executive dysfunction. *Trends Cognitive Science, 10*, 117–123.

Ciliax, B. J., Drash, G. W., Staley, J. K., Haber, S., Mobley, C. J., Miller, G. W. et al. (1999). Immunocytochemical localization of the dopamine transporter in human brain. *Journal of Comparative Neurology, 409*(1), 38–56.

Clarke, A. R., Barry, R. J., McCarthy, R., & Selikowitz, M. (2001). EEG-defined subtypes of children with attention-deficit/hyperactivity disorder. *Clinical Neurophysiology, 112*, 2098–2105.

Clarke, A. R., Barry, R. J., McCarthy, R., & Selikowitz, M. (2002a). Children with attention-deficit/hyperactivity disorder and comorbid oppositional defiant disorder: An EEG analysis. *Psychiatry Research, 11*, 181–190.

Clarke, A. R., Barry, R. J., McCarthy, R., & Selikowitz, M. (2002b). EEG analysis of children with attention-deficit/hyperactivity disorder and comorbid reading disabilities. *J Learning Disabilities, 35*(3), 276–285.

Clarke, A. R., Barry, R. J., McCarthy, R., & Selikowitz, M. (2002c). EEG differences between good and poor responders to methylphenidate and dexamphetamine in children with attention-deficit/hyperactivity disorder. *Clinical Neurophysiology, 113*, 194–205.

Clarke, A. R., Barry, R. J., Bond, D., McCarthy, R., & Selikowitz, M. (2002). Effects of stimulant medications on the EEG of children with attention-deficit/hyperactivity disorder. *Psychopharmacology, 164*, 277–284.

Clarke, A. R., Barry, R. J., McCarthy, R., Selikowitz, M., & Croft, R. J. (2002). EEG differences between good and poor responders to methylphenidate in boys with the inattentive type of attention-deficit/hyperactivity disorder. *Clinical Neurophysiology, 113*, 1191–1198.

Clarke, A. R., Barry, R. J., McCarthy, R., Selikowitz, M., Brown, C. R., & Croft, R. J. (2003). Effects of stimulant medications on the EEG of children with attention-deficit/hyperactivity disorder predominantly inattentive type. *International Journal of Psychophysiology, 47*, 129–137.

Clarke, A. R., Barry, R. J., McCarthy, R., Selikowitz, M., Johnstone, S. J., Abbott, I. et al. (2005). Effects of methylphenidate on EEG coherence in attention-deficit/hyperactivity disorder. *International Journal of Psychophysiology, 58*, 4–11.

Comings, D. E., Gonzalez, N., Wu, S., Gade, R., Muhleman, D., Saucier, G. et al. (1999). Studies of the 48 bp repeat polymorphism of the DRD4 gene in impulsive, compulsive, addictive behaviors: Tourette syndrome, ADHD, pathological gambling, and substance abuse. *American Journal of Medical Genetics, 88*, 358–368.

Cook, E. H., Jr., Stein, M. A., Krasowski, M. D., Cox, N. J., Olkon, D. M., Kieffer, J. E. et al. (1995). Association of attention-deficit disorder and the dopamine transporter gene. *American Journal of Human Genetics, 56*(4), 993–998.

Curran, S., Purcell, S., Craig, I., Asherson, P., & Sham, P. (2005). The serotonin transporter gene as a QTL for ADHD. *American Journal of Medical Genetics – B Neuropsychiatry Genetics, 134*(1), 42–47.

Chabot, R. J., & Serfontein, G. (1996). Quantitative electroencephalographic profiles of children with attention deficit disorder. *Biological Psychiatry, 40*, 951–963.

Chabot, R., Orgill, A., Crawford, G., Harris, M., & Serfontein, G. (1999). Behavioural and electrophysuological predictors of treatment response to stimulants in children with attention disorders. *Journal of Child Neurology, 14*, 343–351.

Chang, F. M., Kidd, J. R., Livak, K. J., Pakstis, A. J., & Kidd, K. K. (1996). The world-wide distribution of allele frequencies at the human dopamine D4 receptor locus. *Human Genetics, 98*, 91–101.

Cheuk, D. K., Li, S. Y., & Wong, V. (2006a). Exon 3 polymorphisms of dopamine D4 receptor (DRD4) gene and attention deficit hyperactivity disorder in Chinese children. *American Journal of Medical Genetics – B Neuropsychiatry Genetics, 141*, 907–911.

Cheuk, D. K., Li, S. Y., & Wong, V. (2006b). No association between VNTR polymorphisms of dopamine transporter gene and attention deficit hyperactivity disorder in Chinese children. *American Journal of Med Genet – B Neuropsychiatry Genetics, 141*, 123–125.

Demirlap, T., Herrmann, C. S., Erdal, M. E., Ergeonglu, T., Keskin, Y. H., Ergen, M. et al. (2007). DRD4 and DAT1 polymorphisms modulate human gamma band responses. *Cerebral Cortex, 17*, 1007–1019.

Dougherty, D. D., Bonab, A. A., Spencer, T. J., Rauch, S. L., Madras, B. K., & Fischman, A. J. (1999). Dopamine transporter density in patients with attention deficit hyperactivity disorder. *Lancet, 354*(9196), 2132–2133.

Duffy, F. H., Denckla, M. B., Bartels, P. H., & Sandini, G. (1980). Dyslexia: Regional differences in brain electric activity by topographic mapping. *Annals of Neurology, 7*, 412–420.

Duffy, F. H., Denckla, M. B., Bartels, P. H., Sandini, G., & Kiessling, L. S. (1980). Dyslexia: Automated diagnosis by computerized classification of brain electric activity. *Annals of Neurology, 7*, 421–428.

Durston, S. (2003). A review of the biological bases of ADHD: What have we learned from imaging studies? *Mental Retardation and Development Disabilities Research Review, 9*(3), 184–195.

Durston, S., Fossella, J. A., Casey, B. J., Hulshoff Pol, H. E., Galvan, A., Schnack, H. G. et al. (2005). Differential effects of DRD4 and DAT1 genotypes on fronto-striatal gray matter volumes in a sample of subjects with attention deficit hyperactivity disorder, their unaffected siblings, and controls. *Molecular Psychiatry, 10*, 678–685.

Ernst, M., Kimes, A. S., London, E. D., Matochik, J. A., Eldreth, D., Tata, S. et al. (2003). Neural substrates of decision making in adults with attention deficit hyperactivity disorder. *American Journal of Psychiatry, 160*(6), 1061–1070.

Fan, J., McCandliss, B. D., Sommer, T., Raz, A., & Posner, M. I. (2002). Testing the efficiency and independence of attentional networks. *Journal of Cognitive Neuroscience, 14*, 340–347.

Fan, J., Byrne, J., Worden, M. S., Guise, K. G., McCandliss, B. D., Fossella, J. et al. (2007). The relation of brain oscillations to attentional networks. *Journal of Neuroscience, 27(23)*, 6197–6206.

Faraone, S. V., Biederman, J., & Monuteaux, M. C. (2000). Toward guidelines for pedigree selection in genetic studies of attention deficit hyperactivity disorder. *Genetic Epidemiology, 18*(1), 1–16.

Faraone, S. V., Doyle, A. E., Mick, E., & Biederman, J. (2001). Meta-analysis of the association between the 7-repeat allele of the dopamine D(4) receptor gene and attention deficit hyperactivity disorder. *American Journal of Psychiatry, 158*(7), 1052–1057.

Faraone, S. V., Sergeant, J. A., Gillberg, C., & Biederman, J. (2003). The worldwide prevalence of ADHD: Is it an American condition? *World Psychiatry, 2*, 104–113.

Faraone, S. V., Perlis, R. H., Doyle, A. E., Smoller, J. W., Goralnick, J. J., Holmgren, M. A. et al. (2005). Molecular genetics of attention-deficit/hyperactivity disorder. *Biological Psychiatry, 57*(11), 1313–1323.

Faraone, S. V., & Khan, S. A. (2006). Candidate gene studies of attention-deficit/hyperactivity disorder. *Journal of Clinical Psychiatry, 67*(Suppl 8), 13–20.

Fisher, B., & Beckley, R. (1999). *Attention deficit disorder: Practical coping methods*. Boca Raton: CRC.

Fisher, S. E., Francks, C., McCracken, J. T., McGough, J. J., Marlow, A. J., MacPhie, I. L. et al. (2002). A genomewide scan for loci involved in attention-deficit/hyperactivity disorder. *American Journal of Human Genetics, 70*(5), 1183–1196.

Hawi, Z., Lowe, N., Kirley, A., Gruenhage, F., Nothen, M., Greenwood, T. et al. (2003). Linkage disequilibrium mapping at DAT1, DRD5 and DBH narrows the search for ADHD susceptibility alleles at these loci. *Molecular Psychiatry, 8*(3), 299–308.

Henríquez, M., López-Calderón, J., López, V., Rotthammer, P., Zamorano, F., Rotthammer, F., & Aboitiz, F. (2007). Confirmation of Genotypic Interaction between DRD4 and DAT1 loci as Risk Factor for Attention- Deficit and Hyperactivity Disorder. *Proceedings of the 39 Danube Symposium and 1st International Congress on ADHD, Wurzburg, Germany. 2–5 June*.

Huang-Pollock, C. L., & Nigg, J. T. (2003). Searching for the attention deficit in attention deficit hyperactivity disorder: The case of visuospatial orienting. *Clinical Psychology Review, 23*, 801–830.

Jensen, P. S., Koretz, D., Locke, B. Z., Schneider, S., Radke-Yarrow, M., Richters, J. E. et al. (1993). Child and adolescent psychopathology research: Problems and prospects for the 1990s. *Journal of Abnormal Child Psychology, 21*(5), 551–580.

Jensen, O., & Tesche, C. D. (2002). Frontal theta activity in humans increases with memory load in a working memory task. *European Journal of Neuroscience, 15*, 1395–1399.

Johnstone, S. J., Barry, R. J., &Anderson, J. W. (2001). Topographic distribution and developmental timecourse of auditory event-related potentials in two subtypes of attention-deficit hyperactivity disorder. *International Journal of Psychophysiology, 42*, 73–94.

Jonkman, L. M., Kemmer, C., Verbaten, M. N., Koelga, H. S., Camfferman, G., Van der Gaag, R. J. et al. (1997). Event-related potentials and performance of attention-deficit hyperactivity disorder: Children and normal controls in auditory and visual selective attention tasks. *Biological Psychiatry, 41*, 595–611.

Jonkman, L. M., Van Melis, J. J. M., Kemner, C., & Markus, C. R. (2007). Methylphenidate improves deficient error evaluation in children with ADHD: An event-related brain potential study. *Biological Psychology, 76*, 217–229.

Joseph, J. (2002). Adoption study of ADHD. *Journal of American Academy of Child and Adolescent Psychiatry, 41*(12), 1389–1390; author reply 1390–1381.

Jucaite, A., Fernell, E., Halldin, C., Forssberg, H., & Farde, L. (2005). Reduced midbrain dopamine transporter binding in male adolescents with attention-deficit/hyperactivity disorder: Association between striatal dopamine markers and motor hyperactivity. *Biological Psychiatry, 57*(3), 229–238.

Kahana, M. J. (2006). The cognitive correlates of human brain oscillations. *Journal of Neuroscience, 26*(6), 1669–1672.

Karayanidis, F., Robaey, P., Bourassa, M., De Koning, D., Geoffroy, G., & Pelletier, G. (2000). ERP differences in visual attention processing between attention deficit hyperactivity disorder and control boys in absence of performance differences. *Psychophysiology, 37*, 319–333.

Kent, L., Doerry, U., Hardy, E., Parmar, R., Gingell, K., Hawi, Z. et al. (2002). Evidence that variation at the serotonin transporter gene influences susceptibility to attention deficit hyperactivity disorder (ADHD): analysis and pooled analysis. *Molecular Psychiatry, 7*(8), 908–912.

Kim, B. N., Lee, J. S., Shin, M. S., Cho, S. C., & Lee, D. S. (2002). Regional cerebral perfusion abnormalities in attention deficit/hyperactivity disorder. Statistical parametric mapping analysis. *European Archives of Psychiatry and Clinical and Neuroscience, 252*(5), 219–225.

Kim, Y. S., Leventhal, B. L., Kim, S. J., Kim, B. N., Cheon, K. A., Yoo, H. J. et al. (2005). Family-based association study of DAT1 and DRD4 polymorphism in Korean children with ADHD. *Neuroscience Letters, 390*, 176–181.

Kirov, R., Kinkelbur, J., Banaschewski, T., & Rothenberger, A. (2007). Sleep patterns in children with attention-deficit/hyperactivity disorder, tic disorder and comorbidity. *Journal of Child Psychology and Psychiatry, 48*(6), 561–570.

Kopeckova, M., Paclt, I., & Goetz, P. (2006). Polymorphisms and low plasma activity of dopamine-beta-hydroxylase in ADHD children. *Neuro Endocrinology Letters, 27*(6), 748–754.

Koschack, J., Kunert, H. J., Derichs, G., Weniger, G., & Irle, E. (2003). Impaired and enhanced attentional function in children with attention deficit/hyperactivity disorder. *Psychological Medicine, 33*, 481–489.

LaHoste, G. J., Swanson, J. M., Wigal, S. B., Glabe, C., Wigal, T., King, N. et al. (1996). Dopamine D4 receptor gene polymorphism is associated with attention deficit hyperactivity disorder. *Molecular Psychiatry, 1*(2), 121–124.

Langley, K., Payton, A., Hamshere, M. L., Pay, H. M., Lawson, D. C., Turic, D. et al. (2003). No evidence of association of two 5HT transporter gene polymorphisms and attention deficit hyperactivity disorder. *Psychiatric Genetics, 13*(2), 107–110.

Lazzaro, I., Gordon, E., Whitmont, S., Plahn, M., Li, W., Clarke, S. et al. (1998). Quantitative EEG activity in adolescent attention-deficit/hyperactivity disorder. *Clinical Electroencephalography, 29*, 37–42.

Leung, P. W., Lee, C. C., Hung, S. F., Ho, T. P., Tang, C. P., Kwong, S. L. et al. (2005). Dopamine receptor D4 (DRD4) gene in Han Chinese children with attention-deficit/hyperactivity disorder (ADHD): increased prevalence of the 2-repeat allele. *American Journal of Medical Genetic – B Neuropsychiatry Genetics, 133*, 54–56.

Li, D., Sham, P. C., Owen, M. J., & He, L. (2006). Meta-analysis shows significant association between dopamine system genes and attention deficit hyperactivity disorder (ADHD). *Human Molecual Genetics, 15*(14), 2276–2284.

Li, J., Wang, Y., Zhou, R., Zhang, H., Yang, L., Wang, B. et al. (2007). Association between polymorphisms in serotonin transporter gene and attention deficit hyperactivity disorder in Chinese Han subjects. *American Journal of Medical Genetics – B Neuropsychiatry Genetics, 144*(1), 14–19.

López, V., López-Calderon, J., Ortega, R., Kreither, J., Carrasco, X., Rothhammer, P. et al. (2006). Attention-deficit hyperactivity disorder involves differential cortical processing in a visual spatial attention paradigm. *Clinical Neurophysiology, 117*, 2540–2548.

López, V., Pavez, F., López-Calderón, J., Ortega, R., Sáez, N., Carrasco, X. et al. (2008). Electrophysiological evidences of inhibition deficit in Attention-Deficit/Hyperactivity Disorder during the attentional blink. *The Open Behavioral Science Journal, 2*, 33–40.

Lowe, N., Kirley, A., Hawi, Z., Sham, P., Wickham, H., Kratochvil, C.J. et al. (2004). Joint analysis of the DRD5 marker concludes association with attention-deficit/hyperactivity disorder confined to the predominantly inattentive and combined subtypes. *American Journal of Human Genetics, 74*(2), 348–356.

Maher, B. S., Marazita, M. L., Ferrell, R. E., & Vanyukov, M. M. (2002). Dopamine system genes and attention deficit hyperactivity disorder: A meta-analysis. *Psychiatric Genetic, 12*(4), 207–215.

Mann, C. A., Lubar, J. F., Zimmerman, A. W., Miller, C. A., & Muenchen, R. A. (1992). Quantitative analysis of EEG in boys with attention-deficit-hyperactivity disorder: Controlled study with clinical implications. *Pediatric Neurology, 8*(1), 30–36.

Mill, J., Asherson, P., Browes, C., D'Souza, U., & Craig, I. (2002a). Expression of the dopamine transporter gene is regulated by the 3' UTR VNTR: Evidence from brain and lymphocytes using quantitative RT-PCR. *American Journal of Medical Genetics, 114*(8), 975–979.

Mill, J., Richards, S., Knight, J., Curran, S., Taylor, E., & Asherson, P. (2004). Haplotype analysis of SNAP-25 suggests a role in the aetiology of ADHD. *Molecular Psychiatry, 9*(8), 801–810.

Mill, J., Xu, X., Ronald, A., Curran, S., Price, T., Knight, J. et al. (2005). Quantitative trait locus analysis of candidate gene alleles associated with attention deficit hyperactivity disorder (ADHD) in five genes: DRD4, DAT1, DRD5, SNAP-25, and 5HT1B. *American Journal of Medical Genetics – B Neuropsychiatry Genetics, 133*(1), 68–73.

Missale, C., Nash, S. R., Robinson, S. W., Jaber, M., & Caron, M. G. (1998). Dopamine receptors: From structure to function. *Physiological Review, 78*(1), 189–225.

Monastra, V. J., Lubar, J. F., Linden, M., Van Deusen, P., Green, G., Wing, W. et al. (1999). Assessing attention deficit hyperactivity disorder via quantitative electroencephalography: An initial validation study. *Neuropsychology, 13*(3), 424–433.

Monastra, V. J., Lubar, J. F., & Linden, M. (2001). The development of a quantitative electroencephalographic scanning process for attention deficit-hyperactivity disorder: Reliability and validity studies. *Neuropsychology, 15*(1), 136–144.

Monastra, V. J., Lynn, S., Linden, M., Lubar, J. F., Gruzelier, J., & LaVaque, T. J. (2005). Electroencephalographic biofeedback in the treatment of attention-deficit/hyperactivity disorder. *Applied Psychophysiology and Biofeedback, 30*(2), 95–114.

Muenke et al. (2007). Abstracts – 39th International Danube Symposium for Neurological Sciences and Continuing Education and 1st International Congress on ADHD, from childhood to adult disease. *Journal of Neural Transmission, 114*(7), XLIII–CXLI.

Murias, M., Swanson, J. M., & Srinivasan, R. (2007). Functional connectivity of frontal cortex in healthy and ADHD children reflected in EEG coherence. *Cerebral Cortex, 17*, 1788–1799.

Niedermeyer, E., & Lopes da Silva, F. (1993). *Electroencephalography: Basic principles, clinical cpplications and related fields,* 4th edn. Baltimore: Williams & Wilkins.

Nigg, J. (2005). Neuropsychologic theory and findings in attentiondeficit/hyperactivity disorder: The state of the field and salient challenges for the coming decade. *Biological Psychiatry, 57*, 1424–1435.

Nunez, P. L., Silberstein, R. B., Shi, Z., Carpenter, M. R., Srinivasan, R., Tucker, D. M. et al. (1999). EEG coherency II: Experimental comparisons of multiple measures. *Clinical Neurophysiology, 110*, 469–486.

Ogdie, M. N., Macphie, I. L., Minassian, S. L., Yang, M., Fisher, S. E., Francks, C. et al. (2003). A genomewide scan for attention-deficit/hyperactivity disorder in an extended sample: Suggestive linkage on 17p11. *American Journal of Human Genetics, 72*(5), 1268–1279.

Paterson, A. D., Sunohara, G. A., & Kennedy, J. L. (1999). Dopamine D4 receptor gene: Novelty or nonsense? *Neuropsychopharmacology, 21*(1), 3–16.

Pogorelov, V. M., Rodriguiz, R. M., Insco, M. L., Caron, M. G., & Wetsel, W. C. (2005). Novelty seeking and stereotypic activation of behavior in mice with disruption of the Dat1 gene. *Neuropsychopharmacology, 30*(10), 1818–1831.

Posner, M. I., Rothbart, M. K., Sheese, B. E., & Tang, Y. (2007). The anterior cingulate gyrus and the mechanism of self-regulation. *Cognitive, Affective, and Behavioral Neuroscience, 7*(4), 391–395.

Purper-Ouakil, D., Wohl, M., Mouren, M. C., Verpillat, P., Ades, J., & Gorwood, P. (2005). Meta-analysis of family-based association studies between the dopamine transporter gene and attention deficit hyperactivity disorder. *Psychiatric Genetics, 15*(1), 53–59.

Raghavachari, S., Kahana, M. J., Rizzuto, D. S., Caplan, J. B., Kirschen, M. P., Bourgeois, B. et al. (2001). Gating of human theta oscillations by a working memory task. *Journal of Neuroscience, 21*(9), 3175–3183.

Raghavachari, S., Lisman, J. E., Tully, M., Madsen, J. R., Bromfield, E. B., & Kahana, M. J. (2006). Theta oscillations in human cortex during a working-memory task: Evidence for local generators. *Journal of Neurophysiology, 95*, 1630–1638.

Raymond, J. E., Shapiro, K. L., & Amell, K. M. (1992). Temporary suppression of visual processing in an RSVP task: An attentional blink? *Journal of Experimental Psychology: Human Perception and Performance, 18*(3), 849–860.

Ribasés, M., Ramos-Quiroga, J. A., Hervas, A., Bosch, R., Bielsa, A., Gastaminza, X. et al. (2007) [Epub ahead of print]. Exploration of 19 serotoninergic candidate genes in adults and children with attention-deficit/hyperactivity disorder identifies association for 5HT2A, DDC and MAOB. *Molecular Psychiatry* [Epub ahead of print].

Roman, T., Schmitz, M., Polanczyk, G., Eizirik, M., Rohde, L. A., & Hutz, M. H. (2001). Attention-deficit hyperactivity disorder: A study of association with both the dopamine transporter gene and the dopamine D4 receptor gene. *American Journal of Medical Genetics, 105*(5), 471–478.

Sadeh, A., Pergamin, L., & Bar-Haim, Y. (2006). Sleep in children with attention-deficit hyperactivity disorder: A meta-analysis of polysomnographic studies. *Sleep Medicine Reviews, 10*, 381–398.

Sagvolden, T., Johansen, E. B., Aase, H., & Russell, V. A. (2005). A dynamic developmental theory of attention-deficit/hyperactivity disorder (ADHD) predominantly hyperactive/impulsive and combined subtypes. *Behavioral and Brain Sciences, 28*, 397–419; discussion 419–368.

Satterfield, J. H., Schell, A. M., Nicholas, T. W., Satterfield, B. T., & Freese, T. E. (1990). Ontogeny of selective attention effects on event-related potentials in attention-deficit hyperactivity disorder and normal boys. *Biological Psychiatry, 28*, 879–903.

Satterfield, J. H., Schell, A. M., & Nicholas, T. W. (1994). Preferential neural processing of attended stimuli in attention-deficit hyperactivity disorder and normal boys. *Psychophysiology, 31*, 1–10.

Schweitzer, J. B., Lee, D. O., Hanford, R. B., Tagamets, M. A., Hoffman, J. M., & Grafton, S. T. et al. (2003). A positron emission tomography study of methylphenidate in adults with ADHD: Alterations in resting blood flow and predicting treatment response. *Neuropsychopharmacology, 28*(5), 967–973.

Shaw, J. (1981). An introduction to the coherence function and its use in EEG signal analysis. *Journal of Medical Engineering Technology, 5*, 279–288.

Sprich, S., Biederman, J., Crawford, M. H., Mundy, E., & Faraone, S. V. (2000). Adoptive and biological families of children and adolescents with ADHD. *Journal of American Acadamy of Child Adolescent Psychiatry, 39*(11), 1432–1437.

Srinivasan, R., Nunez, P. L., & Silberstein, R. B. Spatial filtering and neocortical dynamics: Estimates of EEG coherence. (1998). *IEEE Transaction on Biomedical Engineering, 45*, 814–826.

Srinivasan, R., Winter, W. R., Ding, J., & Nunez, P. L. (2007). EEG and MEG coherence: Measures of functional connectivity at disctint spatial scales of neocortical dynamics. *Journal of Neuroscience Methods, 166*, 41–52.

Steriade, M., Gloor, P., Llinás, R. R., Lopes da Silva, F. H., & Mesulam, M. M. (1990). Basic mechanisms of cerebral rhytmic activities (Report of IFCN Committee on Basic Mechanisms). *Electroencephalography Clinical Neurophysiology, 76*, 481–508.

Strauss, A. A., & Lehtinen, V. (1947). Psychopathology and education of the brain injured child. New York: Grune and Stratton.

Strandburg, R. J., Marsch, J. T., Brown, W. S., Asarnow, R. F., Higa, J., Harper, R. et al. (1996). Continous processing related event-related potentials in children with attention deficit hyperactivity disorder. *Biological Psychiatry, 40*, 964–980.

Swanson, J., Flodman, P., Kennedy, J., Spence, M. A., Moyzis, R., Schuck, S. et al. (2000). Dopamine genes and ADHD. *Neuroscience and Biobehavioral Reviews, 24*, 21–25.

Swanson, J. M., Kinsbourne, M., Nigg, J., Lanphear, B., Stefanatos, G. A., Volkow, N. (2007). Etiologic Subtypes of Attention-Deficit/Hyperactivity Disorder: Brain Imaging, Molecular Genetic and Environmental Factors and the Dopamine Hypothesis. *Neuropsychological Review, 17*, 39–59.

Tang, Y., Buxbaum, S. G., Waldman, I., Anderson, G. M., Zabetian, C. P., Köhnke, M. D. et al. (2006). A single nucleotide polymorphism at DBH, possibly associated with attention-deficit/hyperactivity disorder, associates with lower plasma dopamine beta-hydroxylase activity and is in linkage disequilibrium with two putative functional single nucleotide polymorphisms. *Biological Psychiatry, 60*(10), 1034–1038.

Thatcher, R., Krause, P., & Hrybyk, M. (1986). Cortico-cortical associations and EEG coherence: A two compartamental model. *Electroencephalography Clinical Neurophysiology, 64*, 123–143.

Thatcher, R. W., Biver, C. J., & North, D. (2007). Spatial-temporal current source correlations and cortical connectivity. *Clinical EEG Neuroscience, 38*(1), 35–48.

Van Baal, G. C. M., Boomsma, D. I., & De Geus, E. J. C. (2001). Longitudinal genetic analysis of EEG coherence in young twins. *Behavior Genetics, 31*(6), 637–651.

Van der Heijden, K. B., Smits, M. G., & Gunning, W. B. (2005). Sleep-related disorders in ADHD: A review. *Clinical Pediatric, 44*(3), 201–210.

Van Meel, C. S., Oosterlaan, J., Heslenfeld, D. J., & Sergeant, J. A. (2005). Telling good from bad news: ADHD differentially affects processing of positive and negative feedback during guessing. *Neuropsychologia, 43*, 1946–1954.

Van Meel, C. S., Heslenfeld, D. J., Oosterlaan, J., & Sergeant, J. A. (2007). Adaptive control deficits in attention-deficit/hyperactivity disorder (ADHD): The role of error processing. *Psychiatry Research, 151*, 211–220.

VanNess, S. H., Owens, M. J., & Kilts, C. D. (2005). The variable number of tandem repeats element in DAT1 regulates in vitro dopamine transporter density. *BMC Genetics, 6*, 55.

Verleger, R. (1988). Tge true P3 is hard to see: Some comments on Kok's (1986) paper on degraded stimuli. *Biological Psychology, 27*(1), 45–50.

Volkow, N. D., Wang, G. J., Fowler, J. S., Gatley, S. J., Logan, J., Ding, Y. S. et al. (1998). Dopamine transporter occupancies in the human brain induced by therapeutic doses of oral methylphenidate. *American Journal of Psychiatry, 155*(10), 1325–1331.

Waldman, I. D., & Gizer, I. R. (2006). The genetics of attention deficit hyperactivity disorder. *Clinical Psychological Review, 26*(4), 396–432.

Wohl, M., Purper-Ouakil, D., Mouren, M. C., Ades, J., & Gorwood, P. (2005). Meta-analysis of candidate genes in attention-deficit hyperactivity disorder. *Encephale, 31*(4 Pt 1), 437–447.

Xu, X., Mill, J., Chen, C. K., Brookes, K., Taylor, E., & Asherson, P. (2005). Family-based association study of serotonin transporter gene polymorphisms in attention deficit hyperactivity disorder: No evidence for association in UK and Taiwanese samples. *American Journal of Medical Genetics – B Neuropsychiatric Genetics, 139*(1), 11–13.

Yang, B., Chan, R. C., Jing, J., Li, T., Sham, P., & Chen, R. Y. (2007). A meta-analysis of association studies between the 10-repeat allele of a VNTR polymorphism in the 3'-UTR of dopamine transporter gene and attention deficit hyperactivity disorder. *American Journal of Medical Genetics – B Neuropsychiatric Genetics, 144*(4), 541–550.

Zametkin, A. J., Nordahl, T. E., Gross, M., King, A. C., Semple, W. E., Rumsey, J. et al. (1990). Cerebral glucose metabolism in adults with hyperactivity of childhood onset. *New England Journal of Medicine, 323*(20), 1361–1366.

Chapter 15
The Aberrant Connectivity Hypothesis in Schizophrenia

P.A. Gaspar, C. Bosman, S. Ruiz, and F. Aboitiz

Abstract "The disconnection hypothesis" in schizophrenia emphasizes that the core dysfunction of this disease corresponds to a global decrease of neuronal connectivity involving widespread neuronal networks, or in other words to an alteration of the functional connectivity of the brain. Synchronization of EEG activity at high oscillatory frequencies (20–100 Hz) has been proposed to reflect the degree of functional connectivity. Using this approach, a growing number of studies have shown a decrease in neuronal synchrony in schizophrenics, which correlates with perceptual and cognitive deficits in these patients. However, if impaired brain synchronization is a general feature of the schizophrenic brain, then multiple methods to assess functional connectivity should be expected to find similar results. Nevertheless, other techniques, such as functional resonance magnetic imaging (fMRI), have failed to completely fulfill this prediction. While in some brain regions hemodynamic activity is inferior to that of control subjects, other areas evidence an increased activation relative to controls. Moreover, on measuring the covariances in fMRI signals a significant increase in neuronal connectivity is observed in some areas, concomitant with a decrease of connectivity in other areas. Considering these results, we suggest that localized excesses of connectivity could also be involved in the functional changes accompanying schizophrenia. Thus, a decreased functional connectivity could be part of a more general "aberrant connectivity" phenomenon, whose signature is a disbalanced connectivity, some regions displaying an decrease and others an excess of connectivity; the latter arising in part as a compensatory mechanism attempting to restore the damaged sensorimotor circuits.

P.A. Gaspar

Clínica Psiquiátrica Universitaria, Hospital Clínico de la Universidad de Chile. Av. La Paz 1003, Casilla 70014, Santiago, Chile, e-mail: pgaspar@neuro.med.uchile.cl

F. Aboitiz and D. Cosmelli (eds.) *From Attention to Goal-Directed Behavior.*
© Springer-Verlag Berlin Heidelberg 2009

Abbreviations

DISC1   Disrupted in schizophrenia 1
DLPFC   Dorsolateral prefrontal cortex
FA   Fractional anisotropy
fMRI   Functional magnetic resonance imaging
FS-PV   Fast-spiking, parvalbumin-expressing
GAD67   Glutamic acid decarboxylase 67
GluR   Glutamate receptor
mRNA   Messenger RNA
NMDA N-Methyl-D-aspartate
NRG1   Neuregulin-1

Introduction

At the beginning of the twentieth century, Bleuler coined the term "schizophrenia" ("split mind") to describe a mental disease characterized by cognitive disintegration, hallucinations, delusions, disorganized speech, and thinking. In his definition, Bleuler made emphasized "mind disintegration" as a central core of the disease, claiming that a failure in the integration and/or association of ideas, perceptions, and emotions is a common feature of all symptoms of the disease (Bleuler, 1950/1911). During recent years, advances in different fields of neuroscience have allowed the formulation of theoretical and empirical models for the functional changes accompanying schizophrenia that support Bleuler's idea. Mind disintegration has been reformulated as a failure in the coordinated processing of information over multiple brain areas, and a disruption of neuronal connectivity has been postulated as the phenomenon underlying this uncoordinated processing of information (Friston & Frith, 1995; Lewis & Gonzalez-Burgos, 2000; Selemon & Goldman-Rakic, 1999). Thus, relevant to the understanding of the functional changes accompanying schizophrenia is the study of mechanisms that allow an effective communication between different brain areas and how these mechanisms could be impaired.

Questions about mechanisms involved in effective neuronal communication are still open (Roskies, 1999). The most common model is based on the assumption that brain information converges from sensory neurons into relatively few groups of neurons located in association cortices. These groups of neurons are able to combine the received inputs from neurons that are anatomically connected. Thus, the communication structure of the network in this model is limited by the pattern of anatomical connectedness. In this model, the flexibility required to create new routes of communication is provided by neuronal plasticity mechanisms that allow new anatomically defined routes (Abbott & Nelson, 2000). However, sensory, motor, and cognitive functions require highly flexible processing mechanisms to produce appropriate responses over short periods of time. As a consequence, neuronal plasticity mechanisms might not be able to provide a fast, flexible, and highly dynamical flow of neuronal communication that correctly traces the processing route of these multiple functions. Considering this, alternative models that take advantage of the functional organization in established neural networks have been proposed. In this perspective, neurons are arranged in widespread ensembles covering large regions of the brain, both motor and sensory. These ensembles are characterized by their oscillatory activities, either spontaneously generated or evoked after stimulation. Thus, specific frequencies of oscillations (e.g., gamma frequency) have been associated with attention, stimulus encoding, discrimination, and motor behavior (Engel, Konig, Kreiter, & Singer, 1991; Engel, Kreiter, Konig, & Singer, 1991; Gray, Konig, Engel, & Singer, 1989; Stopfer, Bhagavan, Smith, & Laurent, 1997). Furthermore, neurons are able to synchronize their oscillatory responses at different frequency ranges, providing a cohesion mechanism that organizes the network as a unit (Fries, Nikolic, & Singer, 2007). In this hypothesis, phase synchronization of activity oscillations within neuronal assemblies allows the temporal coincidence of highly excitable states between neurons (Fries, 2005). As a consequence, neuronal transmission is facilitated by means of adjusting the coherence of activities between presynaptic neurons, inducing a better output firing in postsynaptic neurons. Because neither of these hypotheses assumes the anatomical limitation of the participating neuronal network, neuronal synchronization provides a highly dynamical mechanism of effective communication of neurons. Thus, the dynamics of these connectivity patterns over time configures the response of the brain to a certain perturbation (Salinas & Sejnowski, 2001).

In this chapter, we will provide a general review of the mechanisms underlying effective and functional communication between neurons, and how disturbances of these mechanisms could allow the emergence of perceptual and cognitive alterations observed in schizophrenia. Following previous ideas of Friston and Frith (1995), recent reviews (Ford, Krystal, & Mathalon, 2007; Stephan, Baldeweg, & Friston, 2006; Uhlhaas & Singer, 2006) have emphasized the role of neuronal disconnectivity as a major pathophysiological failure in schizophrenia. These reviews have pointed out that corticocortical and thalamocortical synaptic failure in several brain regions is one possible mechanism behind the establishment of neuronal disconnectivity. This prediction has been confirmed by several studies using electroencephalography (see later) especially when assessing gamma-band oscillations during attentional binding tasks. However, some experiments have assessed functional connectivity by measuring the covariances in functional magnetic resonance imaging (fMRI) signals during the execution of specific cognitive tasks (Shergill et al., 2007; Whitford et al., 2007a; Whitford et al., 2007b). In these studies, concomitant with a decrease of connectivity pattern in particular areas, a significant increase in neuronal connectivity was also observed in different areas. In our opinion, this observation has profound implications in the understanding of the functional changes accompanying schizophrenia. Thus, functional disconnection could be considered as a particular case of a more general "aberrant connectivity hypothesis," where there may be both pathological increases and decreases in neuronal connectivity in different brain regions.

So, we intend to discuss the aberrant connectivity hypothesis and the evidence supporting it. For this, we will address how several neurodevelopmental dysfunctions that have been related to schizophrenia could lead to aberrant connectivity in the adult brain. Then, we will review evidence supporting the hypothesis of disconnection in schizophrenia, discuss the crucial role of inhibitory interneurons in the generation of neuronal synchronization and as a key node for desynchronization in schizophrenia, and finally present evidence that supports the notion of aberrant connectivity in schizophrenia. As a conclusion, we discuss the specificity of aberrant connectivity as a dysfunction in schizophrenia and other diseases.

Abnormal Brain Development in Schizophrenia: Pathways to Aberrant Connectivity

One of the currently dominant etiopathogenic models in schizophrenia considers this condition as a neurodevelopmental dysfunction (Rapoport, Addington, Frangou, & Psych, 2005; Weinberger, 1987). This hypothesis proposes that a mixture of genetic, epigenetic, and environmental factors would induce early anatomical brain lesions in circuits critical for cognition and emotion. These brain insults might affect the normal positioning of neurons and the establishment of connectivity in development, as well as the synaptic pruning through childhood and adolescence, configuring the onset of the clinical syndrome years later (Selemon & Goldman-Rakic, 1999). Studies of genome expression and molecular markers have shown that most of the genes and gene products involved in schizophrenia are directly or indirectly related to the organization and operation of the synaptic machinery (Frankle, Lerma, & Laruelle, 2003). For instance, complementary DNA microarray techniques have shown that the majority of the transcripts related to presynaptic secretory machinery, especially those involved in excitatory glutamatergic neuro-transmission and inhibitory GABAergic neurotransmission were consistently decreased in schizophrenic patients compared with healthy subjects (Mirnics, Middleton, Marquez, Lewis, & Levitt, 2000). This evidence suggests that genes whose expression is affected in schizophrenia are involved in synaptic connectivity (Lewis, Hashimoto, & Volk, 2005). Moreover, inhibitory GABAergic neurotransmission plays an important role in the generation of oscillatory gamma synchronizations in normal conditions (Bartos et al., 2002), suggesting a possible mechanism by which alterations in synaptic connectivity might generate dynamic alterations leading to cognitive dysfunctions in this disease.

On the other hand, several genetic polymorphisms have been associated with schizophrenia. The most widely acknowledged at this point relate to the loci disrupted in schizophrenia 1 (*DISC1*), dysbindin, neuregulin-1 (*NRG1*), chatecol-*O*-methyltransferase, and D-amino acid oxidase (Stephan et al., 2006). The aim of this chapter is not to make an exhaustive revision of all the genes implicated in

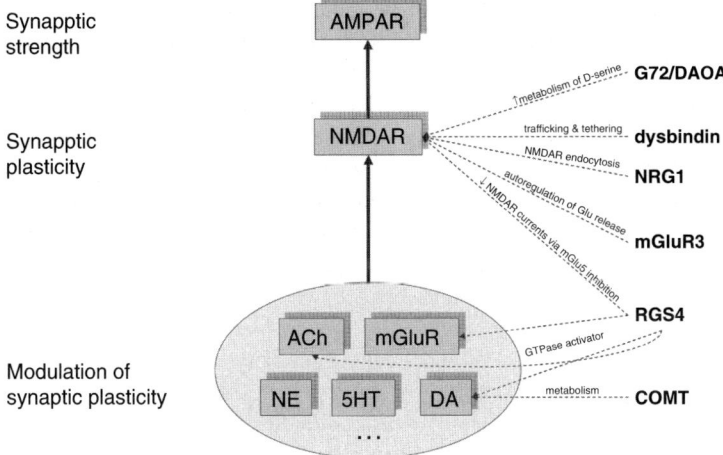

Fig. 15.1 Interrelations of major genetic markers postulated in schizophrenia and glutamatergic neurotransmission. Note that all the genetic markers in this disease are directly and indirectly related to glutamatergic synaptic failure. *AMPAR* α-amino-3-hydroxy-5-methylisoxazole-4-propionic acid receptor, *NDMAR* N-methyl-D-aspartate receptor, *ACh* acetylcholine, *COMT* chatecol-*O*-methyltransferase, *DA* dopamine, *mGluR* metabotropic glutamate receptor, *5HT* 5-hydroxytryptamine, *NE* norepinephrine, *NRG1* neuregulin-1. (Modified from Stephan et al., 2006 with permission from Elsevier Ltd, license number 1893241223905)

schizophrenia; for a complete revision see Harrison & Weinberger (2005). We rather intend to illustrate some examples of how genetic alterations could cause disruptions in the development of critical cortical and subcortical brain circuits that are related to the abovementioned mechanisms of neuronal communication previously described (Fig. 15.1) (see also Chap. 14 by Carrasco et al.).

DISC1 and Abnormal Synaptic Plasticity in Schizophrenia

Loss of genetic control of the dendritic growth process could explain part of the altered connectivity observed in schizophrenia. Since the first description of an association between schizophrenia and a balanced translocation in *DISC1*, this gene has received considerable attention owing to its profound implication in synaptic plasticity (Millar et al., 2000). *DISC1* has also been associated with schizoaffective disorder (Hamshere et al., 2005), unipolar and bipolar disease (Blackwood, Fordyce, Walker, St Clair, Porteous, 2001; Palo et al., 2007), and is now considered to be an ubiquitous marker of several major mental illnesses (Hennah, Thomson, Peltonen, & Porteous, 2006). *DISC1* participates in the normal dendritic growth and maintenance, both during development and in the adult brain (Mackie, Millar, & Porteous, 2007). It has been shown that the interactions of the DISC1 protein with other proteins involved in the presynaptic machinery, such as the NudE-like–lissencephaly-1 complex, fasciculation and elongation protein zeta 1, F-actin, and the zinc-finger binding protein, are crucial to control dendritic growth in neurodevelopment (Kirkpatrick, Xu, Cascella, Ozeki, Sawa, 2006; Lipska et al., 2006; Roberts, 2007). Therefore, it seems that functional mutations in *DISC1* are able to provoke strong impairments in the dendritic growth process. This observation fits the neuropil hypothesis in schizophrenia that predicts a decrease of the dendritic tree as a basis of disconnectivity in the schizophrenic brain. Other gross anatomical and functional alterations have been also associated with *DISC1* in schizophrenia, including a decreased temporal cortex size, impaired response in behavioral tasks, and decreased fMRI activation during working memory paradigms in schizophrenics compared with healthy controls (Callicott et al., 2005), and working memory dysfunctions in first-degree relatives (Hennah, Thomson, Peltonen, & Porteous, 2005) and twins not affected (Gasperoni et al., 2003). This line of evidences implies that alterations in *DISC1* could be related to dysfunctions in the cognitive domain of symptoms in schizophrenia.

NRG1 and Altered Timing of Nervous Impulse

Another example of how genetic alterations may produce disconnectivity is through disruption of the formation and maintenance of myelin sheets. NRG1 and its family of epidermal growth factor receptors are genetically linked with susceptibility to schizophrenia (Stefansson et al., 2002) and bipolar disorders (Green et al., 2005). Association with 8p21–22 haplotypes involving this gene has been observed in samples of patients from Iceland (Stefansson et al., 2002), Scotland (Stefansson, Thorgeirsson, Gulcher, & Stefansson, 2003), the UK (Williams et al., 2003) and other countries Corvin et al., 2004; Yang et al., 2003). NRG1 proteins (neuregulins) have been demonstrated to play important roles during the development of the nervous system. Neuregulins interact with transmembrane tyrosine kinase receptors, and through the activation of intracellular signaling cascades induce proliferation, migration, differentiation, and survival or apoptosis. It has been considered that this signaling pathway is essential to control the proliferation, maturation, and myelination in oligodendrocytes. Recently it has been suggested that epidermal growth factor receptor signaling, throughout expression of NRG1, regulates oligodendrocyte maturation and myelin production in the central nervous system (Roy et al., 2007). Thus, alterations of the *NRG1* gene might produce a decrease in conduction velocity of the nervous impulses, leading to a functional disconnectivity between larger areas, owing to the incapacity of the cells to send precisely timed impulses through large distances in the brain (Aboitiz, Lopez, & Montiel, 2003). However, there are contradictory results with respect of the expression of messenger RNA *NRG1* gene in prefrontal cortex of adult schizophrenic brains Bertram et al., 2007; Hashimoto, Straub,

Weickert, Hyde, Kleinman, 2004). Larger samples of patients must be tested to confirm the relevance of *NRG1* to functional disconnection.

Reelin and Malformation of Cortical Layers in Schizophrenia

Reelin is an extracellular matrix glycoprotein widely distributed in the brain (Quattrocchi et al., 2002). Reelin has important functions during the formation of cortical layers, because it regulates the organization of radial and tangential migration processes during brain development (Aboitiz, 1999). These processes are at the origins of future GABAergic interneuron and glutamatergic neuron populations (Aboitiz, Morales, & Montiel, 2003). In the adult brain, this protein is exclusively secreted by GABAergic interneurons. Although the role of reelin in the adult is still under discussion, it has been postulated that an important function could be related to cortical plasticity (Niu, Renfro, Quattrocchi, Sheldon, & D'Arcangelo, 2004; Weeber et al., 2002). An increasing number of studies point to a strong relation between alterations of the reelin gene and schizophrenia. Postmortem studies have shown a 50% reduction of reelin expression in schizophrenic patients in prefrontal, temporal, and occipital cortices, and in the caudate nucleus and cerebellum (Fatemi, Earle, & McMenomy, 2000; Guidotti et al., 2000; Impagnatiello et al., 1998), but not in mood disorders (Fatemi, Cuadra, El-Fakahany, Sidwell, & Thuras, 2000). Besides, inoculation of the neurotrophic influenza virus in rats during critical periods of prebirth neurodevelopment induces a decrease in reelin expression in temporal and hippocampal cortex, a decrease of brain volume in these same areas, and cortical layer disorganization (Fatemi et al., 1999). These alterations strongly resemble the postmortem brain alterations in schizophrenics. In addition, reelin has been postulated to play a critical role in the renewal and maintenance of dendritic spines in cortical interneurons (Quattrocchi et al., 2002). Furthermore, a failure in the expression of reelin might also have consequences in the stability of neuronal networks involved in the maintenance of gamma oscillations in the brain (Lewis, Hashimoto, & Volk, 2005), underlying a possible origin of the observed gamma dysfunction in schizophrenia.

   In summary, alterations in genes associated with neurodevelopment, but also with important functions in the adult brain, such as synaptic communication (*DISC1*), the normal maintenance of the myelin sheet (neuregulins), and cortical layer formation and synaptic plasticity (reelin), may produce disruptions in anatomical connectivity between neuronal populations and, as a consequence, a functional disconnectivity in schizophrenia. It has been proposed that schizophrenia might be a synaptic disease (Frankle, Lerma, & Laruelle, 2003; Stephan, Baldeweg, & Friston, 2006), taking all this evidence into consideration. In the following sections we will discuss evidence indicating anatomical and functional disconnectivity in schizophrenia.

The Hypothesis of Disconnection in Schizophrenia: Implications for Neuronal Synchronization

The disconnection hypothesis implies that the core of this disease is a global decrease of neuronal connectivity (anatomical and functional) involving networks that support different cognitive processes (Friston & Frith, 1995). As we stated previously, dysfunctional connectivity is likely to be a consequence of an alteration of mechanisms that support synaptic communication (Frankle, Lerma, & Laruelle, 2003). Furthermore, neuronal synchronization critically depends on the integrity of neuroanatomical networks and synaptic communication. In this sense, the study of neuronal synchronization in the brain might shed light on the degree of functional connectivity in the living brain, but could also provide clues about the different mechanisms involved in synaptic disconnectivity. We will discuss the underlying mechanisms of neuronal synchronization, and findings based on noninvasive electromagnetic recordings that support the notion of disconnectivity in schizophrenia. Interestingly, normal synchronization depends on many mechanisms that have been considered to be disturbed in schizophrenia (Friston & Frith, 1995).

Neural Synchrony and Perceptual Deficits in Schizophrenia

Neuronal synchrony is a ubiquitous phenomenon that has been studied on different spatial scales and with different functional implications (Aboitiz et al., 2003; Engel, Fries, & Singer, 2001; Singer & Gray, 1995; Varela, Lachaux, Rodriguez, & Martinerie, 2001). For instance, short-range neuronal synchrony involves the coincident firing of two neurons, generally in the same cortical column (Gray, Konig, Engel, & Singer, 1989; Gray & Singer, 1989; Maldonado & Gray, 1996). In visual areas, neurons of different receptive fields fire synchronously when one stimulus passes through both receptive fields, but not when two separated stimuli stimulate the same receptive fields (Singer & Gray, 1995). Thus, short-range synchrony involves local computations between neurons. In contrast, large-scale neuronal synchrony refers to the coincident phases of two or more groups of neurons whose activities oscillate at a particular frequency of interest. Neuronal synchronization (normally recognized as phase synchronization) has been studied between small groups of neurons (Fries, Roelfsema, Engel, Konig, & Singer, 1997; Roelfsema, Engel, Konig, & Singer, 1997) or in humans, using noninvasive electrophysiological recordings (Rodriguez, George, Lachaux, Martinerie, Renault, 1999; Tallon-Baudry & Bertrand, 1999) and it has been consistently related to several perceptual and cognitive tasks. In humans, induced oscillatory activity in the gamma range (Engel, Fries, & Singer, 2001; Tallon-Baudry & Bertrand, 1999) has been postulated as a binding element of distributed brain activity in relation to the integration of sensory features, reflecting top-down brain processes and temporal integration of functional neural networks (Gray et al., 1989; von der Malsburg, 1995). A growing number of studies support this hypothesis (Raghavachari, Lisman, Tully, Madsen, Bromfield, 2006; Tallon-Baudry, 2003). High-contrast images and visual illusory figures elicit an increase of high-frequency oscillations (gamma band) in healthy subjects (Rodriguez et al., 1999).

Considering that Gestaltic-like stimuli involve the binding of individual characteristics of the objects (color, shape) into a single percept, these kinds of stimuli are frequently tested in schizophrenic patients. Under these stimulation conditions, deficits in large-scale, high-frequency rhythmic neural synchrony have been observed in these patients, compared with normal controls (Ford, Roach, Faustman, & Mathalon, 2007a; Spencer, Nestor, Niznikiewicz, Salisbury, Shenton, 2003; Uhlhaas et al., 2006). Specifically, a lower than normal degree of phase synchrony in gamma (30–100 Hz) (Spencer et al., 2003; Spencer et al., 2004) and beta (15–30 Hz) frequencies (Uhlhaas et al., 2006) has consistently been observed in a sample of chronic medicated patients after the presentation of Gestalt stimuli. Moreover, these perceptual abnormalities showed a direct correlation with the severity of the symptoms (Spencer et al., 2004).

Altered Synchrony and Cognitive Dysfunctions in Schizophrenia

Studies of coherent long-range dysfunctions have not only focused on perceptual processes but have also been performed using tasks that evaluate cognitive functions. Cognitive dysfunctions have been postulated as core symptoms in schizophrenia (Green & Nuechterlein, 1999; Lee, Williams, Breakspear, & Gordon, 2003; van der Stelt, Belger, & Lieberman, 2004), motivating the study of neuronal correlates of cognitive tasks in these patients. A recent study has indirectly tested synchrony alterations in the context of cognitive dysfunctions in schizophrenic patients. In one study, a large cohort of schizophrenic patients ($n = 100$) displayed reduced gamma-band synchronization while click trains were delivered at 40 Hz, but not at other rates of stimulation. Cognitive status was also evaluated in these patients using several cognitive tests, there being a significant correlation with gamma synchrony alterations only in the case of performance in working memory tasks (Light et al., 2006). It is important to address correlations with working memory alterations because these behavioral alterations have been observed in prodromical phases of the disease, in first-episode patients, and also in populations with high risk of developing schizophrenia (Green & Nuechterlein, 1999). Alterations in executive functions have also been related to cognitive dysfunctions in schizophrenia. Recently, a decrease in induced gamma band activity was demonstrated in schizophrenic patients in a paradigm that evaluates inhibition of executive functions (Cho, Konecky, & Carter, 2006). In this study, subjects received a color cue that indicated if they responded with the same or the opposite hand to the direction of an arrow. Error rates, reaction time, and topographical spectral power were measured. Schizophrenic patients had more errors and longer reaction times in the condition in which they responded with the opposite hand with respect to the direction of the arrow, reflecting a deficit in the cognitive control of impulsive responses. Concomitant to this deficit, a significant decrease in gamma-band power compared with normal controls was found in frontal regions. Finally, it has been postulated that a disorganized speech in schizophrenia would reflect the disorganized thinking underlying this disease. In a recent study, phase synchrony of neural oscillations preceding speech onset was examined during a simple vocalization task (Ford, Roach, Faustman, & Mathalon, 2007b). The authors showed a significant inverse correlation between the severity of hallucinations and the phase-locked neuronal synchrony of preceding speech, and significant differences in the phase-locking value of synchrony were found between groups (patients vs. controls). These deficits may underlie the poor capacity of schizophrenic patients to engage different areas involved in the generation of a sensory motor activity, thus contributing to the symptom of disorganized speech typically observed in these patients (Ford, Roach et al., 2007a).

Taken together, these pieces of evidence suggest an association between a global decrease of long-range synchrony and perceptual and cognitive dysfunctions in schizophrenia. Moreover, these results are compatible with the relative incapacity of the schizophrenic brain to bind object features. Thus, insights into the mechanisms that support neuronal synchronization must be relevant to illuminate the mechanisms of disease in schizophrenia.

Impairment of Neuronal Synchronization as a Result of Inhibitory Neuron Dysfuntions

Although, in general, functional connectivity can be impaired by several mechanisms in schizophrenia (i.e., by synaptic malfunction at different levels or by abnormal axonal conduction; see earlier), the specific alteration producing a decrease in phase synchronization is not yet clear. Here we will focus on one possible mechanism for this phenomenon, which considers the role of inhibitory neurons in driving synchronic firing of groups of excitatory neurons.

It is important to recall that oscillatory activity of neuronal networks provides windows of excitability for particular neurons, thus controlling the probability of concurrent spiking (see Chap. 1 by Bosman and Womelsdorf). In other words, if two neuronal populations synchronize their local field potentials in such a way that both groups present a temporally coincident excitability period, then the probability of common firing between both groups is enhanced. In contrast, if two neuronal groups are not able to synchronize their local field potentials, the probability of common firing between the neurons decreases (Fries, 2005). Thus, phase locking of oscillatory patterns between two neuronal groups provides effective conditions for optimal communication between neurons, while the absence of phase synchronization prevents the communication between groups (Fries, 2005).

Oscillatory activity depends on the proper connectivity of interneuronal networks (Bartos, Vida, & Jonas, 2007). Synaptic coupling across fast-spiking, parvalbumin-expressing (FS-PV) interneurons and prefrontal pyramidal neurons is critical for establishing the basic circuit of neuronal synchronization in high- and low-frequency rhythms (Fries, 2005). FS-PV interneurons represent the 70% of total interneurons in the cerebral cortex, and are conspicuous in regions where synchronization has been observed (Bartos et al., 2007). This suggests that these cells provide an important means to modulate neuronal oscillatory activity via powerful axosomatic inhibition, at least in prefrontal and temporal cortices. Furthermore, FS-PV interneurons are connected among themselves by gap junctions, mediated by connexin 36, forming extended neuronal networks that fire in a highly synchronous fashion. Malfunction of FS-PV interneurons might disrupt the glutamatergic pyramidal output. Reduced levels of glutamic acid decarboxylase-67 (GAD67, one of the main enzymes that synthesizes GABA in inteneurons) mRNA and protein in prefrontal (Knable, Barci, Bartko, Webster, & Torrey, 2002) and temporal (Torrey et al., 2005) cortices are among the most consistent findings in postmortem studies of schizophrenic patients. Evidence that neuronal synchrony might be disrupted by altering FS-PV interneurons has been recently assessed in animal models of schizophrenia. For instance, GAD67 mutant ferrets exhibit altered local synchrony mediated by dysfunctional FS-PV interneurons in prefrontal cortex (Gao, 2007). Besides, acute induction of clozapine (one of the most effective antipsychotic drugs) in GAD67 mutants reduces the synchrony firing events through the same mechanism (Fig. 15.2) (Gao, 2007).

FS-PV interneuron function can be also altered by disruption of the synaptic machinery that regulates these neurons. In mammalian cortex, one of the main modulators of FS-PV interneuron activity is metabotropic glutamatergic receptors located on the soma of these neurons (Huang, Di Cristo, & Ango, 2007). Different types of noncompetitive (ketamine, MK-801) and competitive (phencyclidine) antagonists of the N-methyl-D-aspartate (NMDA) receptor induce schizophreniform psychosis, hallucinations, and cognitive deficits in healthy subjects, and exacerbate symptoms in schizophrenic patients (Adler et al., 1998; Krystal et al., 1994; Newcomer et al., 1999). In this sense, although psychosis-like symptoms might be evoked by other neurochemical systems such as dopaminergic agonists, only the

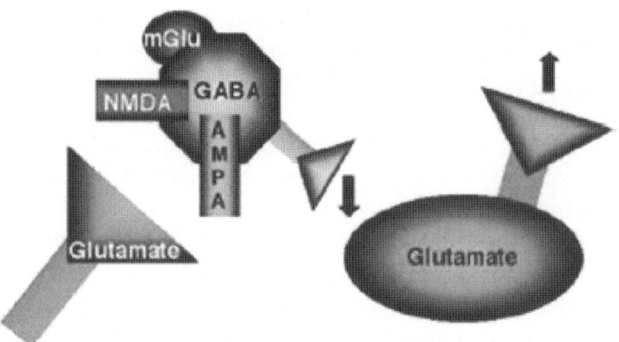

Fig. 15.2 Interrelations between GABAergic interneurons and glutamatergic neurotransmission in schizophrenia. Alterations or blockade of metabotropic receptors on the soma of GABA inteneurons decreases the activity of that neuron. The reduction of GABA release increases the activity of glutamatergic neurotransmission

symptoms elicited by NMDA antagonists occur in a physiological range (for instance, subanesthesic doses) (Krystal et al., 1994). This evidence has focused the attention of schizophrenia research on glutamatergic receptors as a possible therapeutic target (Moghaddam, 2003; Mouri, Noda, Enomoto, & Nabeshima, 2007). Evidence of alterations in molecular components of NMDA receptors and glutamate receptors (GluR) supports this hypothesis. Protein and mRNA subunits of NMDA receptors are frequently diminished in the neocortex of first-episode patients and in animal models of schizophrenia (Mouri et al., 2007). Besides, transgenic mice for GluR subunit genes exhibit neuroanatomic and physiological alterations resembling those of schizophrenic patients. In conclusion, hypofunction of glutamatergic neurotransmission and especially NMDA alterations may have pivotal importance in schizophrenia (Mori & Zhang, 2007). Considering that FS-PV interneurons are considered to be crucial to the generation of synchronic ensembles, we hypothesize that alterations of FS-PV interneuron activity (through NMDA receptors and other GluR) and their relation to pyramidal glutamatergic circuits could be considered a key node of desynchronization postulated in this disease.

In this context, recent reports imply participation of these receptors in the normal function of neural synchrony. For instance, gamma power and spiking coincidence (a measure of synchrony) were strongly decreased in the hippocampus of transgenic GluR-D and GluR-A knockout mice. Interestingly, these mutants display impairments in spatial working memory and novel object recognition, among other behavioral deficits (Fuchs et al., 2007). Phencyclidine, a common antagonist of NMDA receptors, induces in these animals reduced synchrony in delta frequency and psychosis-like symptoms that are reversed by haloperidol and clozapine (Kargieman, Santana, Mengod, Celada, & Artigas, 2007). Although these are observations in animal models, the impairments resemble the positive and cognitive dysfunctions in schizophrenia.

Taken together, these pieces of evidence provide a physiological and neuroanatomical basis to local neural synchronization, where the interplay between FS-PV interneurons and glutamatergic pyramidal neurons plays a key role to establish a normal short-range synchrony in the mammalian cerebral cortex. Alterations in key aspects of FS-PV interneurons, together with the modulation that occurs in their interaction with glutamatergic neurons, could disrupt neuronal synchronization in animal models of schizophrenia.

The Aberrant Connectivity Hypothesis in Schizophrenia

Until this part of the chapter, we have been discussing the well-known implications of anatomical disconnectivity and failure in neuronal coherence as a functional counterpart of the observed anatomical disconnectivity. Moreover, the search for key molecular mechanisms that could explain the failure of schizophrenia are also in agreement with these statements. However, our purpose is to draw attention to additional studies that support the opposite effect, i.e., an excess of connectivity, at least in some brain regions. Alterations in cortical circuit development may result in decreased synaptic connectivity in some regions, but in other brain areas there may be an excess of neuronal connectivity. In other words, we suggest that instead of there being a generalized decrease of connectivity, in the schizophrenic brain there is a disbalance in the establishment of appropriate connectivity, resulting both in regions with a deficit of functional connectivity and in regions with an excess of functional connectivity. The latter condition might partly be caused by a compensatory mechanism attempting to restore the operation of sensorimotor circuits. We will term this alternative the "aberrant connectivity hypothesis." Although this is not strictly a new concept (Stephan et al., 2006), we consider that the emphasis of other authors has been placed on synaptic disconnectivity, perhaps neglecting other forms of aberrant connectivity that might lead to synaptic overfunctioning.

Three main lines of evidence support an aberrant connectivity model in this disease. First, there are controversies about the integrity of white matter in schizophrenia. Intra- and interhemispheric disturbance has been postulated to play a key role in this disease (Crow, 1998; Hoffman & McGlashan, 1998). Considering that some characteristics of white matter may represent an accurate reflection of normal or abnormal neuronal fiber connectivity, it is important to discuss this topic. Multiple kinds of techniques are being developed to map white matter in a noninvasive way. Magnetic resonance diffusion-weighted imaging is thought to be an indicator of the integrity, organization, and density of fiber bundles in white matter regions of the brain. It is among the newest and most promising in vivo methods that have made it possible to investigate white matter abnormalities (Mori & Zhang, 2006). Multiple studies have reported inconsistent results in chronic and first-episode schizophrenic patients. Although decreased fractional anisotropy (FA), a measure of white matter integrity, has been observed in the corpus callosum and the fiber tracts connecting temporal to frontal and occipital lobes of schizophrenics, others have not reported alterations in FA, and even describe an increase in FA in schizophrenic brains (Fig. 15.3) (Hubl et al., 2004; Shergill et al., 2007; Whitford et al., 2007b).

On the basis of the hypothesis that disruption of the lateral arcuate fasciculus connecting auditory to speech cortex is associated with auditory hallucinations, fiber tracts in the Sylvian fissure white matter were examined using a line-scan diffuse tensor imaging sequence (Hubl et al., 2004). A particular pattern of FA in the schizophrenic sample with demonstrated history of auditory hallucinations was observed. An increased FA in the lateral process of the arcuate fasciculus but a decreased FA in the medial process of this structure was demonstrated, especially in the left hemisphere. This pattern was not observed in schizophrenics without history of auditory hallucinations (Hubl et al., 2004). In a study using a whole-brain method and recruiting a larger number of subjects than in previous studies (80 in both groups), a decrease of FA relative to healthy subjects was observed in the superior longitudinal fasciculus and genu of corpus callosum (Shergill et al., 2007), but interestingly an increased FA in tapetum, anterior cingulum, and in the lateral process of the superior longitudinal fasciculus was observed associated with propensity of auditory hallucinations. Moreover, in first-episode patients, a two-pattern alteration of white matter has recently been reported. A decreased connectivity was demonstrated between

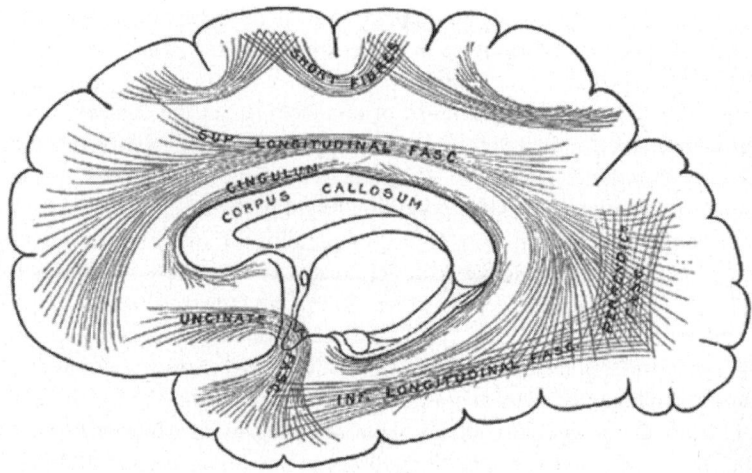

Fig. 15.3 The principal systems of association fibers in the cerebrum. Most of them are disrupted in schizophrenia, as evaluated by tractography techniques

dorsolateral prefrontal cortex (DLPFC) and parietal regions, but there was also an increased connectivity from DLPFC to temporal cortex (Whitford et al., 2007b). The follow-up of these subjects indicated a progressive decrease of connectivity years later (Whitford et al., 2007a). Although some methodological issues still remain to be improved in diffusion-weighted imaging (Kanaan et al., 2005), it seems that in the schizophrenic brain there is the coexistence of overconnected and under-connected brain regions, in accordance with the aberrant connectivity hypothesis (Fig. 15.4).

The second line of evidence that support the aberrant connectivity hypothesis arises from estimations of functional connectivity using fMRI methods evaluating local blood perfusion. Certain regions of prefrontal cortex are the main substrates of several higher-order cognitive processes such as executive functions and working memory (Aboitiz, 1995; Baddeley, 1992; Goldman-Rakic, 1996; Smith & Jonides, 1998). Because schizophrenic patients have strong disabilities in these tasks, the prefrontal cortex has received plenty of attention in this disease (Shad, Tamminga, Cullum, Haas, & Keshavan, 2006). The apparent inability of schizophrenic patients to significantly increase DLPFC perfusion during some cognitively demanding tasks (Jansma, Ramsey, Slagter, & Kahn, 2001; Perlstein, Dixit, Carter, Noll, & Cohen, 2003) contrasts with the overperfusion detected in other prefrontal regions using similar tasks. For instance, in a recent meta-analysis, over 12 imaging studies based on different working memory load tasks revealed an increased activation in ventrolateral prefrontal cortex, anterior cingulate, and left frontal polar regions concomitant with DLPFC activation reductions (Glahn et al., 2005). Although several studies have replicated this finding, other authors claim a normal activation and even hyperactivation of DLPFC in the same tasks. Such increases in perfusion

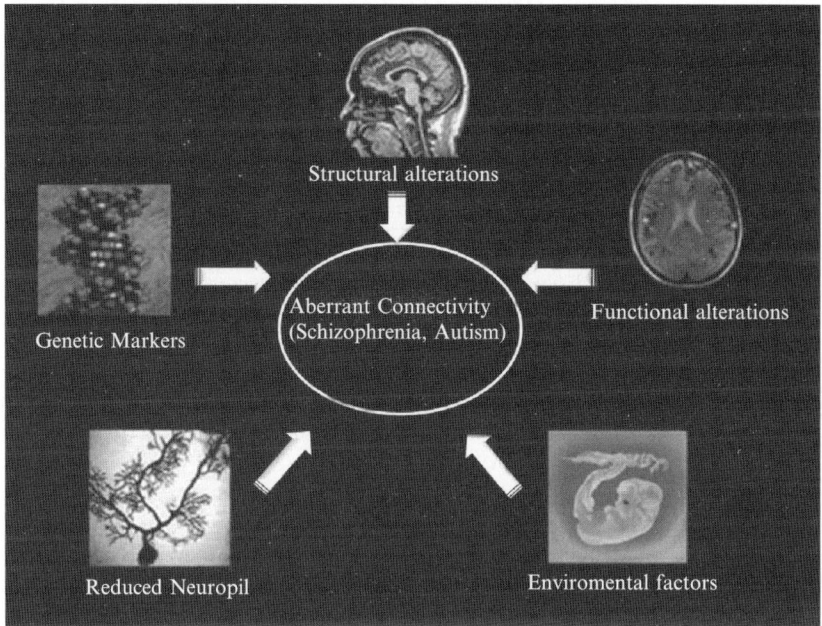

Fig. 15.4 Lines of evidence supporting the aberrant connectivity hypothesis in schizophrenia

changes have been interpreted as an inefficient effort of the schizophrenic brain to compensate for the cognitive and behavioral alterations of this disease (Callicott et al., 2003), which is consistent with the concept that the excess of connectivity in some regions may be viewed as a compensatory mechanism oriented to restore normal function.

The last evidence to be discussed comes from fMRI studies testing functional connectivity. Initially, the theoretical concept of disconnectivity in schizophrenia arose from the statistical association of activity changes in different brain regions using fMRI. Like the previous lines of evidence, the first reports observed a decreased functional connectivity between prefrontal and temporal and parietal cortices in schizophrenic patients. For instance, frontotemporal disconnectivity has been observed in chronic and first-episode patients, both during working memory processes (Barch & Csernansky, 2007; Meyer-Lindenberg et al., 2001) and associated with auditory hallucinations (Lawrie et al., 2002). Frontotemporal disconnectivity has also been associated with genetic risk of schizophrenia (Winterer, Coppola, Egan, Goldberg, & Weinberger, 2003). Positron emission tomography imaging techniques have also reported frontoparietal disconnectivity during the maintenance period of working memory tasks (Kim et al., 2003). Nevertheless, a contrasting fMRI pattern of functional connectivity in resting conditions has recently been observed in a sample of first-episode patients (Zhou et al., 2007). In this study, DLPFC showed a bilateral reduction in functional connectivity to parietal, thalamus, and posterior cingulate cortex; but there was also an enhanced connectivity between DLPFC and the medial portion of the temporal lobe and paralimbic regions. In summary, most studies have placed emphasis on the disconnection of cell assemblies as an etiological factor in schizophrenia, perhaps overlooking structural and functional evidence suggesting that in schizophrenia there could be not only disconnectivity, but also an overconnectivity in different parts of the brain.

Discussion: The Specificity of Disconnectivity in Schizophrenia

A wealthy extent of literature has considered schizophrenia as a disconnection disease. In this chapter, we consider the notion of aberrant connectivity as an extension of the disconnectivity hypothesis. This concept extends the notion of abnormality of synaptic connectivity beyond decreasing synaptic strength to consider in more general terms an abnormal pattern of connectivity during development. This notion includes the anatomical and functional abnormalities previously described by disconnectivity, but it is also in agreement with the results of studies that show overconnectivity in some brain regions. Aberrant connectivity has also been observed in epilepsy, autism, Parkinson's disease, and Alzehimer's disease. This observation suggests that connectivity alterations might represent a more general mechanism of brain damage (Uhlhaas & Singer, 2006). Since autism spectrum disorders have been among the most studied in this context, we will review some evidence that links schizophrenia and autistic disorders in the context of aberrant connectivity.

Autism is currently considered a neurodevelopmental syndrome that starts during early childhood. Like schizophrenics, these patients have an impaired capacity to express emotions and also display cognitive dysfunctions (Goldstein, Minshew, Allen, & Seaton, 2002) and language and speech disabilities (Pardo & Eberhart, 2007). More importantly, most autism-predisposing genetic polymorphisms are also shared with schizophrenia (Rzhetsky, Wajngurt, Park, & Zheng, 2007). We have already discussed the importance of these genes in the establishment of neural networks during the brain neurodevelopment (for instance, reelin). Autistic disorders also present impairments in neuronal coherence during the performance of perceptual and cognitive tasks (Grice et al., 2001). As a consequence of that, a disconnection mechanism has been invoked to explain the core of autism spectrum disorders (Belmonte et al., 2004; Brock, Brown, Boucher, & Rippon, 2002).

If those statements are correct, we should be able to detect evidence of aberrant communication in both schizophrenia and autism. In a recent fMRI study (Just, Cherkassky, Keller, & Minshew, 2004), an impaired functional connectivity in language temporal networks of high-functioning autistics was found during the execution of a sentence comprehension task. Interestingly, a concomitance of high and low activations in different brain regions was found, according to aberrant connectivity assumptions (Just et al., 2004). The localization of the dissimilar activation was related to cortical language processing regions. A reduced activation in Broca's area (left inferior frontal gyrus) but enhanced activation in the left posterior superior temporal gyrus compared with the control group was found (Just et al., 2004). With use of a working memory task, different patterns of activation were observed in an autistic group compared with the control sample. Left hemispheric activation was observed when healthy subjects performed the task, but right predominant hemispheric activation was found in the autistic sample (Just et al., 2004). This evidence suggests that autism might be also considered to result from an aberrant connectivity established during early neuronal development.

Finally, this chapter was intended to expand the concept of reduced connectivity in schizophrenia, suggesting that there is a general alteration in the normal development of neuronal connectivity. This results in several domains of hypoconnectivity, but in some instances connectivity is hyperdeveloped, resulting in abnormal function as well. This aspect might be partly associated with a decreased number and activity of inhibitory processes and with the abnormal development of certain fiber tracts, and could relate to some of the positive symptoms observed in these patients. At this point it is difficult to determine which networks display a hyperconnectivity in the schizophrenic brain, which is a matter for future research. In general, a disbalance in the development of normal connectivity may result in profound alterations in sensorimotor integration, leading to compensatory mechanisms which, although oriented to minimize dysfunction, produce overall disbalances in networks involved in cognitive processing, emotional expression, and behavioral control. Thus, there is an overall distortion of the appropriate balance between different networks that regulate goal-directed and social behavior. Thus, we propose that not only decreases in connectivity, but also pathological increases in connectivity in certain regions should be searched for in future studies of this disease.

Acknowledgments This work was supported by a MECESUP grant to the Pontificia Universidad Católica de Chile, the Millenium Nucleus for Integrative Neuroscience, and CONICYT.

References

Abbott, L. F., & Nelson, S. B. (2000). Synaptic plasticity: Taming the beast. *Nature Neuroscience*, *3*(Suppl), 1178–1183.

Aboitiz, F. (1995). Working memory networks and the origin of language areas in the human brain. *Medical Hypotheses*, *44*(6), 504–506.

Aboitiz, F. (1999). Evolution of isocortical organization. A tentative scenario including roles of reelin, p35/cdk5 and the subplate zone. *Cerebral Cortex*, *9*(7), 655–661.

Aboitiz, F., Lopez, J., & Montiel, J. (2003). Long distance communication in the human brain: Timing constraints for inter-hemispheric synchrony and the origin of brain lateralization. *Biological Research*, *36*(1), 89–99.

Aboitiz, F., Morales, D., & Montiel, J. (2003). The evolutionary origin of the mammalian isocortex: Towards an integrated developmental and functional approach. *Behavioral and Brain Sciences*, *26*(5), 535–552; discussion 552–585.

Adler, L. E., Olincy, A., Waldo, M., Harris, J. G., Griffith, J., Stevens, K. et al. (1998). Schizophrenia, sensory gating, and nicotinic receptors. *Schizophrenia Bulletin*, *24*(2), 189–202.

Baddeley, A. (1992). Working memory. *Science*, *255*(5044), 556–559.

Barch, D. M., & Csernansky, J. G. (2007). Abnormal parietal cortex activation during working memory in schizophrenia: Verbal phonological coding disturbances versus domain-general executive dysfunction. *Americal Journal of Psychiatry*, *164*(7), 1090–1098.

Bartos, M., Vida, I., Frotscher, M., Meyer, A., Monyer, H., Geiger, J. R. et al. (2002). Fast synaptic inhibition promotes synchronized gamma oscillations in hippocampal interneuron networks. *Proceedings of the National Academy of Science USA*, *99*(20), 13222–13227.

Bartos, M., Vida, I., & Jonas, P. (2007). Synaptic mechanisms of synchronized gamma oscillations in inhibitory interneuron networks. *Nature Reviews Neuroscience*, *8*(1), 45–56.

Belmonte, M. K., Allen, G., Beckel-Mitchener, A., Boulanger, L. M., Carper, R. A., & Webb, S. J. (2004). Autism and abnormal development of brain connectivity. *Journal of Neuroscience*, *24*(42), 9228–9231.

Bertram, I., Bernstein, H. G., Lendeckel, U., Bukowska, A., Dobrowolny, H., Keilhoff, G. et al. (2007). Immunohistochemical evidence for impaired neuregulin-1 signaling in the prefrontal cortex in schizophrenia and in unipolar depression. *Annals of the New York Academy of Sciences*, *1096*, 147–156.

Blackwood, D. H., Fordyce, A., Walker, M. T., St Clair, D. M., Porteous, D. J., & Muir, W. J. (2001). Schizophrenia and affective disorders – cosegregation with a translocation at chromosome 1q42 that directly disrupts brain-expressed genes: Clinical and P300 findings in a family. *American Journal of Human Genetics*, *69*(2), 428–433.

Bleuler, E. (1950/1911). *Dementia praecox or the group of schizophrenias*. New York: International UP.

Brock, J., Brown, C. C., Boucher, J., & Rippon, G. (2002). The temporal binding deficit hypothesis of autism. *Development and Psychopathology*, *14*(2), 209–224.

Callicott, J. H., Mattay, V. S., Verchinski, B. A., Marenco, S., Egan, M. F., & Weinberger, D. R. (2003). Complexity of prefrontal cortical dysfunction in schizophrenia: More than up or down. *American Journal of Psychiatry*, *160*(12), 2209–2215.

Callicott, J. H., Straub, R. E., Pezawas, L., Egan, M. F., Mattay, V. S., Hariri, A. R. et al. (2005). Variation in DISC1 affects hippocampal structure and function and increases risk for schizophrenia. *Proceedings of the National Academy of Sciences USA*, *102*(24), 8627–8632.

Cho, R. Y., Konecky, R. O., & Carter, C. S. (2006). Impairments in frontal cortical gamma synchrony and cognitive control in schizophrenia. *Proceedings of the National Academy of Sciences USA*, *103*(52), 19878–19883.

Corvin, A. P., Morris, D. W., McGhee, K., Schwaiger, S., Scully, P., Quinn, J. et al. (2004). Confirmation and refinement of an 'at-risk' haplotype for schizophrenia suggests the EST cluster, Hs.97362, as a potential susceptibility gene at the Neuregulin-1 locus. *Molecular Psychiatry*, *9*(2), 208–213.

Crow, T. J. (1998). Schizophrenia as a transcallosal misconnection syndrome. *Schizophrenia Research*, *30*(2), 111–114.

Engel, A. K., Fries, P., & Singer, W. (2001). Dynamic predictions: Oscillations and synchrony in top-down processing. *Nature Reviews Neuroscience*, *2*(10), 704–716.

Engel, A. K., Konig, P., Kreiter, A. K., & Singer, W. (1991). Interhemispheric synchronization of oscillatory neuronal responses in cat visual cortex. *Science*, *252*(5010), 1177–1179.

Engel, A. K., Kreiter, A. K., Konig, P., & Singer, W. (1991). Synchronization of oscillatory neuronal responses between striate and extrastriate visual cortical areas of the cat. *Proceedings of the National Academy of Sciences USA*, *88*(14), 6048–6052.

Fatemi, S. H., Cuadra, A. E., El-Fakahany, E. E., Sidwell, R. W., & Thuras, P. (2000). Prenatal viral infection causes alterations in nNOS expression in developing mouse brains. *Neuroreport*, *11*(7), 1493–1496.

Fatemi, S. H., Earle, J. A., & McMenomy, T. (2000). Reduction in Reelin immunoreactivity in hippocampus of subjects with schizophrenia, bipolar disorder and major depression. *Molecular Psychiatry*, *5*(6), 654–663, 571.

Fatemi, S. H., Emamian, E. S., Kist, D., Sidwell, R. W., Nakajima, K., Akhter, P. et al. (1999). Defective corticogenesis and reduction in Reelin immunoreactivity in cortex and hippocampus of prenatally infected neonatal mice. *Molecular Psychiatry*, *4*(2), 145–154.

Ford, J. M., Krystal, J. H., & Mathalon, D. H. (2007). Neural synchrony in schizophrenia: From networks to new treatments. *Schizophrenia Bulletin, 33*(4), 848–852.

Ford, J. M., Roach, B. J., Faustman, W. O., & Mathalon, D. H. (2007a). Out-of-Synch and Out-of-Sorts: Dysfunction of Motor-Sensory Communication in Schizophrenia. *Biological Psychiatry*.

Ford, J. M., Roach, B. J., Faustman, W. O., & Mathalon, D. H. (2007b). Synch before you speak: Auditory hallucinations in schizophrenia. *American Journal of Psychiatry, 164*(3), 458–466.

Frankle, W. G., Lerma, J., & Laruelle, M. (2003). The synaptic hypothesis of schizophrenia. *Neuron, 39*(2), 205–216.

Fries, P. (2005). A mechanism for cognitive dynamics: Neuronal communication through neuronal coherence. *Trends in Cognitive Sciences, 9*(10), 474–480.

Fries, P., Nikolic, D., & Singer, W. (2007). The gamma cycle. *Trends Neurosci, 30*(7), 309–316.

Fries, P., Roelfsema, P. R., Engel, A. K., Konig, P., & Singer, W. (1997). Synchronization of oscillatory responses in visual cortex correlates with perception in interocular rivalry. *Proceedings of the National Academy of Sciences USA, 94*(23), 12699–12704.

Friston, K. J., & Frith, C. D. (1995). Schizophrenia: A disconnection syndrome? *Clinical Neuroscience, 3*(2), 89–97.

Fuchs, E. C., Zivkovic, A. R., Cunningham, M. O., Middleton, S., Lebeau, F. E., Bannerman, D. M. et al. (2007). Recruitment of parvalbumin-positive interneurons determines hippocampal function and associated behavior. *Neuron, 53*(4), 591–604.

Gao, W. J. (2007). Acute clozapine suppresses synchronized pyramidal synaptic network activity by increasing inhibition in the ferret prefrontal cortex. *Journal of Neurophysiology, 97*(2), 1196–1208.

Gasperoni, T. L., Ekelund, J., Huttunen, M., Palmer, C. G., Tuulio-Henriksson, A., Lonnqvist, J. et al. (2003). Genetic linkage and association between chromosome 1q and working memory function in schizophrenia. *American Journal of Medical Genetics – B Neuropsychiatric Genetics, 116*(1), 8–16.

Glahn, D. C., Ragland, J. D., Abramoff, A., Barrett, J., Laird, A. R., Bearden, C. E. et al. (2005). Beyond hypofrontality: A quantitative meta-analysis of functional neuroimaging studies of working memory in schizophrenia. *Human Brain Mapping, 25*(1), 60–69.

Goldman-Rakic, P. S. (1996). Regional and cellular fractionation of working memory. *Proceedings of the National Academy of Sciences USA, 93*(24), 13473–13480.

Goldstein, G., Minshew, N. J., Allen, D. N., & Seaton, B. E. (2002). High-functioning autism and schizophrenia: A comparison of an early and late onset neurodevelopmental disorder. *Archives of Clinical Neuropsychology, 17*(5), 461–475.

Gray, C. M., Konig, P., Engel, A. K., & Singer, W. (1989). Oscillatory responses in cat visual cortex exhibit inter-columnar synchronization which reflects global stimulus properties. *Nature, 338*(6213), 334–337.

Gray, C. M., & Singer, W. (1989). Stimulus-specific neuronal oscillations in orientation columns of cat visual cortex. *Proceedings of the National Academy of Sciences USA, 86*(5), 1698–1702.

Green, E. K., Raybould, R., Macgregor, S., Gordon-Smith, K., Heron, J., Hyde, S. et al. (2005). Operation of the schizophrenia susceptibility gene, neuregulin 1, across traditional diagnostic boundaries to increase risk for bipolar disorder. *Archives of General Psychiatry, 62*(6), 642–648.

Green, M. F., & Nuechterlein, K. H. (1999). Should schizophrenia be treated as a neurocognitive disorder? *Schizophrenia Bulletin, 25*(2), 309–319.

Grice, S. J., Spratling, M. W., Karmiloff-Smith, A., Halit, H., Csibra, G., de Haan, M. et al. (2001). Disordered visual processing and oscillatory brain activity in autism and Williams syndrome. *Neuroreport, 12*(12), 2697–2700.

Guidotti, A., Auta, J., Davis, J. M., Di-Giorgi-Gerevini, V., Dwivedi, Y., Grayson, D. R. et al. (2000). Decrease in reelin and glutamic acid decarboxylase67 (GAD67) expression in schizophrenia and bipolar disorder: A postmortem brain study. *Archives of General Psychiatry, 57*(11), 1061–1069.

Hamshere, M. L., Bennett, P., Williams, N., Segurado, R., Cardno, A., Norton, N. et al. (2005). Genomewide linkage scan in schizoaffective disorder: Significant evidence for linkage at 1q42 close to DISC1, and suggestive evidence at 22q11 and 19p13. *Archives of General Psychiatry*, *62*(10), 1081–1088.

Harrison, P. J., & Weinberger, D. R. (2005). Schizophrenia genes, gene expression, and neuropathology: On the matter of their convergence. *Molecular Psychiatry*, *10*(1), 40–68; image 45.

Hashimoto, R., Straub, R. E., Weickert, C. S., Hyde, T. M., Kleinman, J. E., & Weinberger, D. R. (2004). Expression analysis of neuregulin-1 in the dorsolateral prefrontal cortex in schizophrenia. *Molecular Psychiatry*, *9*(3), 299–307.

Hennah, W., Thomson, P., Peltonen, L., & Porteous, D. (2006). Genes and schizophrenia: Beyond schizophrenia: The role of DISC1 in major mental illness. *Schizophrenia Bulletin*, *32*(3), 409–416.

Hennah, W., Tuulio-Henriksson, A., Paunio, T., Ekelund, J., Varilo, T., Partonen, T. et al. (2005). A haplotype within the DISC1 gene is associated with visual memory functions in families with a high density of schizophrenia. *Molecular Psychiatry*, *10*(12), 1097–1103.

Hoffman, R. E., & McGlashan, T. H. (1998). Reduced corticocortical connectivity can induce speech perception pathology and hallucinated 'voices'. *Schizophrenia Research*, *30*(2), 137–141.

Huang, Z. J., Di Cristo, G., & Ango, F. (2007). Development of GABA innervation in the cerebral and cerebellar cortices. *Nature Reviews Neuroscience*, *8*(9), 673–686.

Hubl, D., Koenig, T., Strik, W., Federspiel, A., Kreis, R., Boesch, C. et al. (2004). Pathways that make voices: White matter changes in auditory hallucinations. *Archives of General Psychiatry*, *61*(7), 658–668.

Impagnatiello, F., Guidotti, A. R., Pesold, C., Dwivedi, Y., Caruncho, H., Pisu, M. G. et al. (1998). A decrease of reelin expression as a putative vulnerability factor in schizophrenia. *Proceedings of the National Academy of Sciences USA*, *95*(26), 15718–15723.

Jansma, J. M., Ramsey, N. F., Slagter, H. A., & Kahn, R. S. (2001). Functional anatomical correlates of controlled and automatic processing. *Journal of Cognitive Neuroscience*, *13*(6), 730–743.

Just, M. A., Cherkassky, V. L., Keller, T. A., & Minshew, N. J. (2004). Cortical activation and synchronization during sentence comprehension in high-functioning autism: Evidence of underconnectivity. *Brain*, *127*(Pt 8), 1811–1821.

Kanaan, R. A., Kim, J. S., Kaufmann, W. E., Pearlson, G. D., Barker, G. J., & McGuire, P. K. (2005). Diffusion tensor imaging in schizophrenia. *Biol Psychiatry*, *58*(12), 921–929.

Kargieman, L., Santana, N., Mengod, G., Celada, P., & Artigas, F. (2007). Antipsychotic drugs reverse the disruption in prefrontal cortex function produced by NMDA receptor blockade with phencyclidine. *Proceedings of the National Academy of Sciences USA*, *104*(37), 14843–14848.

Kim, J. J., Kwon, J. S., Park, H. J., Youn, T., Kang, D. H., Kim, M. S. et al. (2003). Functional disconnection between the prefrontal and parietal cortices during working memory processing in schizophrenia: a[15(O)]H2O PET study. *Am J Psychiatry*, *160*(5), 919–923.

Kirkpatrick, B., Xu, L., Cascella, N., Ozeki, Y., Sawa, A., & Roberts, R. C. (2006). DISC1 immunoreactivity at the light and ultrastructural level in the human neocortex. *J Comp Neurol*, *497*(3), 436–450.

Knable, M. B., Barci, B. M., Bartko, J. J., Webster, M. J., & Torrey, E. F. (2002). Molecular abnormalities in the major psychiatric illnesses: Classification and Regression Tree (CRT) analysis of post-mortem prefrontal markers. *Mol Psychiatry*, *7*(4), 392–404.

Krystal, J. H., Karper, L. P., Seibyl, J. P., Freeman, G. K., Delaney, R., Bremner, J. D. et al. (1994). Subanesthetic effects of the noncompetitive NMDA antagonist, ketamine, in humans. Psychotomimetic, perceptual, cognitive, and neuroendocrine responses. *Archives of General Psychiatry*, *51*(3), 199–214.

Lawrie, S. M., Buechel, C., Whalley, H. C., Frith, C. D., Friston, K. J., & Johnstone, E. C. (2002). Reduced frontotemporal functional connectivity in schizophrenia associated with auditory hallucinations. *Biological Psychiatry*, *51*(12), 1008–1011.

Lee, K. H., Williams, L. M., Breakspear, M., & Gordon, E. (2003). Synchronous gamma activity: A review and contribution to an integrative neuroscience model of schizophrenia. *Brain Research –Brain Research Reviews*, *41*(1), 57–78.

Lewis, D. A., & Gonzalez-Burgos, G. (2000). Intrinsic excitatory connections in the prefrontal cortex and the pathophysiology of schizophrenia. *Brain Research Bulletin, 52*(5), 309–317.

Lewis, D. A., Hashimoto, T., & Volk, D. W. (2005). Cortical inhibitory neurons and schizophrenia. *Nature Reviews Neuroscience, 6*(4), 312–324.

Light, G. A., Hsu, J. L., Hsieh, M. H., Meyer-Gomes, K., Sprock, J., Swerdlow, N. R. et al. (2006). Gamma band oscillations reveal neural network cortical coherence dysfunction in schizophrenia patients. *Biololgical Psychiatry, 60*(11), 1231–1240.

Lipska, B. K., Peters, T., Hyde, T. M., Halim, N., Horowitz, C., Mitkus, S. et al. (2006). Expression of DISC1 binding partners is reduced in schizophrenia and associated with DISC1 SNPs. *Human Molecular Genetics, 15*(8), 1245–1258.

Mackie, S., Millar, J. K., & Porteous, D. J. (2007). Role of DISC1 in neural development and schizophrenia. *Current Opinion in Neurobiology, 17*(1), 95–102.

Maldonado, P. E., & Gray, C. M. (1996). Heterogeneity in local distributions of orientation-selective neurons in the cat primary visual cortex. *Visual Neuroscience, 13*(3), 509–516.

Meyer-Lindenberg, A., Poline, J. B., Kohn, P. D., Holt, J. L., Egan, M. F., Weinberger, D. R. et al. (2001). Evidence for abnormal cortical functional connectivity during working memory in schizophrenia. *American Journal of Psychiatry, 158*(11), 1809–1817.

Millar, J. K., Wilson-Annan, J. C., Anderson, S., Christie, S., Taylor, M. S., Semple, C. A. et al. (2000). Disruption of two novel genes by a translocation co-segregating with schizophrenia. *Human Molecular Genetics, 9*(9), 1415–1423.

Mirnics, K., Middleton, F. A., Marquez, A., Lewis, D. A., & Levitt, P. (2000). Molecular characterization of schizophrenia viewed by microarray analysis of gene expression in prefrontal cortex. *Neuron, 28*(1), 53–67.

Moghaddam, B. (2003). Bringing order to the glutamate chaos in schizophrenia. *Neuron, 40*(5), 881–884.

Mori, S., & Zhang, J. (2006). Principles of diffusion tensor imaging and its applications to basic neuroscience research. *Neuron, 51*(5), 527–539.

Mouri, A., Noda, Y., Enomoto, T., & Nabeshima, T. (2007). Phencyclidine animal models of schizophrenia: Approaches from abnormality of glutamatergic neurotransmission and neurodevelopment. *Neurochemistry International, 51*(2–4), 173–184.

Newcomer, J. W., Farber, N. B., Jevtovic-Todorovic, V., Selke, G., Melson, A. K., Hershey, T. et al. (1999). Ketamine-induced NMDA receptor hypofunction as a model of memory impairment and psychosis. *Neuropsychopharmacology, 20*(2), 106–118.

Niu, S., Renfro, A., Quattrocchi, C. C., Sheldon, M., & D'Arcangelo, G. (2004). Reelin promotes hippocampal dendrite development through the VLDLR/ApoER2-Dab1 pathway. *Neuron, 41*(1), 71–84.

Palo, O. M., Antila, M., Silander, K., Hennah, W., Kilpinen, H., Soronen, P. et al. (2007). Association of distinct allelic haplotypes of DISC1 with psychotic and bipolar spectrum disorders and with underlying cognitive impairments. *Human Molecular Genetics, 16*(20), 3517–3528.

Pardo, C. A., & Eberhart, C. G. (2007). The neurobiology of autism. *Brain Pathology, 17*(4), 434–447.

Perlstein, W. M., Dixit, N. K., Carter, C. S., Noll, D. C., & Cohen, J. D. (2003). Prefrontal cortex dysfunction mediates deficits in working memory and prepotent responding in schizophrenia. *Biol Psychiatry, 53*(1), 25–38.

Quattrocchi, C. C., Wannenes, F., Persico, A. M., Ciafre, S. A., D'Arcangelo, G., Farace, M. G. et al. (2002). Reelin is a serine protease of the extracellular matrix. *Journal of Biological Chemistry, 277*(1), 303–309.

Raghavachari, S., Lisman, J. E., Tully, M., Madsen, J. R., Bromfield, E. B., & Kahana, M. J. (2006). Theta oscillations in human cortex during a working-memory task: Evidence for local generators. *Journal of Neurophysiology, 95*(3), 1630–1638.

Rapoport, J. L., Addington, A. M., Frangou, S., & Psych, M. R. (2005). The neurodevelopmental model of schizophrenia: Update 2005. *Molecular Psychiatry, 10*(5), 434–449.

Roberts, R. C. (2007). Schizophrenia in translation: Disrupted in schizophrenia (DISC1): Integrating clinical and basic findings. *Schizophrenia Bulletin*, *33*(1), 11–15.

Rodriguez, E., George, N., Lachaux, J. P., Martinerie, J., Renault, B., & Varela, F. J. (1999). Perception's shadow: Long-distance synchronization of human brain activity. *Nature*, *397*(6718), 430–433.

Roelfsema, P. R., Engel, A. K., Konig, P., & Singer, W. (1997). Visuomotor integration is associated with zero time-lag synchronization among cortical areas. *Nature*, *385*(6612), 157–161.

Roskies, A. L. (1999). The binding problem. *Neuron*, *24*(1), 7–9, 111–125.

Roy, K., Murtie, J. C., El-Khodor, B. F., Edgar, N., Sardi, S. P., Hooks, B. M. et al. (2007). Loss of erbB signaling in oligodendrocytes alters myelin and dopaminergic function, a potential mechanism for neuropsychiatric disorders. *Proceedings of the National Academy of Sciences USA*, *104*(19), 8131–8136.

Rzhetsky, A., Wajngurt, D., Park, N., & Zheng, T. (2007). Probing genetic overlap among complex human phenotypes. *Proceedings of the National Academy of Sciences USA*, *104*(28), 11694–11699.

Salinas, E., & Sejnowski, T. J. (2001). Correlated neuronal activity and the flow of neural information. *Nature Reviews Neuroscience*, *2*(8), 539–550.

Selemon, L. D., & Goldman-Rakic, P. S. (1999). The reduced neuropil hypothesis: A circuit based model of schizophrenia. *Biological Psychiatry*, *45*(1), 17–25.

Shad, M. U., Tamminga, C. A., Cullum, M., Haas, G. L., & Keshavan, M. S. (2006). Insight and frontal cortical function in schizophrenia: A review. *Schizophrenia Research*, *86*(1–3), 54–70.

Shergill, S. S., Kanaan, R. A., Chitnis, X. A., O'Daly, O., Jones, D. K., Frangou, S. et al. (2007). A diffusion tensor imaging study of fasciculi in schizophrenia. *American Journal of Psychiatry*, *164*(3), 467–473.

Singer, W., & Gray, C. M. (1995). Visual feature integration and the temporal correlation hypothesis. *Annual Review of Neuroscience*, *18*, 555–586.

Smith, E. E., & Jonides, J. (1998). Neuroimaging analyses of human working memory. *Proceedings of the National Academy of Sciences USA*, *95*(20), 12061–12068.

Spencer, K. M., Nestor, P. G., Niznikiewicz, M. A., Salisbury, D. F., Shenton, M. E., & McCarley, R. W. (2003). Abnormal neural synchrony in schizophrenia. *Journal of Neurosci*, *23*(19), 7407–7411.

Spencer, K. M., Nestor, P. G., Perlmutter, R., Niznikiewicz, M. A., Klump, M. C., Frumin, M. et al. (2004). Neural synchrony indexes disordered perception and cognition in schizophrenia. *Proceedings of the National Academy of Sciences USA*, *101*(49), 17288–17293.

Stefansson, H., Sigurdsson, E., Steinthorsdottir, V., Bjornsdottir, S., Sigmundsson, T., Ghosh, S. et al. (2002). Neuregulin 1 and susceptibility to schizophrenia. *American Journal of Human Genetics*, *71*(4), 877–892.

Stefansson, H., Thorgeirsson, T. E., Gulcher, J. R., & Stefansson, K. (2003). Neuregulin 1 in schizophrenia: Out of Iceland. *Molecular Psychiatry*, *8*(7), 639–640.

Stephan, K. E., Baldeweg, T., & Friston, K. J. (2006). Synaptic plasticity and dysconnection in schizophrenia. *Biological Psychiatry*, *59*(10), 929–939.

Stopfer, M., Bhagavan, S., Smith, B. H., & Laurent, G. (1997). Impaired odour discrimination on desynchronization of odour-encoding neural assemblies. *Nature*, *390*(6655), 70–74.

Tallon-Baudry, C. (2003). Oscillatory synchrony and human visual cognition. *Journal of Physiology – Paris*, *97*(2–3), 355–363.

Tallon-Baudry, C., & Bertrand, O. (1999). Oscillatory gamma activity in humans and its role in object representation. *Trends in Cognitive Sciences*, *3*(4), 151–162.

Torrey, E. F., Barci, B. M., Webster, M. J., Bartko, J. J., Meador-Woodruff, J. H., & Knable, M. B. (2005). Neurochemical markers for schizophrenia, bipolar disorder, and major depression in postmortem brains. *Biol Psychiatry*, *57*(3), 252–260.

Uhlhaas, P. J., Linden, D. E., Singer, W., Haenschel, C., Lindner, M., Maurer, K. et al. (2006). Dysfunctional long-range coordination of neural activity during Gestalt perception in schizophrenia. *Journal of Neurosciences*, *26*(31), 8168–8175.

Uhlhaas, P. J., & Singer, W. (2006). Neural synchrony in brain disorders: Relevance for cognitive dysfunctions and pathophysiology. *Neuron*, *52*(1), 155–168.

van der Stelt, O., Belger, A., & Lieberman, J. A. (2004). Macroscopic fast neuronal oscillations and synchrony in schizophrenia. *Proceedings of the National Academy of Sciences USA*, *101*(51), 17567–17568.

Varela, F., Lachaux, J. P., Rodriguez, E., & Martinerie, J. (2001). The brainweb: Phase synchronization and large-scale integration. *Nature Reviews Neuroscience*, *2*(4), 229–239.

von der Malsburg, C. (1995). Binding in models of perception and brain function. *Current Opinion in Neurobiology*, *5*(4), 520–526.

Weeber, E. J., Beffert, U., Jones, C., Christian, J. M., Forster, E., Sweatt, J. D. et al. (2002). Reelin and ApoE receptors cooperate to enhance hippocampal synaptic plasticity and learning. *Journal of Biological Chemistry*, *277*(42), 39944–39952.

Weinberger, D. R. (1987). Implications of normal brain development for the pathogenesis of schizophrenia. *Archives of General Psychiatry*, *44*(7), 660–669.

Whitford, T. J., Farrow, T. F., Rennie, C. J., Grieve, S. M., Gomes, L., Brennan, J. et al. (2007a). Longitudinal changes in neuroanatomy and neural activity in early schizophrenia. *Neuroreport*, *18*(5), 435–439.

Whitford, T. J., Grieve, S. M., Farrow, T. F., Gomes, L., Brennan, J., Harris, A. W. et al. (2007b). Volumetric white matter abnormalities in first-episode schizophrenia: A longitudinal, tensor-based morphometry study. *American Journal of Psychiatry*, *164*(7), 1082–1089.

Williams, N. M., Preece, A., Spurlock, G., Norton, N., Williams, H. J., Zammit, S. et al. (2003). Support for genetic variation in neuregulin 1 and susceptibility to schizophrenia. *Molecular Psychiatry*, *8*(5), 485–487.

Winterer, G., Coppola, R., Egan, M. F., Goldberg, T. E., & Weinberger, D. R. (2003). Functional and effective frontotemporal connectivity and genetic risk for schizophrenia. *Biological Psychiatry*, *54*(11), 1181–1192.

Yang, J. Z., Si, T. M., Ruan, Y., Ling, Y. S., Han, Y. H., Wang, X. L. et al. (2003). Association study of neuregulin 1 gene with schizophrenia. *Molecular Psychiatry*, *8*(7), 706–709.

Zhou, Y., Liang, M., Jiang, T., Tian, L., Liu, Y., Liu, Z. et al. (2007). Functional dysconnectivity of the dorsolateral prefrontal cortex in first-episode schizophrenia using resting-state fMRI. *Neuroscience Letters*, *417*(3), 297–302.

Index

Printing: Krips bv, Meppel, The Netherlands
Binding: Stürtz, Würzburg, Germany